SCIENCE AND IDEOLOGY

Does science work best in a democracy? Was "Soviet" or "Nazi" science fundamentally different from science in the USA? These questions have been passionately debated in the recent past. Particular developments in science took place under particular political regimes, but they may or may not have been directly determined by them.

Science and Ideology brings together a number of comparative case studies to examine the relationship between science and the dominant ideology of a state. Cybernetics developments in the USA are compared to that of France and the Soviet Union. Postwar Allied science policy in occupied Germany is juxtaposed to that in Japan. The essays are narrowly focused, yet cover a wide range of countries and ideologies. The collection provides a unique comparative history of scientific policies and practices in the twentieth century.

Mark Walker teaches modern European history and the history of science and technology at Union College in Schenectady, NY. He has published several books and articles on science under National Socialism, including *German National Socialism and the Quest for Nuclear Power, 1939–1949* (1989) and *Nazi Science: Myth, Truth, and the German Atom Bomb* (1995).

ROUTLEDGE STUDIES IN THE HISTORY OF SCIENCE, TECHNOLOGY AND MEDICINE
Edited by John Krige, CRHST, Paris

Routledge Studies in the History of Science, Technology and Medicine aims to stimulate research in the field, concentrating on the twentieth century. It seeks to contribute to our understanding of science, technology and medicine as they are embedded in society, exploring the links between the subjects on the one hand and the cultural, economic, political and institutional contexts of their genesis and development on the other. Within this framework, and while not favouring any particular methodological approach, the series welcomes studies which examine relations between science, technology, medicine and society in new ways, for example, the social construction of technologies, large technical systems and so on.

SCIENCE AND IDEOLOGY

A comparative history

Edited by Mark Walker

Routledge
Taylor & Francis Group

LONDON AND NEW YORK

First published 2003
by Routledge
2 Park Square, Milton Park, Abingdon, Oxon OX14 4RN

Simultaneously published in the USA and Canada
by Routledge
711 Third Avenue, New York, NY 10017

Routledge is an imprint of the Taylor & Francis Group, an informa business

Typeset in Goudy by Taylor & Francis Books Ltd

British Library Cataloguing in Publication Data
A catalogue record for this book is available from the British Library

Library of Congress Cataloging in Publication Data
A catalogue record for this book has been requested

ISBN 0–415–27122–3 (hbk)
ISBN 0–415–27999–2 (pbk)

I WOULD LIKE TO THANK DR. FREDERICK
SEITZ AND THE RICHARD LOUNSBERY
FOUNDATION FOR THEIR GENEROUS
SUPPORT.

THIS BOOK IS DEDICATED TO MY FAMILY,
CHRISTOPHER, KERRY, AND LINDA.

CONTENTS

CONTENTS

CONTRIBUTORS

Richard H. Beyler received his doctorate in history of science from Harvard University. He teaches history of science and modern European history at Portland State University, and has recently been a guest researcher at the Presidential Commission of the Max Planck Society for the History of the Kaiser Wilhelm Society under National Socialism. In addition to the political relations of science in Weimar, Nazi, and post-Second World War Germany, his research has focused on the history of physics and biophysics in the early twentieth century.

Burghard Ciesla did his Ph.D. in economic history and specializes in twentieth-century history of science and technology and German economic and social history after 1945. He is a research fellow at the Research Center for Contemporary History in Potsdam. His publications include *Technology Transfer Out of Germany After 1945* (1996); *"Sterben für Berlin?" Berliner Krisen 1948: 1958. Politik, Gesellschaft und Kultur im Kalten Krieg* (2000); and *Vertreibung-Neuanfang-Integration. Vertriebene in der brandenburgischen Nachkriegsgesellschaft* (2001).

Slava Gerovitch is a Dibner/Sloan Postdoctoral Researcher at the Dibner Institute for the History of Science and Technology. He has two Ph.Ds, one in 1992 from the Russian Academy of Sciences, the second in 1999 from MIT. In 1997 he established the Virtual Guide to the History of Russian and Soviet Science and Technology on the World Wide Web, and is currently working on the Dibner/Sloan project, "History of Recent Science and Technology on the Web." His publications include *From Newspeak to Cyberspeak: A History of Soviet Cybernetics* (2002); "'Mathematical Machines'? of the Cold War: Soviet Computing, American Cybernetics and Ideological Disputes in the Early 1950s," *Social Studies of Science*, 31, 2 (April 2001); and "'Russian Scandals'? Soviet Readings of American Cybernetics in the Early Years of the Cold War," *The Russian Review*, 60, 4 (October 2001).

Michael Gordin is a historian of the physical sciences, with a particular emphasis on science in Russia and the Soviet Union. He received his doctorate from the History of Science Department at Harvard in 2001, is currently a Junior Fellow at the Harvard Society of Fellows, and in 2003 will begin an assistant professorship in the History Department at Princeton University. Gordin is currently finishing a book manuscript on a cultural history of Russian chemist D.I. Mendeleev in Imperial St. Petersburg. His publications include "The Importation of Being Earnest: The Early St. Petersburg Academy of Sciences," *Isis*, 91 (2000); "Loose and Baggy Spirits: Reading Dostoevskii and Mendeleev," *Slavic Review*, 60 (2001); "The Organic Roots of Mendeleev's Periodic Law," *Historical Studies in the Physical and Biological Sciences* (2002); and "The Anthrax Solution: The Sverdlovsk Incident and the Resolution of a Biological Weapons Controversy," *Journal of the History of Biology*, 30 (1997).

Walter Grunden received his Ph.D. from the University of California, Santa Barbara in 1998 and is an Assistant Professor at Bowling Green State University, Ohio. His research and teaching focus on modern Japan and East Asia, history of science and technology, the Second World War, and public policy. He has received fellowships and grants from the Japan Society for the Promotion of Science (2001–2), the Association for Asian Studies' Northeast Asia Council (2001, 1993), as well as the US Department of Education for Foreign Language and Area Studies, and the University of Michigan's Center for Japanese Studies. He is the author of "Hungnam and the Japanese Atomic Bomb: Recent Historiography of a Postwar Myth," *Intelligence and National Security* (1998). He is presently working on a manuscript, "Japanese Secret Weapons of World War II: The Mobilization of Science for Advanced Research and Why It Failed."

Dieter Hoffmann studied physics and the history of science at the Humboldt University in Berlin. He has been a research scholar at the East German Academy of Sciences, the Federal German Physical-Technical Institute, and since 1996 at the Max Planck Institute for the History of Science in Berlin. Hoffmann is a lecturer at the Humboldt University, where he has also worked as a visiting professor. His research interests focus on the history of physics in the nineteenth and twentieth centuries, especially Max Planck, the institutional and experimental history of quantum theory, and modern metrology. He also studies the history of science in East Germany. His publications include biographies of Erwin Schrödinger (1984), Robert Havemann (1990), Ernst Mach (1991), Hermann Helmholtz (1995) and Max Planck (1997); the German version of the Farm Hall Papers, *Operation Epsilon* (1993); and *Science under Socialism* (1999).

Uwe Hoßfeld is a historian of the biological sciences. His fields of interest include the history of evolutionary biology, morphology and anthropology in the nineteenth and twentieth centuries; the reception of Ernst Haeckel in the Third Reich, and the history of University of Jena. Hossfeld received his Ph.D. from the Biology Department at the University of Jena. He has worked as a postdoctoral fellow at the Universities of Tübingen and Göttingen. He catalogued the Haeckel correspondence at the University of Jena as part of a German Research Council project, and is now a research fellow at the Ernst-Haeckel-House and in the "Senate Commission for the History of the University of Jena in the Twentieth Century." His publications include *Gerhard Heberer (1901–1973) – Sein Beitrag zur Biologie im 20. Jahrhundert* (1997); *Evolutionsbiologie von Darwin bis heute* (2000); *Darwinismus und/als Ideologie* (2001); and *Die Entdeckung der Evolution. Eine revolutionäre Theorie und ihre Geschichte* (2001).

Jürgen John is professor at the Historical Institute of the Friedrich-Schiller-University Jena. He studied history and art history at the universities of Jena and Halle. John worked at the film and television federation in Berlin and the Academy of Sciences in Berlin before coming to the University of Jena. His special fields of research include German history, regional history, urban history, cultural history and university history of the nineteenth and twentieth centuries. His publications include *Die Wiedereröffnung der Friedrich-Schiller-Universität Jena 1945* (1998); *Weimar 1930. Politik und Kultur im Vorfeld der NS-Diktatur* (1998); *Das Dritte Weimar. Klassik und Kultur im Nationalsozialismus* (1999); and *"Mitteldeutschland". Begriff-Geschichte-Kultur* (2001).

Paul Josephson teaches history at Colby College. He is the author of four books, most recently *Red Atom*, and is currently writing a study of big technology, the state and environmental issues. His publications include *Physics and Politics in Revolutionary Russia* (1991); *New Atlantis Revisited: Akademgorodok, the Siberian City of Science* (1997); *Totalitarian Science and Technology* (1996); and *Red Atom: Russia's Nuclear Power Program from Stalin to Today* (2000).

Morris F. Low is senior lecturer in Asian Studies at the University of Queensland in Brisbane, Australia. He is an historian of Japanese science. His publications include *Science, Technology and Society in Contemporary Japan* (1999); *Beyond Joseph Needham: Science, Technology, and Medicine in East and South East Asia* (1999); *The Politics of Knowledge: Science and Evolution in Asia and the Pacific* (1999); and *Science, Technology and R&D in Japan* (2001).

David Mindell is Dibner Associate Professor in the History of Engineering and Manufacturing at the Massachusetts Institute of Technology, and the founder and director of MIT's "DeepArch" research group in technology, archaeology, and the deep sea. He has degrees in Literature and in Electrical Engineering from Yale University, and received his doctorate in the history of technology from MIT in 1996. His research interests include technology policy (historical and current), the history of automation in the military, the history of electronics and computing, new theories of engineering systems, and deep ocean robotic archaeology. His publications include *War,*

Technology, and Experience aboard the USS Monitor (2000) and *Between Human and Machine: Feedback, Control, and Computing before Cybernetics* (2002).

Elena Z. Mirskaya is a professor of sociology and head of the Department of the Sociology of Science at the Institute for History of Science and Technology, Russian Academy of Science. Her main field of research is social studies of science, including: science and society, scientific community, scientific communications, social production of knowledge, and ethics of science. Her current interests focus on topics of Soviet and post-Soviet science. Her publications include "From East to West: New Patterns of International Relations," *East European Academies in Transition* (1998); "Russian Academic Science Today," *Social Studies of Science*, 25, 4 (1995); "Stalinismus und Wissenschaft, Die 30-er Jahre: Präeludium des grossen Terrors," in *Kultur im Stalinismus* (1994).

Yakov M. Rabkin is a professor of the history of science at the University of Montreal and has received over twenty research awards and fellowships from several countries. His publications include comparative studies of science in totalitarian societies, investigations of cultural aspects of science in Africa and in North America, studies of non-Western research cultures, and of relations between science, cultures and traditions. Books include *Science Between the Superpowers: A Study of Soviet-American Relations in Science and Technology* (1988); *The Interaction of Scientific and Jewish Cultures in Modern Times* (1995); and *Diffusion of New Technologies in Post-Communist World* (1997).

Jérôme Segal is assistant professor of history of science and epistemology at the IUFM in Paris. He spent two years at the Max Planck Institute for History of Science in Berlin (Germany) as a post-doctoral researcher. A specialist in the history of information theory, he recently became interested in the history of protein folding theories. His publications include "Protein Structure, From Model Kits to Computer Screens," in *Displaying the Third Dimension: Models in the Sciences, Technology and Medicine* (forthcoming); "Kybernetik in der DDR – Begegnung mit der marxistischen Ideologie," *Dresdner Beiträgen zur Geschichte der Technik und der Technikwissenschaften*, 27 (2001); "The Pigeon and the Predictor Miscarriage of a Cyborg," in *American Foundations and Large Scale Research: Construction and Transfer of Knowledge* (2001); and "Hermann Schmidt (1894–1968) et la théorie générale de la régulation: Une cybernétique allemande en 1940?" *Annals of Science*, 54 (1997).

Rüdiger Stutz earned his Ph.D. in 1985 from the Friedrich-Schiller-University of Jena. Since 1993, he has worked in the Historical Institute in Jena and currently is a research fellow in a German Research Council project at the universities of Halle and Jena examining social revolutions after the collapse of the GDR. His publications include "Im Schatten von Zeiss. Die NSDAP in Jena," in *Nationalsozialismus in Thüringen* (1995); *Macht und Milieu. Jena zwischen Kriegsende und Mauerbau* (2000); and *Studien zur Geschichte der Universität Jena im Nationalsozialismus* (2002).

Helmuth Trischler is Director of Research of the Deutsches Museum and Acting Head of the Munich Center for the History of Science and Technology. He studied history and literature at the University of Munich. Since 1990 he has worked at the Deutsches Museum in Munich. His publications include *Luft- und Raumfahrtforschung in Deutschland 1900-1970. Politische Geschichte einer Wissenschaft* (1992) and "Aeronautical Research under National Socialism: Big Science or Small Science?" in *Science in the Third Reich* (2001).

Mark Walker teaches history at Union College. His publications include *German National Socialism and the Quest for Nuclear Power, 1939–1949* (1989); *Science, Technology, and National Socialism* (1993); and *Nazi Science: Myth, Truth, and the German Atom Bomb* (1995).

Zuoyue Wang received a B.S. in physics from Henan Normal University, Xinxiang, a M.S. in history of science from the Chinese Academy of Sciences, Beijing, and in 1994 a Ph.D. from the University of California, Santa Barbara. He has taught history of science and technology and US and Chinese history at UCSB, UC Berkeley and, since 1999, California State Polytechnic University, Pomona. He is currently finishing two books: a history of the US President's Science Advisory Committee during the Cold War and a collection of biographical studies of Asian American scientists and engineers. His publications include "Saving China Through Science: The Science Society of China, Scientific Nationalism, and Civil Society in Republican China," *Osiris*, 17 (2002); "Between the Devil and the Deep Sea: C.K. Tseng, Ocean Farming, and the Politics of Science in Modern China," *Isis*, 91 (2000); and "US-China Scientific Exchange: A Case Study of State-Sponsored Scientific Internationalism During the Cold War and Beyond," *Historical Studies in the Physical and Biological Sciences*, 30, 1 (1999).

Thomas Zeller teaches history at the University of Maryland in College Park and is a research Fellow at the German Historical Institute in Washington, D.C.. After attending universities in the USA and Germany, Zeller received his Ph.D. from the University of Munich in 1999 and has been working in the USA since then. He is particularly interested in the growing intersections between technological and environmental history. His publications include "'The Landscape's Crown': Landscape, Perceptions, and Modernizing Effects of the German Autobahn System, 1934-1941," in *Technologies of Landscape: Reaping to Recycling* (1999); *Straße, Bahn, Panorama. Verkehrswege und Landschaftsveränderung in Deutschland 1930 bis 1970* (2002); *Germany's Nature* (co-editor, forthcoming); and *How Green Were the Nazis?* (co-editor, forthcoming).

1

INTRODUCTION
Science and ideology

Mark Walker[1]

How does ideology affect science? Does it hurt or help, or is this dichotomy too simplistic? Science is normally portrayed as ideology-free, and ideology is usually accused of interfering with or distorting science. As the contributions to this book demonstrate, however, the interaction of science and ideology is more subtle, complex, and interesting than this model of science and ideology as separate spheres would imply.

The discussion of science and ideology during the first three decades since the end of the Second World War has been dominated by the legacy of National Socialism and the pressure exerted by the Cold War. As this introduction will show, this dominance has had a long-lasting influence on our understanding of how ideology and science interact. This literature also makes clear that ideology can affect science in two very different ways: (1) ideological pressure on scientists (as on everyone) for political conformity; and (2) ideological interference in the practice of science itself. It is important to keep this distinction in mind. The former case was quite widespread and is easy to locate, but may tell us little about scientists under a given ideology that would not hold equally well for other social groups. The latter case would be more interesting and significant, but is more difficult to determine.

When the editor of this book began studying science under National Socialism in the early 1980s, one fundamental question stood out: what was the specific and unique effect of National Socialism on science? By the late 1980s, first of all a trickle, then a flood of work on many different aspects and subjects of science, medicine, and technology under National Socialism appeared. The problem with our historical understanding of science under Hitler is no longer that we do not know what happened, rather that it is becoming difficult to see the forest for the trees. Despite the wealth of information and analysis now available, we arguably have come no closer to answering this question. What difference did National Socialism really make?

The break-up of the Soviet Union has both liberated a wealth of new source materials on Soviet science and encouraged historians from the former Soviet Union to be more independent. This has, of course, also generated more work on Lysenkoism (see below),[2] but this has been accompanied by a wealth of studies covering many different areas of science under Soviet Communism.[3] However, like the situation for science and National Socialism, this increase in detail and information about Soviet science has arguably not provided a more profound understanding of the subject.

A more modest, but significant amount of work has appeared on Communist China. Scholars are waiting, not always patiently, for Charles C. Gillispie's major work on science and the French Revolution.[4] Surprisingly, only recently have studies on American science during the McCarthy period begun to go beyond the Oppenheimer Affair. Some topics have fairly recently become feasible subjects of study, like science in East Germany[5] and the other Eastern Bloc states since the collapse of communism in 1989. Finally, there are other subjects that remain relatively inaccessible, like science in Fascist Italy.[6] In other words, much remains to be done.

Science and the Cold War

The historiography of ideology and science has been profoundly influenced, if not distorted by the pressures of the Cold War. But even before the end of the Second World War, contemporary observers like Robert Merton and J.D. Bernal showed a keen interest in the subject of ideology and science through their analyses of the practice of science under extreme political regimes. In the late 1930s and early 1940s, Merton had set out a vision of a free democratic society in which science would play an important role by encouraging the spread of liberal democracy.[7] The scientist and leftist Bernal joined Merton in being sharply critical of science under fascism, but was much less objective with regard to science in the Soviet Union, thereby revealing the blinkered perspective of some leftist scholars during the Stalinist years.

Thus in 1938, the sociologist Merton wrote with regard to the National Socialist regime that the conflict between the totalitarian state and the scientist was the result of the "incompatibility between the ethic of science and the new political code which is imposed upon all, irrespective of occupational creed." Science required that "theories or generalizations be evaluated in terms of their logical consistency and consonance with facts," but the National Socialist state had introduced the "hitherto irrelevant criteria of the race or political creed of the theorist.[8] Merton's comments were a response to the so-called "Aryan Science" movements like the "Aryan Physics" of Philipp Lenard and Johannes Stark or "Aryan Mathematics" of Ludwig Bieberbach, and the attacks on "Jewish Physics" or "White Jews in Science" that accompanied them.

In a subsequent paper, Merton also examined the Soviet Union and its leaders' emphasis on "Russian nationalism" and insistence on the "'national' character of science." Merton argued that this confused two distinct issues: (1) "the cultural context in any given nation or society may predispose scientists to focus on certain problems, to be sensitive to some and not other problems on the frontier of science"; but (2) "the criteria of validity of claims to scientific knowledge are not matters of national taste and culture. Sooner or later, competing claims to validity are settled by universalistic criteria."[9] What Merton is calling "national" or "nationalism" (a term the Soviets would not have used) is Lysenkoism and its effects on Soviet science in general and on genetics in particular.

In 1939, on the eve of the Second World War, Bernal had the following to say about science in the Third Reich. The National Socialists had "reversed the values on which not only liberal but Christian society is ostensibly founded." Since the "ideals of blood and soil" which had taken their place had "no scientific backing," science had to be distorted in order to provide one. The National Socialist state had demanded that "old prejudices should be revived and put in the place of new discoveries." Perhaps most disturbing was the National Socialists' anti-Semitic science policy. Not only Jews, but also Jewish ideas were persecuted. But if "in logic, mathematics, or physics everything a Jew did must necessarily have been wrong," then the "whole edifice of science" would have to be rebuilt with "diminished and incongruous materials." Finally, Bernal perceptively noted that militarism had also affected German science. Biology, psychology, and the social sciences had been distorted in order to provide a scientific basis for the "great Nazi myth of race superiority and the necessity for military struggle."[10]

In contrast to his sharp rejection of racial science and militarism in Germany, in a footnote to the same 1939 book Bernal notes that there had been a "very important controversy on the subject of the foundations of genetics" between Vavilov and Lysenko, but that outside the Soviet Union, this controversy had been "magnified out of all proportion."[11] A decade later, in his book *The Freedom of Necessity*, he could still approvingly quote Lysenko: "Men are not born in the Soviet Union; organisms may be born but men are made."[12] Bernal's support of Soviet science policy in general and Lysenko in particular was not due to ignorance. In contrast to Bernal, in the early 1950s the émigré biologist Theodosius Dobzhansky painted a very different picture of science in the Soviet Union and thereby made clear that information about the excesses of Lysenkoism under Stalin was available.

In the 1920s and 1930s, Dobzhansky noted, the USSR had been a leading country in genetics, second perhaps only to the United States. But from about 1935 on, genetics had been attacked by a "group of ambitious men, spearheaded by the agronomist T.D. Lysenko." Their attacks were encouraged by the Communist Party hierarchy, and vice versa. In August of 1940 the leading geneticist N.I. Vavilov was arrested and subsequently died. At least six other internationally known geneticists also vanished without a trace. In the summer of 1948 a "sort of a popular tribunal" convoked in Moscow to "judge the sins of genetics." When Lysenko proclaimed at this meeting that "The Central Committee of the Party has examined my report and approved it," it was clear that science and ideology were not always separate in the Soviet Union.[13]

Thus the two obvious case studies for the influence of ideology upon science in the postwar world were science under Hitler and under Stalin. But whereas the excesses of science and medicine under National Socialism could easily be condemned and often the scientific work itself dismissed (see below), the case of the Soviet Union was more complicated. On one hand, Lysenkoism was well known and had been influential for many years. However, it was only after the end of the Second World War that, with the help of Stalin, it essentially achieved the banning of modern genetics in the Soviet Union.

On the other hand, the Soviet atomic bomb in 1949, its first fusion bomb a few years later, and the launch of the Sputnik satellite into space in 1957 made it difficult to belittle Soviet science. Even the critics who pointed to the Soviet use of espionage and German scientists and engineers left over from Hitler's regime did not dispute that the Soviet scientific potential was profound, if not dangerous. If the National Socialists had perverted and ruined German science – and even here there were always exceptions noted like aeronautic and rocket research – then the Soviet Union had managed to distort but still exploit its science. Thus it is clear here that ideology was working at different levels. Stalin's regime could simultaneously nurture an irrational doctrine like Lysenkoism within agricultural science and genetics while generously supporting sciences and scientists who could contribute to nuclear weapons research and development.

The practical examples of science under Hitler and Stalin provided an important theme, the interaction of ideology and science, precisely when the academic discipline of the history of science was becoming established in the United States. Arguably it is no coincidence that, in the ideologically charged climate of the 1950s, several young historians of science who would later emerge as some of the most important founders of their discipline in the United States, Charles C. Gillispie, L. Pearce Williams, and Henry Guerlac, all tackled the subject of science and ideology head on. They did not study National Socialist or Soviet science, but instead a historically more remote, but equally important episode, science in the French Revolution.[14] The questions they articulated shortly after Sputnik and before the Berlin Wall and Cuban Missile Crisis are still significant.

Gillispie's analysis of the "Jacobin philosophy of science" is particularly important because he was one of the first to ask an important question: can ideology affect not only the practice of science, but also its "hard" content? In their "year of exaltation" the Jacobins not only tried to "change human nature," but both the leaders and the rank and file in particular tried to replace "the image of nature with which science confronts humanity" with a different one, "one sympathetic to the ordinary man." The technological side of Jacobin science would be a "docile servant of humanity," while its conceptual side would be a "simple extension of consciousness to nature, the seat of virtue, attainable by any instructed citizen through good will and moral insight."[15] What Gillispie is describing here was an attempt to produce what can be termed "ideologically-correct-science," which is the subject of one of the essays in this volume.

There is a second, equally important innovation in Gillispie's analysis. In the spirit of the Cold War, Gillispie went on to compare the French case with science under Stalin. The interesting thing about the Jacobin philosophy of science, he notes, was not that it was "futile...Nature is not like that," rather "that it happened at all." The parallels to Lysenkoism in the Soviet Union were obvious for a contemporary like Gillispie. However, the "analogy with Marxism flags if it does not altogether fail." It was inconceivable for Gillispie that "anything of the sort could happen today." The Lysenko controversy "never went so deep" and "it did not really matter to science that it happened then. It mattered only to scientists."[16] With the benefit of hindsight, Gillispie appears to have been too optimistic. Lysenkoism had a profound effect on Soviet science, although this was not so clear in the late 1950s. Gillispie's distinction between the fate of science and scientists has some merit, for Soviet science did eventually throw off Lysenko's influence while individual scientists had their careers ruined, but it is not so clear that these two fates can be so easily separated, or what such a separation would mean.

Scholars began to study Soviet science in the 1960s, notably David Joravsky and Loren Graham. From the very beginning, the Lysenko Affair cast a long shadow over the historiography of Soviet science.[17] By this time, Lysenko's long career had finally ended together with Khruschev's tenure as Soviet leader and the rehabilitation of genetics in the Soviet Union had begun. Like Gillispie, Joravsky also contemplated a comparison of science in the French and Russian Revolutions. His book "grew out of an interest in the intellectual history of the Russian Revolution, out of a desire to understand the modern analogues to Marat and Lavosier in an earlier revolution."[18]

Joravsky left no doubt of the damage done by Soviet ideology to science. During the "great break" the new Soviet Union was making with its past, the "primitive zealots who assumed control of Bolshevik ideology" had "razed the walls of academic autonomy" without addressing the "substantive issues of natural sciences."[19] This was ironic, for as Graham noted, Marxism was founded on the "scientific" study of society and professed great respect for science. The Soviet revolutionaries had "subscribed in some degree to the belief that culture is a derivative superstructure above the economic base and that a modification of the base inevitably results in a transformation of culture." Since science was one of the layers of culture, perhaps the highest layer, it would also acquire "unique characteristics," These characteristics would include "a new theory of the place of science in society, a more fertile economic environment for technological growth, unprecedented governmental support for research, a superior organizational scheme of research interests, and a methodology for the planning of science." A "socialist science" would emerge, a science "superior to all its capitalist competitors."[20] In practice, "a transformation of quantity had occurred, but the goal had been a transformation of quality." The unique Soviet science never appeared.[21]

Joravsky went further, noting the fundamental hypocrisy of Bolshevik science policy. When the Bolsheviks forced intellectuals to "shout encouragement in fervent unison" and scientists to "profess Marxism at gun point," they were "doing violence to their own assurance that scientists would spontaneously recognize Marxism as the logical extension of science into human affairs." By pursuing recalcitrant scientists into their special disciplines and demanding the "reconstruction" of these disciplines as proof of the scientists' conversion, the Bolsheviks were "casting doubt on their own faith that dialectical materialism formulates the methods that have brought success to scientists in their cognition of the world."[22]

Graham agreed that great harm was done to science in the Soviet Union, particularly to genetics, by the "wedding of centralized political control to a system of philosophy with claims to universality." When Lysenko's views of biology were officially approved in August 1948, it soon became clear that other scientific fields were also objects of ideological attack. Soviet scientists were generally fearful that each field would produce its own Lysenko.[23] But Graham added that "Science is not shattered by a few blows. Like an injured plant, it grows around its wounds, seeks other paths to its goals, and continues its development so long as it can find minimum sustenance."[24] In contrast with Joravsky, Graham also argued that all aspects of Soviet ideology should not be tarred with the same brush: "Lysenkoism had nothing to do with dialectical materialism."[25] In fact, Graham has made a case for dialectical materialism benefiting Soviet science.[26] However, dialectical materialism or any Marxist influence on science has been discredited since the collapse of the Soviet Union.

In addition to Soviet science, other examples of ideology affecting science have arisen since the publication of Gillispie's essay, including science in Communist China. Of course the McCarthy period in the United States, including the infamous and well-publicized "Oppenheimer Affair" (whereby the Atomic Energy Commission publicly stripped the physicist of his security clearance because of his opposition to the hydrogen bomb) was another candidate for the study of ideology and science, but it was too close at hand for historians in the 1950s. As described below, the writer Robert Jungk, however, had no such inhibitions. Indeed, with the exception of Joseph Haberer (examined below), scholars have only recently begun to examine this ambivalent chapter of American science.[27]

It is perhaps most striking that Gillispie and Joravsky make no mention of science in the Third Reich. Scholars of the "generation" that institutionalized the history of science, as well as the one that immediately followed, generally appear to have held "science under Hitler" at arm's length. Loren Graham compared eugenics in the Soviet Union and Germany, but limited himself to the Weimar Republic.[28] Even Paul Forman, who explicitly addressed the question of an ideologically charged environment influencing science, limited himself to the study of science in Weimar Germany.[29] It is true that Soviet

Communism represented a real threat in the 1950s and 1960s, while a divided Germany mainly provoked bad memories, but this explanation alone appears insufficient. Perhaps the answer lies in the fact that, at least in theory, Marxism was conducive to science, and indeed claimed to be especially so, while science and National Socialism, with its blatant anti-intellectualism and racism, seemed irreconcilable.

Science and National Socialism

Along with the effect of the Cold War, the historiography of ideology and science has also been profoundly influenced by the horrific legacy of National Socialism. After the fall of the Third Reich and the defeat of Japan the subject of science and ideology was transformed by revelations of the crimes of National Socialism. Here the emphasis has been on the collaboration of scientists, physicians, engineers, and other intellectuals with National Socialist policies. If Lysenkoism overshadowed historical work on Soviet science, the study of science under National Socialism in the postwar era has also been dominated by a few examples, including the "German atomic bomb" and the so-called "Aryan Physics" movement.

Perhaps the first postwar study of science under National Socialism was Max Weinreich's *Hitler's Professors*, a passionate work written in the immediate aftermath of the Second World War. Weinreich is primarily concerned with the responsibility that Germany's "intellectual leaders" had for the horrors of National Socialism. Working together with Germany's political and military leaders, they "declared Germany the final judge of her own acts and then renounced accepted morality," "arrogated to themselves the right to dispose of millions of people for their own and their fatherland's greater glory," and "prepared, instituted, and blessed the program of vilification, disfranchisement, dispossession, expatriation, imprisonment, deportation, enslavement, torture, and murder."[30]

In 1949 Alexander Mitscherlich, head of the German Medical Commission serving at the Nuremberg Trials, and Fred Mielke published *Doctors of Infamy*. This documentation of medical atrocities committed under National Socialism was a courageous indictment of the German physicians, who under Hitler had become "licensed murderer[s]" and "publicly appointed torturer[s]." Mitscherlich and Mielke also warned of an inhumane future, with its "transmogrification of subject into object, of man into a thing against which the destructive urge may wreak its fury without restraint." Finally, they also condemned "high-ranking men of science" who had perhaps "themselves committed no culpable act," but nevertheless had taken an "objective interest" in the atrocities.[31]

Both Weinreich on one hand and Mitscherlich and Mielke on the other were mainly concerned with, if not overwhelmed by the practical horrors of the role "intellectual leaders" had played in the Holocaust. One of the harshest postwar critics of German scientists under National Socialism, Samuel Goudsmit, also turned his attention to the effect of National Socialism on science itself. A physicist who had emigrated from Holland to the United States, Goudsmit worked on radar during the war, lost his parents in Auschwitz, and led a scientific intelligence-gathering unit sent to Europe in the wake of the advancing Allied armies.[32] This "Alsos Mission" was first of all charged with assessing and neutralizing German progress on nuclear weapons, but also investigated biological and chemical weapons.

Goudsmit wanted the United States in the postwar period to learn some lessons from science under Hitler. "Too many of us," he wrote, still assumed that "totalitarianism gets things done where democracy only fumbles along" and that "the Nazis were able to cut all corners and proceed with ruthless and matchless efficiency." Nothing was further from the truth. The German scientists were too complacent. Politicians interfered in the affairs of science, and "party hacks" were appointed to important administrative posts. Dogma, "whether it be political, scientific, or religious," was dangerous. The "stubborn blindness of dogma" and the "free inquiring spirit of science" did not mix, nor was there any such thing as "Jewish science" or "Aryan science."[33]

Secrecy in science was a particular concern of many scientists in the U.S.A. in the postwar years. They were justifiably concerned that the model of military control of scientific research exemplified by the Manhattan Project would be carried over to the postwar peace, including restrictions on the publication of scientific work. Once again, Goudsmit argued that science under Hitler provided an important lesson. If certain basic scientific discoveries were kept hidden from other scientists because of the need for secrecy, then the result would prove "more disastrous than the prohibition of teaching Einstein's work in Germany." When it came to secrecy in scientific matters, the scientists were the best judges of "what to keep secret – and when."[34]

For Goudsmit, science flourished best in a democracy with a great deal of intellectual freedom and a minimum of political and ideological control. Where Merton had argued that science would encourage liberal democracy, Goudsmit believed that liberal democracy was the best political system for science. In his view, science under National Socialism, and in particular the German uranium project, were sobering examples of how political ideology can ruin science and arguments to use in the contemporary debate over how much control the State should exert over science.

Goudsmit's articles and book were also a part of an ongoing debate with Werner Heisenberg over how to judge the German work on nuclear energy and nuclear weapons during the war.[35] Heisenberg's most influential advocate was the writer Robert Jungk's book *Brighter than a Thousand Suns*.[36] Like Goudsmit, Jungk compared German and American science, but while the former saw the United States and democracy as ideologically and morally superior, the latter juxtaposed Werner Heisenberg's work under Hitler with Robert Oppenheimer's victimization at the hand of the Atomic Energy Commission.

Perhaps Jungk's harshest criticism came in a comparison of German and American efforts to harness nuclear fission. It seemed "paradoxical" that German scientists, living under a "saber-rattling dictatorship, obeyed the voice of conscience" and attempted to "prevent the construction of atomic bombs," while their professional colleagues in the democracies, who had "no coercion to fear," with very few exceptions "concentrated their whole energies" on the production of nuclear weapons.[37] Shortly before his death, Jungk renounced this interpretation, but this myth of resistance to Hitler by denying him nuclear weapons nevertheless remains potent to the present day.[38]

In 1967, David Irving published the first scholarly study of nuclear weapons research under National Socialism. In the end, he also compared the German and American efforts. One of the main reasons why the German project did not keep pace with the United States was because "the project was directed by scientists throughout its history, and not by military commanders as in America." Whereas in the United States General Groves reined in and harnessed the scientists, and in particular, the "theoreticians" in order to create the first nuclear weapons, the behavior of the leading German scientists demonstrated that "during war, science cannot be left to the scientists." Germany's nuclear scientists failed to win the confidence of their government, and were "left stranded on the shores of the atomic age."[39]

In one of the first and most perceptive scholarly examinations of science under Hitler, the political scientist Joseph Haberer argued in 1969 that in National Socialist Germany the scientific leaders were unable to "beat political leaders at their own game." Instead of using the political order for their own purposes, it was the scientific leadership that was used. Moreover, championing a policy "founded on expediency and compliance" in such a time of crisis was to "acquiesce in the threat to the existence of the community" and to "confuse means as ends." The "obligations of leadership" were betrayed by permitting without opposition the "victimization of members of the community" and the "destruction of fundamental institutions and values." In the final analysis, this collaboration brought about the very conditions which made the "eventual collapse of the community itself more likely."[40]

Like Jungk, Haberer juxtaposed German scientists like Max Planck and Heisenberg under Hitler with American scientists like Oppenheimer and Edward Teller under McCarthy. Noting that "on the surface" Oppenheimer himself "appeared to accept the social responsibility of science," on closer examination Oppenheimer's statements invariably dissolved into "elusive ambiguity." Even Oppenheimer's well-known phrase after the first test of an atomic bomb, that "the physicists have known sin" contained "no admission of sin, only knowledge of it."[41]

The "Oppenheimer Affair" actually produced very little public comment by scientists. "Public discussion of the matter by scientists evaporated so soon after the verdict had been delivered that it is difficult to substantiate the claim that its effect on the community of science was traumatic." Indeed the scientists' public response on behalf of Oppenheimer was "weak, nonactive, and tenuous."[42] Thus Haberer finds parallels in Hitler's Germany and McCarthy's America that echo those of Jungk, but are more critical and disturbing. In both cases, if for different reasons, the scientific community acquiesced in a politically and ideologically motivated purge.

One of the first scholarly studies done in Germany on science and technology under National Socialism was Karl-Heinz Ludwig's analysis of technology and engineers during the Third Reich. His analysis and moral criticism of engineers was compatible with Haberer's judgment of scientists. The National Socialists' promise to apply technology in a socially responsible way both influenced the "genesis of the Third Reich" and later also "helped to stabilize it." At the same time, many engineers hoped that a stronger state would constrain capitalism and its domination of technology. These engineers were disappointed by the National Socialists' policies after 1933 and "perverted" by the overwhelming and "destructive effect of rearmament." Finally, during the war years the engineers' "techno-political" program "atrophied into a purely occupational professionalism."[43]

Alan Beyerchen's 1977 book *Scientists under Hitler*,[44] the next substantial scholarly work on science under National Socialism, focused first of all on the physics community, and secondly on the "Aryan Physics" movement. During the National Socialist period the members of the physics community cared most about the "protection of their autonomy against political encroachment." The vast majority of the scientists under Hitler were "neither anti-Nazis nor Nazis," but rather were "committed solely to independence in the conduct of professional affairs." Thus Beyerchen agrees with Haberer, although he is perhaps more forgiving. According to Beyerchen, "Aryan Physics" failed because of their "lack of success in obtaining backing from political sources," and their "inability to win the support of the professional physics community."[45]

Along with Ludwig's book on engineers, a collection of essays edited by Herbert Mehrtens and Steffen Richter was the path-breaking book in German on the subject of science and National Socialism. This volume devoted a great deal of space to the various "Aryan" sciences, but Herbert Mehrtens also offered a general analysis. The sciences under the "German Fascist regime" offered a "complex" and "opaque" picture. Individual disciplines, theories, and many scientists were clearly closely connected with the "deeply reactionary and cynical ideology and rule" of the National Socialists. Most scientists probably rejected this ideology, the "anti-Semitic terror," and war, and "fought for the conservation of scientific autonomy and rationality." However, they also compromised themselves through their collaboration with big business and industry, with "men like Carl Krauch and concerns like IG Farben." The end result of their efforts was accommodation to Fascism.[46]

Enough time has now passed since Hitler's suicide and the Soviet conquest of Berlin that most German scientists, indeed most Germans, cannot bear personal responsibility for the horrors of the Third Reich. It is probably no coincidence that recently German industrial firms and banks, scientific institutions and universities have begun to show more enthusiasm for dealing openly with their National Socialist past.

Comparative studies of science and ideology

During the Cold War, studies of science and ideology stressed how ideology was bad for science by focusing on, or at least emphasizing extreme examples: the Reign of Terror, Lysenkoism, medical experiments with concentration camp inmates, "Aryan science," the "German atomic bomb," and the "Oppenheimer Affair." Only Graham stands out as even asking whether ideology might have facilitated scientific research. The overwhelming majority of these authors did not consider the possibility that science was helped, or even not really harmed by ideology.

But the question remains: what difference did National Socialism, or Soviet communism, or any of these other ideologies make for science? Of course, logically this question cannot be answered by studies of science affected by any one ideology. One way to answer this question is to do what Gillispie and Joravsky contemplated, and what Jungk, Graham, and Haberer did: use comparisons of science under different ideological regimes to determine significant differences and similarities and thereby draw conclusions about the relationship between science and ideology.

Unfortunately, the criminal legacy of National Socialism and the polarizing effect of the Cold War have discredited comparative history. This can perhaps be best and most succinctly illustrated by the German historian Ernst Nolte's disingenuous use of the argument, "there are no forbidden questions," and his comparison of Cambodian, Soviet, and German genocide in order to whitewash the Holocaust.[47] Thus it is understandable why comparative history has sometimes been greeted with skepticism, if not hostility. However, despite Nolte, it is striking that the historiography of science and ideology in the first three decades since the Second World War is not dominated by comparisons between science in the Third Reich and the Soviet Union, as one might have expected, but instead between the French Reign of Terror and Stalin's regime on one hand, and National Socialism and McCarthyism on the other.

"Comparison" need not be a dirty word. It is not the act of comparison that is problematic, rather how it is done and subsequently used. Indeed, strictly speaking, one can only assert that two things are incomparable after a comparison has been attempted. Only by means of comparisons with science under different political regimes could one determine what difference National Socialism or any other ideology made, or to put it another way, to determine what might or might not have happened anyway, even if Hitler had never been appointed Reich Chancellor in 1933, or if Stalin had failed to succeed Lenin, or if McCarthy had never become powerful.

This book takes up the challenge of comparative history in order to break new ground. Rather than having one essay on science under Hitler, another on science under Stalin, etc., the usual format that leaves it up to the reader to make the comparison, each essay in this book includes a comparison of science influenced by different ideologies. Furthermore, in order to provide the daunting range of competencies such comparisons require, each of these essays is co-authored by two or more historians. The result is a first step towards a fuller understanding of the interaction between science and ideology. This book does not provide the definitive answer to the question, does ideology matter for science? However, it does confront this question head on.

Yakov Rabkin's and Elena Mirskaya's essay uses the term "totalitarianism" to investigate science under ideological pressure, thereby arguing that this concept has not yet outlived its usefulness. Focusing on science in National Socialist Germany, Fascist Italy, and the Soviet Union, the authors reject the claim that science and freedom are necessarily linked, instead noting historically that science has been made to support any social or political claim. Rather than science protecting humanity, Rabkin and Mirskaya believe that moral, ethical and religious convictions are needed.

The second essay in this collection introduces the concept of "ideologically-correct-science" (ICS) to examine science and scientists during periods of political, social, and ideological upheaval. The four authors, Michael Gordin, Walter Grunden, Mark Walker, and Zuoyue Wang, collectively compare and contrast science during the French Revolution, the Soviet Union, National Socialism, Second World War Japan, the early Cold War in the United States, and Communist China under Mao. There are clearly differences in the relationship between science and the state in these cases, but no one ideology of political system has shown itself to be either clearly superior at harnessing science, or incapable of it. War and the preparation for it have a great effect on these relationships. Finally, in general the reaction by scientists to these pressures was to accommodate themselves to the regime, not to resist or try to transform it.

Dave Mindell, Jérôme Segal, and Slava Gerovitch compare the post-Second World War development of information theory and cybernetics in the United States, France, and Soviet Union, thereby showing how these highly technical theories acquired ideological baggage and what scientists in these countries did about it. Information theory and cybernetics were first developed in the United States, initially were greeted with hostility in the Soviet Union, and were taken up by scientists in France at least in part, precisely because of their apparent ideological implications. The authors conclude that the main difference between American, Soviet, and French cybernetics lies not in mathematical applications or models, but instead in the political and cultural meanings attached to cybernetic ideas.

The fourth chapter compares postwar science policy in West Germany and Japan, the two losers in the Second World War. Richard Beyler and Morris Low demonstrate how, in both countries, debates over science policy played an important role in both economics and ideology. Discussions of the political significance of science, and the social role of scientists, served as a forum for alternative visions for ideological reconstruction. The basic American occupation strategy consisted of democratization, stabilization, and economic reconstruction, but during the Cold War the occupation authorities also became concerned about the economic and political deployment of science. On one hand, science could be used to stabilize and internationally integrate the occupied countries, but on the other hand, the occupation powers were also fearful of the ideological uses that science might be put to by suspect intellectuals. As the authors describe in their essay, the subsequent relationship between scientific expertise and public policy was sometimes turbulent.

Paul Josephson and Thomas Zeller compare and contrast the role played by science and engineering in the transformation of the environment under Hitler in National Socialist Germany and under Stalin in the Soviet Union. Taking advantage of their closed political systems, the National Socialists and Soviets used coercion, arrest, and execution in their attempts to reconstruct their countries' economic, social, and cultural systems into unbeatable world powers. In both systems scientists and engineers were involved in this political and ideological work, in the Nazi case in the elaboration of a notion of *völkisch* nature, and in the Soviet case a kind of "proletarian" nature. Nazi and Soviet scientists and engineers accepted that economic and political considerations might overrule their environmental concerns, complied with state programs, and accepted the research funding offered them. Nature itself remained an adversarial object in both countries.

In the sixth chapter, Burghard Ciesla and Helmuth Trischler follow rocket and aeronautic research in Germany from the Third Reich to the postwar United States. After working for the National Socialist state during rearmament and the Second World War, these scientists and engineers were brought to the United States to continue their research during the Cold War. The German "specialists" not only brought concrete scientific and technological knowledge; they also had experience with the organization of research in complex fields like aeronautics and rocket technology. Thus Big Science was transferred as an organizational model of science from the Third Reich to the U.S.A. Despite the differences in the political and social systems in Germany and America, these experts legitimated themselves by their military-relevant research results and technical innovations.

The history of science at the University of Jena from the Weimar Republic to the postwar East German state provides a case study for the contrast between institutional continuity and ideological upheaval. Uwe Hoßfeld, Jürgen John, and Rüdiger Stutz note the tensions and contradictions inherent in this study: on one hand the university was often described as a "laboratory of modernity"; on the other hand, it also supported the infrastructure of terror in the nearby infamous Buchenwald concentration camp. The university also experienced radical nationalism during and after the First World War, including a virulent anti-Semitism. During the Third Reich, it was considered a "model" National Socialist university, and under the postwar Communist regime it transformed itself to serve the Socialist state.

The final essay in this collection follows the career of the talented physicist Friedrich Möglich from the last years of the Weimar Republic, through the Third Reich and Second World War, to the communist German Democratic Republic. In contrast to the well-known "Aryan physicists" like Johannes Stark, Möglich was a different type of "Nazi physicist": he supported and contributed to modern physics while at the same time taking an active role in the National Socialist political movement. However, Möglich fell from grace during the Third Reich when he ran afoul of the Nuremberg Race Laws, in time coming to see the National Socialists as his enemy. When the Second World War was over, Möglich successfully made the transition to serving communism and the new East German state.

Collectively, these essays shed new light on the interaction of science and ideology, and demonstrate the power of the comparative method in history.

Notes

1 Department of History, Union College; the author would like to thank Yakov Rabkin and Jessica Wang for their helpful comments on an earlier draft of this introduction.
2 Nikolai L. Krementsov, *Stalinist Science* (Princeton, NJ: Princeton University Press, 1997).
3 For example, see Alexei Kojevnikov, "Rituals of Stalinist Culture at Work: Science and the Games of Intraparty Democracy circa 1948," *Russian Review*, 57, 1 (1998): 25–52.
4 James McClellan, "En attendant Charles Gillispie," in *Sciences et techniques autour de la Revolution française* (Paris: Société des études robespierristes, 2000), 219–224.
5 Kristie Macrakis and Dieter Hoffmann (eds), *Science Under Socialism: East Germany in Comparative Perspective* (Cambridge, MA: Harvard University Press, 1999).
6 See *Scienza e razza nell'Italia fascista* (Bologna: El Mulino, 1998) and *Le interdizione del Duce* (Torino: Albert Meynier, 1988).
7 Jessica Wang, "Merton's Shadow: Perspectives on Science and Democracy since 1940," *Historical Studies in the Physical and Biological Sciences*, 30, 1 (1999): 279–306, here 279.
8 Robert Merton, *The Sociology of Science: Theoretical and Empirical Investigations* (Chicago: University of Chicago Press, 1973), 258; also see Wang, "Merton's Shadow."
9 Merton, 271, footnote 6.
10 J.D. Bernal, *The Social Function of Science* (New York: Macmillan, 1939), 213–19.
11 Bernal, *The Social Function of Science*, 237, endnote 26.
12 J.D. Bernal, *The Freedom of Necessity* (London: Routledge & Kegan Paul, 1949), 59.
13 Theodosius Dobzhansky, "Russian Genetics," in Ruth Christman (ed.), *Soviet Science* (Washington, DC: AAAS, 1952), 1–7, here 1, 7.

14 See their respective contributions to Marshall Clagett (ed.), *Critical Problems in the History of Science* (Madison, WI: University of Wisconsin Press, 1959), for example.

15 Charles Coulston Gillispie, "The *Encyclopédie* and the Jacobin Philosophy of Science: A Study in Ideas and Consequences," in Clagett, 255–289, here 279.

16 Gillispie, "The *Encyclopédie* and the Jacobin Philosophy of Science," 281.

17 David Joravsky, *The Lysenko Affair* 2nd edn (Chicago: University of Chicago Press, 1986).

18 David Joravsky, *Soviet Marxism and Natural Science 1917–1932* (New York: Columbia University Press, 1961), ix.

19 Joravsky, *Soviet Marxism and Natural Science*, 248.

20 Loren Graham, *The Soviet Academy of Sciences and the Communist Party, 1927–1932* (Princeton, NJ: Princeton University Press, 1967), 190.

21 Graham, *The Soviet Academy of Sciences*, 191.

22 Joravsky, *Soviet Marxism and Natural Science*, 312.

23 Loren Graham, *Science and Philosophy in the Soviet Union* (New York: Knopf, 1972), 8, 19–20.

24 Graham, *The Soviet Academy of Sciences*, 208.

25 Graham, *Science and Philosophy in the Soviet Union*, 6.

26 See Loren R. Graham, *Science in Russia and the Soviet Union: A Short History* (Cambridge: Cambridge University Press, 1993), chapter 5.

27 See Joseph Haberer, *Politics and the Community of Science* (New York: Van Nostrand Reinhold, 1969) and Jessica Wang, *American Science in an Age of Anxiety: Scientists, Anti-Communism, and the Cold War* (Chapel Hill, NC: University of North Carolina Press, 1998).

28 Loren Graham, "Science and Values: The Eugenics Movement in Germany and Russia in the 1920's," *American Historical Review*, 82, 5 (1977): 1133–64.

29 Paul Forman, "Weimar Culture, Causality, and Quantum Theory, 1918–1927: Adaptation by German Physicists and Mathematicians to a Hostile Intellectual Environment," *Historical Studies in the Physical Sciences*, 3 (1971): 1–115.

30 Max Weinreich, *Hitler's Professors: The Part of Scholarship in Germany's Crimes against the Jewish People* 2nd edn (New Haven, CT: Yale University Press, 1999).

31 Alexander Mitscherlich and Fred Mielke, *Doctors of Infamy: The Story of the Nazi Medical Crimes* (New York: Henry Schuman, 1949), 152.

32 For Goudsmit, see Mark Walker, *German National Socialism and the Quest for Nuclear Power, 1939–1949* (Cambridge: Cambridge University Press, 1989), 204–21.

33 Samuel Goudsmit, *Alsos* 2nd edn (Los Angeles: Tomash, 1983), xxvii–xxviii.

34 Goudsmit, 244–5.

35 See Walker, *German National Socialism and the Quest for Nuclear Power*, 204–21.

36 Robert Jungk, *Brighter than a Thousand Suns* (New York: Harcourt, Brace, and Co., 1958).

37 Jungk, *Brighter than a Thousand Suns*, 105.

38 Mark Walker, *Nazi Science* (New York: Plenum, 1995), 243–68.

39 David Irving, *The German Atomic Bomb* 2nd edn (New York: DaCapo, 1983), 297–303.

40 Haberer, 181.

41 Haberer, 259–60.

42 Haberer, 292–3.

43 Karl-Heinz Ludwig, *Technik und Ingenieure im Dritten Reich* (Düsseldorf: Droste, 1979), 515.

44 Alan Beyerchen, *Scientists under Hitler* (New Haven, CT: Yale University Press, 1977).

45 Beyerchen, 199.

46 Herbert Mehrtens, "Das 'Dritte Reich' in der Naturwissenschaftsgeschichte. Literaturbericht und Problemskizze," in Herbert Mehrtens and Steffen Richter (eds), *Naturwissenschaft, Technik und NS-Ideologie* (Frankfurt am Main: Suhrkamp Taschenbuch, 1980), 15–87, here 69.
47 See "'*Historikerstreit*": *Die Dokumentation der Kontroverse um die Einzigartigkeit der nationalsozialistischen Judenvernichtung* (Munich: Piper, 1987).

2

SCIENCE AND TOTALITARIANISM

Lessons for the twenty-first century

Yakov M. Rabkin and Elena Z. Mirskaya

Introduction

"Science made of us not only singers but it made us into citizens who perform on stage. What we do in music is caused by our being scientists," remarked a duet of Russian musicians, former scientists, performing in Montreal in the late 1990s.[1] Such regard for science as the all encompassing formative influence on values, art and morality is not unusual in totalitarian societies that marked the history of this century. Science and scientism offer an important, albeit not exclusive, angle to understand totalitarianism which has been explained in artistic, not only scientistic, terms.[2] Literature offers another explanatory approach according to which totalitarian societies are organized along the lines similar to those used by fiction writers. The formal aspect of art, the total organization of material in a literary work, its reduction to a simplified, stylized form offer a model for the totalitarian organization of social and political life.[3] The content, the result remain secondary to the imperatives of a rigidly organized process and ritual which were, indeed, remarkably similar in various totalitarian societies. The spectacle of totalitarianism replaced reality and common sense.

Science and totalitarianism may be two of the most characteristic traits of the twentieth century. Science, in the comprehensive sense that encompasses most organized knowledge, has attracted much scholarly attention, particularly in the aftermath of Auschwitz and Hiroshima. There have been many images of science in this century: a noble, selfless activity in search of new knowledge and a pathological obsession with controlling the world; a means of improving the material quality of life and a threat to the very physical survival of the planet; a relentless progress on the path to greater enlightenment and a politically manipulated attempt to enslave humankind; selfless pursuit of objective truth and routine manufacturing of socially constructed scientific facts; a new way of understanding God's work and an implacable antithesis to religion; an activity grounded in personal freedom and a basis for negating humans their free will.

In this chapter, "totalitarian" refers to societies that lived through attempts at "total control of human behavior" in the course of the twentieth century.[4] The term "totalitarianism" is controversial, and has been often used polemically. Initially used with pride by Mussolini and his circle to describe Fascist Italy, the term quickly acquired a pejorative and polemical connotation outside the totalitarian realm. Germany, Italy, China, the Soviet Union, their dependencies and occupied territories, are examples of cruel and oppressive regimes that exhibited many totalitarian traits. These regimes attempted to establish total control of society and total mobilization of its citizens. Human and technological imperfection prevented such total controls from becoming a reality, and certain aspects of life escaped state intrusion. However, the ideal was totalitarian, and it is this ideal that distinguishes these regimes from other oppressive regimes (Franco's Spain or Pinochet's Chile) often referred to as authoritarian. As is shown elsewhere in this book, this distinction becomes even more pronounced when totalitarian experiences are compared with periods of political tension (such as the late eighteenth century in France and the early 1950s in the United States).

An essential similarity among the three political revolutions (in Italy, Germany and the USSR) was noticed quite early; possibly one of the first to do so was Michael Florinsky, a Russian historian.[5] Mussolini became aware of a gradual convergence between Soviet socialism ard fascism as early as 1933, just after Hitler came to power and a few years after the Great Break which solidified Stalin's grip on power in the course of forced collectivization and industrialization.[6] From a communist perspective, Leon Trotsky, by then exiled by Stalin from the USSR, also deemed Stalinism and fascism to be "deadly similar."[7] While these observations are certainly polemical, their authors cannot be denied intimate knowledge of the phenomenon under discussion.

However, many deem it sacrilegious to compare German Nazism and Soviet socialism.[8] In this respect, it is quite instructive to review *The Social Function of Science* by J.D. Bernal, a pioneer of social studies of science. He is one of the first to discuss "Science and Fascism" and "Science and Socialism" in the same chapter as early as 1939.[9] Bernal decries links between ideological uses of science in Nazi Germany: "The ideas of blood and soil have no scientific basis; it is therefore necessary that science should be distorted to give them one". Yet, he finds it natural to legitimate national policies with scientific arguments, and writes in a rather different vein about the "Marxists [who] have always conceived of a society thoroughly permeated by science, one in which science becomes the corner-stone of education and culture."

"The first duty of the scientist ceases to be the discovery of truth or the well-being of the mankind and becomes the service of his nation in peace and war," writes Bernal about science in Germany. He is appalled by the German scientists' "combination of benevolence and docility, especially in relation to the State". Yet, these same faults become virtues when Bernal turns to Soviet science. The link with industry, which he finds detrimental to science in Italy and Germany, becomes praiseworthy when Soviet "industry serves to present science with new and original problems." He categorically refutes criticism of Soviet science: "accusations of the detractors of the Soviet Union – that Marxism is dogma imposed on and distorting the findings of science – is [sic] palpably absurd, as anyone who bothers to look into the writings of Marx, Engels, or Lenin can see in a moment."

Relations between science and totalitarianism were often perceived in diametrically opposing manners. Some regard science and scientists as innocent victims of totalitarian oppression.[10] Others see science as an essential ingredient of totalitarian ideologies, and scientists as "willing executioners" at the employ of totalitarian tyrants.[11] Most accounts consider "normal" science devoid of ideology, and therefore view interactions of dominant totalitarian ideologies with scientific knowledge as sad perversions of science.

The earliest work concerned with the interface between science and totalitarianism is, perhaps, Zamiatin's *We*, a dystopia rather than an historical account.[12] His perspicacity was remarkable: he boldly outlined an entirely mobilized Stalinist society based on brutal uses of science and technology, a society that would begin to emerge only a decade later. An engineer by training, he foresaw the potential of science and technology becoming not only instruments of power but also the means of its legitimization. While fiction writers were impressed – and depressed – by the prospect of a robotic, rational society, early scholarly interpretations of totalitarian regimes emphasized their irrational and impulsive nature. They were often portrayed as the antithesis of progress towards a modern, liberal, rational society. In this perspective, science was seen as a victim of totalitarianism, bound by its ideological constraints, and forced to perform tasks that were essentially alien to the ethos and nature of science. These critics of totalitarianism, just as most internal critics of these regimes, had embraced rationalism and reproached totalitarian leaders for their ideological blindness. They bemoaned insufficient, rather than excessive, reliance on science and scientific rationality.[13]

The recency and the intensity of the totalitarian experiences tend to make the subject emotionally and politically charged. For example, former inmates of the Gulag[14] and a son of the once omnipotent head of Stalin's political police[15] present conflicting first-hand testimonies of their respective experiences with science and totalitarianism. These sources offer useful empirical material, but also complicate the task of the historians in their attempt to present a balanced, trustworthy view of science in totalitarian states, to understand the protagonists' behavior, to evaluate their objectives and to attempt to reconstruct their motivations.

The authors of this chapter share a moral revulsion of totalitarianism, consider human life to be a supreme value, and abhor the use of humans as tools in grand social designs. However, it is also a matter of historical record that Italian trains came to run on time under Fascism, that an economically backward country could launch a Sputnik under communism, and that inventive leaps in rocket design took place under National Socialism. In the same vein, one may recall that many widely admired historical artifacts – from the awe-inspiring pyramids of Egypt to the breath-taking vistas of imperial St Petersburg – were produced in unsavory conditions of exploitation and cruelty. The purpose of this chapter is to examine interactions between science and totalitarianism rather than pass a moral judgement which, it should be recalled, has nothing to do with achievements in science and technology. Meritocratic criteria are highly praised in professional advancement but are questionable in moral assessments.

Moreover, a moral judgement may be passed only on individuals, not on concepts such as science, totalitarianism, nation or state. Humans, rather than activities they are involved in, make choices and commit acts that carry moral responsibility, whatever the moral framework applied. Indeed, it is important to see science as a human activity rather than an autonomous living organism. Similarly, technology is often presented as a force independent of human control. One of the earliest references is the Golem, a robot reportedly developed by Rabbi Lowe in Prague in the seventeenth century. Technologies and technological artifacts that impose their own control on humans is a common theme in science fiction. The idea of self-propelling technology which absolves humans of personal responsibility has also been quite common.[16]

All too often, science has been portrayed as an entity endowed with its own objectives, preferences, achievements and moral values. Scientists and engineers, generals and heads of state have objectives, preferences, achievements and moral values. It is their behavior and interactions, their ambitions and frustrations, rather than ideological constructs such as "the good of science," that constitute the focus of this chapter. Rejecting animism of scientific activities and institutions ought not imply oblivion of science as a social institution.

Science and scientism in totalitarian ideologies

Clear links exist between the belief in the scientific method and the familiar emphasis on the "one truth," the "one really true doctrine," and the "one true way" one finds in both real totalitarian societies and in totalitarian utopias. Scientific transformation of society, a nineteenth-century belief carried over and reinforced into this century, tempted political leaders, scientists and rank-and-file citizens alike.

The ensuing enthusiasm had dramatic consequences, and was manifest, particularly in the Soviet Union, in the widely spread belief that scientific and technological progress, part of the overall sense of mobilization prevalent in such societies, legitimated human sacrifices. For example, the Soviet film *Nine Days of a Year*[17] portrays nuclear scientists routinely disregarding mortal dangers in order to arrive at scientific results. This technological romanticism was widely spread as an ideal in the Soviet Union and other totalitarian societies: "If only the mankind consisted of Gusevs," remarks a protagonist in the film with respect to Gusev, a physicist who, during an experiment exposes himself to a lethal dose of radiation. This brings to mind a biblical commentary on the destruction of the Tower of Babel. God destroyed it because he was enraged that the builders regretted more the fall of a brick from the top floor than the fall of a man.

While Stalin's forceful policies of industrialization were predicated on scientistic optimism,[18] Hitler's vision of Germany contained a mystical, romantic streak which glorified the relationship between Aryans and Nature.[19] Ideologically, science was elevated to a higher status in the Soviet Union, goaded by its leaders to modernization, than in Nazi Germany where the ideal was in fact antimodernist and bucolic. However, it is in Germany, not in the USSR, that scientific justification was consistently articulated for the formulation and execution of policies with severe consequences for mankind.

While the dominant image of science under totalitarianism is that of a hapless victim, scientific, or rather scientistic, roots of totalitarian ideologies have also been identified. Natural, medical, and exact sciences, on the one hand, and social sciences, on the other, played different roles and, most certainly, experienced different fates in totalitarian societies. While totalitarian regimes tend to instrumentalize social sciences for ideological persuasion, they are particularly censured as an alternative source of knowledge about society, a clear threat to the monopoly in the hands of totalitarian leaders. The disruptive potential of social sciences was well recognized by totalitarian leaders. They were equally wary of religion, particularly of the less centralized and diffuse denominations that offered an alternative moral framework and, on the other hand, remained harder to control than the more established churches. Whenever possible, both social sciences as well as organized religions were instrumentalized but both also remained suspicious in the eyes of totalitarian leaders.

Natural sciences became strategically more important as industrialized states came to depend on scientists employed in key civilian and military industries. Consequently, even totalitarian states allowed scientists to acknowledge a more transnational character of natural sciences and to exhibit a greater identification with scientific internationalism than would be allowed in social sciences. There was another important reason for this tolerance of scientific internationalism, namely, the use of scientific researchers for technological and political espionage. Examples abound, particularly in the case of the Soviet Union, of scientists used as intelligence analysts as well as conduits of sensitive information from other countries. Even among scientists who would later fall victims of Stalinism, there were quite a few who are known to have collaborated with Soviet intelligence, an activity which very few indeed could then eschew with impunity.

Totalitarian leaders understood science in its deterministic, pre-Heisenberg sense, an understanding which was common in the nineteenth and much of the twentieth century. This image of science continues to be widespread, particularly in circles removed from the practice of science. In the context of rapid secularization that swept many European societies in the last two centuries, it is this deterministic image of science that has inspired millions as a moral force, as an alternative and a substitute to traditional systems of morality. For example, the theory of scientific communism engendered a new teleological morale based on class interests as interpreted by the Communist Party; it was to replace the old "bourgeois" morale inherited from the past. Scientific communism and racial science were ideologically incarnated in, respectively, *partiynost'* (the Party Principle) and *Führerprinzip* (the Leadership Principle), each hailed as a higher form of human morality. It is instructive to reflect on the choice of the title for the most important communist newspaper in Russia, *Pravda*, or Truth. The new doctrine could be disputed no more than laws of nature which appeared immutable to the totalitarian doctrinaires.

In the course of the twentieth century, several societies were formed or shaped in a forceful manner, as "a triumph of the will" on the part of their leaders. They overtly requested sacrifices from their citizens in order to bring about a particular ideological vision. Unlike builders of empires in antiquity, the totalitarian regimes, whether in Russia, Italy, Germany, China, or Cambodia, all aimed at an organized and radical transformation of societies and human behavior. The leaders of these countries paternalistically decided what was better for the "people," and, in the name of a radically better future, shaped society in accordance with their designs. Since "the people" often lacked enthusiasm for the new society, the leaders declared their avowed goal to create "a new man." The Nazis used eugenics, practiced mass murder and encouraged human breeding by the SS. The Soviets, the Chinese, and the Cambodians put emphasis on "re-education" that, paradoxically, led to an even wider scale of human destruction than the German policies predicated on genetic determination of human behavior. In spite of differences among them, all these regimes legitimated atrocities by references to science.

Science used to be neglected as a building block of totalitarian ideologies.[20] Yet, science was clearly a constitutive, i.e. more than merely instrumental, element of totalitarian forms of political oppression. While different degrees of rejection of modernity characterized all totalitarian ideologies, they were, at the same time, permeated by scientific discourse. Their common trait is reductionism, i.e. the reduction of complex political and societal issues to a limited number of variables that are then presented as part of a scientific problem. Social and political tensions were reduced to a "problem" for which solutions, preferably final, were enthusiastically sought. The very use of the word "problem" presupposes that a solution exists and should be found. Scientific language was adapted to totalitarian uses, thereby conferring the air of objectivity on value judgements of most serious nature.

Whereas Mussolini and his ideologue philosopher, Giovanni Gentile, denied science any prescriptive or normative value, both National, Soviet and Chinese socialists accepted science, or rather, certain kinds of science, as prescriptive, often coaching legitimation of policies in scientific terms. For example, Marx and the Marxists attributed to economic theories the qualities of laws, "imma-nent laws" that beget specific consequences "with the inexorability of a law of nature." Marx and Engels claimed that they could predict social change with the certainty of "a mathematical or mechanical demonstration."[21]

The transition from descriptive to normative functions of science reflects a reductive definition of individuality and freedom on the basis of sociology, anthropology or genetics. In this context, a substantive identity of private and public interest is a matter of definition. This definition can be found in a broad range of contemporary societies, but it becomes the charter myth only in consciously revolutionary societies whose leaders, to use Marx's words, are not content to explain the world but want to change it. For a society conceived as an equation, Marx postulated that the only independent variables are economic ones. They are by definition also collective variables. All political, social, intel-lectual, or moral diversity is reduced to the status of superstructure that is ultimately dependent on economics.

This belief in science, and in Marx the scientist, endured for decades, and overcame serious empirical inconsistencies with his "laws." It was a truly Messianic belief couched in scientific discourse. In practical terms, Lenin and Stalin cast "the proletariat" in the role of Messiah, Mussolini referred to "prole-tarian nations," while for Hitler the Aryans were genetically predetermined to redeem the imperfect world. Science as a redeeming force is present in all of these messianic strategies.

These uses of science are predicated on reductionism. Traditional scape-goating required no scientific proof: for example, infidels could be pointed out quite easily in many societies. But in such societies, the stigma was not permanent: there was always a way out in the form of conversion, genuine or feigned. In totalitarian systems the culprit was identified "objectively" with the help of *nauka* or *Wissenschaft*. Thus "inferior races" were ostensibly iden-tified by physical anthropology, "class enemies" by "rigorous Marxist analysis," and this "objective" definition made their status inescapable. The "kulak" or a "Jew" were not individuals who acted in a certain manner, and could therefore exercise their free will and mend their ways. These were defined as "objective realities" that no conversion, political or religious, was allowed to alter.

Soviet experience introduced the term "objective enemy" to identify someone who may have committed no crime against the Revolution but, according to the "scientific theory of Marxism" which, like all science, possessed a predictive value, was bound to commit it. Similarly, the term "Jew" was applied according to apparently objective criteria of biological provenance, not to the more arbitrary criteria such as behavior or belief. This constitutes an important departure from the more traditional Christian anti-Jewish persecu-tion which always left the door open to conversion, sincere or not, thereby offering a chance to survive. Modern biological anti-Semitism, a pillar of Nazi ideology, left no such way to save one's life. The *libre arbitre*, or freedom of choice as a condition of personal responsibility disappeared while traditional morality was being replaced with it totalitarian surrogates.

What makes the twentieth century unique is not so much the total number of civilian casualties of organized violence as the fact that this mass violence was brought about by "advanced" or cultured societies, usually with scientific legitimation, and occasionally with the help of science-based technologies. Technological and scientific terms were put to common use as writers became "engineers of human souls" and common people were initiated to "the theory of scientific communism" or "theories of race." The political and social goal of excluding Jews from German society was presented as "the Jewish problem," a term shared by those interested in the welfare of the Aryan nation as well as those intent on forging "a new Hebrew nation" in Palestine. A problem must have a solution, and this is exactly how the National Socialists called their extermination policy: "the final solution." The use of medical terms was also common in Nazi Germany. An individual patient was replaced by the collective concept of "the Nation" whose "health" required surgical intervention such as amputation applied, similarly, to scientifically determined groups of people.

Ideological uses of science in totalitarianism can be better understood against the background of the rapid demise of traditional organized religions. Science as an antithesis to religion was among the most common beliefs in the last century. The twentieth century saw science become an accepted cultural form and invade the place of organized religions in many societies. Science spectacularly expanded the realm of human cognition and, concomitantly, reinforced the illusion of human omnipotence.

This belief in human omnipotence, common to both scientistic and totalitarian ideologies, was able to affect the twentieth century precisely because of the formal secularization of most European societies. The goals of "elimination of the bourgeoisie," of "liquidation of the *kulaks*," of "the final solution of the Jewish question," were all predicated on omnipotence, on the human capacity to change society just as advanced technological means appeared to change, or at least subdue, nature. Dehumanizing scientific and medical metaphors (leech, vermin, cancer) were used in the context of cleansing that would rid the unjustly disdained races (Aryans) or classes (workers and peasants) of such impurities, and thereby elevate and redeem them to the well deserved state of superiority and dominance. These grand programs of social transformation were devised by intellectuals who shared a profound respect for science and learning. In the face of empirical discrepancies with their programs, they were reluctant to admit errors in their ideological programs. Rather, they insisted on a harsher, more consistent application of their radical policies.

In order to understand the specificity of science under totalitarianism, science also ought to be compared with other creative pursuits such as art, literature and, above all, technology. Scientists played a role different from that of the engineers, but in many respects their social status was comparable to that of creative artists and architects. Indeed, the very definition of our theme, science and totalitarianism, may be perceived as a reflection of the old belief that genius and evil are incompatible, a reflection of the antinomy between a once pristine image of science and a uniformly sinister image of totalitarian realities.

It is thus quite telling that corruption in science represents a novelty, nay incongruity, while corruption in politics is perceived as common and unavoidable. Scientists, artists, poets were cast in the category that used to be occupied by clergy: they were expected to embody morality, to become a source of new moral values. Frequent use of metaphors such as "Temple of Reason" (brought about by the French Revolution), "temple of science" or "temple of arts," ritualized language and behavior common to these vocations, were all quite significant in this regard. The same was usually not true for technology and engineering, which, while also admittedly powerful, were not endowed with moral superiority in secularized imagination.

Science was not only expected to improve the moral character of its practitioners; it was also cast in the role of a deity whose main attributes were universality, uniqueness, and jealousy. The goal of science was, with the help of one method which was to be applied universally, to reveal the truth which was believed to be one and the same for all, across the board in all scientific disciplines. The scientific method was deemed unique and superior to other means of human cognition. Finally, science was often perceived as a jealous deity that would tolerate no non-scientific approaches.

Engineering metaphors and, more generally, technocratic thinking were also prominent in totalitarian societies. But it was scientism, in the guise of science, that gave most totalitarian regimes a seemingly rational justification to embark on a radical transformation of man and society. Science-based sociopolitical engineering may not have caused all the atrocities in the twentieth century; it did, however, serve to legitimate most of them. Borrowing the language and norms of science sometimes reached the degree of *imitatio dei*, i.e. the attitudes common in scientific research came to be considered worthy of imitation in political and social life. Thus, emotional detachment from the object of research came to be extended, for example, to the operation of state bureaucracies which cultivated similar detachment from the human object of policies. This trend culminated in totalitarian societies in what later came to be known as "crimes against humanity" apparently perpetrated, and this is these crimes' distinctive trait, with no more emotional involvement than in vivisection of a frog. References to physical anthropology and racial theory neutralized the moral sensitivity of thousands of educated German and other nationals who voluntarily participated in unprecedented genocide.

When ideologies of scientism and totalitarianism offer a temptation of total control, of human omnipotence, they also provide a quasi-religious solace of participation in gigantic and therefore collective projects to a populace suffering from "atomization, impotence and despair."[22] Science was among the most important activities which epitomized the contradictions characteristic of totalitarian cultures. These contradictions could be seen between the technical need to maintain foreign contacts, essential for the conduct of research, with the political need to isolate society from political "contamination"; between the need to develop critical thinking among scientists and the need to prevent this critical thinking from being applied to social and political domains. This dilemma did not have time to mature in Nazi Germany (and was hardly an issue in Fascist Italy), but the Soviet Union, and to a lesser degree, China had to face such political dissidence emerging among the scientific elites. Some dissidents resorted to scientistic arguments to oppose the regime. Thus scientism was used both in the formulation of totalitarian ideologies and in attempts to oppose them.

An interesting debate about science and totalitarianism took place in Germany after the defeat of Nazism. It was argued that human freedom should enable man, albeit of a different nature, to make technological choices without fear of an autonomous demonic power of technology. However, it was science that, once again, was used to support a moral argument: "A crucial feature of modern science is that it does not provide a total world view, because it recognizes that this is impossible. It was science that liberated us from total view of the world, for the first time in history. All previous epochs (and even our own at the level of the average man) have clung to general conceptions of this kind. Because it takes seriously the principles of cogent, universal, and systematic knowledge, science is always aware of its limitations, understands the particularity of its insights, and knows that it nowhere explores Being, but only objects in the world."[23]

Totalitarian ideologies in science

Restrictions on scientific research are known to have been imposed in a variety of political regimes. Restrictions practiced in totalitarian societies may only stand out because of their overt links with an explicitly ambitious political ideology. The magnitude of ideological censorship does not seem to correlate with the intensity of overall political oppression. In fact, one of the more intensive cases of such censorship of science was the opprobrium heaped on researchers interested in studies of gender and race differentials in intellectual achievement in the last three decades of the twentieth century.[24] The very attempt to investigate these issues was opposed by liberal circles in North America: they were concerned about past misuse of such data for formulation of discriminatory policies directed at certain racial groups and at women. Research grants were revoked, scholarly articles were refused, and pressure to fire recalcitrant researchers was brought to bear on university administrators. This censorship, occasionally supported by "progressive" local and federal government agencies, left traces on the scholarship in several fields of psychology. The nature versus nurture debate that underlies this issue was politically decided in favor of nurture, partly as a reaction to the exclusive emphasis on nature made in totalitarian societies a few decades earlier.

There is an interesting paradox between, on the one hand, the totalitarian regimes' use of science as an objective universal truth and as a pillar of the official ideology (such as the Soviet discipline of scientific communism or the Nazis' race sciences), and, on the other hand, their commitment to the idea of total control which should make no exception for science. The commitment to total control offered unusual opportunities to scientists willing to resort to appropriate discursive and political strategies in the advancement of their scientific ideas and careers. Such scientists, some of whom were responsible for totalitarian crimes and abuses, have been often portrayed as exceptions. Many scholars continue to see scientists as victims of totalitarianism. With the development of social studies of science, research on science in totalitarian societies has revealed a degree of symbiosis between scientists and political elites. In some instances, it has been shown that scientists used totalitarian leaders in their quest for power and influence in science.

The Lysenko affair, the story of repression of genetics in the Soviet Union, has been the most researched episode in the history of Soviet science.[25] Lysenko, and his ally philosopher Prezent, resorted to Marxist-Leninist rhetoric to declare genetics "a bourgeois pseudo-science." They succeeded in shutting down institutes, imprisoning prominent geneticists, and purging biology textbooks of references to genetics. Most histories of the Lysenko affair portray scientists as innocent victims of political intervention into science. One tends to forget that it takes a scientifically literate person to be able to effect such intervention. The recent evolution of scholarship on the Lysenko affair shows how scientists, rather than party functionaries, took the initiative and used communist rhetoric to expand their influence and eliminate competition. The picture of Lysenko using Stalin in finalizing the destruction of Soviet genetics in 1948[26] is quite different from the early representations of politics encroaching on science. Moreover, Nikolai Vavilov, the most prominent victim of Lysenkoism, had also used direct access to Stalin to seek the dictator's favor in his belated defense against Lysenko's imperial designs.[27] Without insinuating moral equivalence between Vavilov and Lysenko, it remains a fact that both contestants had sought the dictator's support even though only one, Lysenko, actually received it. Once Stalin overtly supported Lysenko, there was no further recourse left, political or scientific.

Scientists, just like other people, settled accounts according to the possibilities of their times. The official domination of one correct ideology was one of such possibilities. Under Stalin, proving to the authorities that one's own kind of science was ideologically correct was the surest way to defeat an adversary and build an empire. It is in this manner that Moscow University physicists attempted to torpedo their Leningrad colleagues, who, in turn, invoked the strategic importance of physics to deflect the attack.[28] Quite a few intradisciplinary battles were waged with resort to the Communist Party, its functionaries and its ideology by scientists who were not too scrupulous about means to be used.[29] Rather than cases of extraneous intervention, many such instances were internal fights politicized and amplified in the context of a totalitarian regime.

It matters little whether these were opportunistic strategies of "protective coloration" or genuine attempts to develop an "ideologically correct science." Moreover, it is impossible to ascertain the motivations of the protagonists of these attempts to harmonize science with the official ideology. Resort to the politically acceptable discourse was common throughout all totalitarian societies, and scientists, along with other intellectuals, naturally found in their ranks quite a few who could master and manipulate the discourse to derive individual or institutional benefit. Besmirching of scientific opponents and competitors by affixing politically derogatory labels was part of this practice rooted in the totalitarian ambitions of these regimes. Such labeling would have been meaningless, and possibly counterproductive, in societies devoid of an overarching political discourse.

The political certification of the content of science was done in a haphazard and arbitrary manner. By necessity the protagonists were scientists, some of whom had become philosophers of science.[30] They denounced adversaries and appeared oblivious to the often cruel sequels of such denunciations. It is impossible to judge the sincerity of such action, but it appears that opportunism was not always the main reason behind censoring science. For example, the main proponents of "Aryan Physics" (*Deutsche Physik*), two Nobel laureates, had expressed their ideas well before the dawn of Hitler's "One-Thousand-Year Reich."[31] Attempts to build a nationally specific science predate the victory of totalitarianism in Russia and Germany. Some German scientists had long argued for building a spiritually meaningful, holistic science "freed" from the materialistic and mechanistic French influence.[32] Similarly, there were proponents of a nationally specific Russian science, this time "freed" from German influence as early as the nineteenth century.[33]

While unlike Stalin, Hitler is not known to have been drawn into scientists' squabbles, the SS certainly was. Werner Heisenberg, a world renowned physicist who chose to stay on in Nazi Germany, was attacked in the National Socialist newspaper *Völkischer Beobachter* and subsequently in the SS weekly *Das Schwarze Korps*. In the latter case, he was labeled a "white Jew," an admittedly dangerous label in Nazi Germany which designated someone who had been tainted by "Jewish ideas." In self-defense he responded in *Völkischer Beobachter*, and appealed, promptly and successfully, to the head of the SS to clear his name. Moreover, with the war looming on the horizon, he succeeded, with the help of high Nazi officials, in expanding his influence and consolidating the position of modern physics in Germany. Just as in the case of the Soviet Union, the totalitarian regime was mobilized by scientists for purposes of intradisciplinary warfare.

Proponents of "Aryan physics" encouraged open debate and reached the international scientific community via the respectable *Nature* magazine.[34] Lysenko had fewer illusions about the attractiveness of his ideas in the international scientific community, and the best he could hope for would be for his doctrine to be adopted by scientists in communist countries as well as communist scientists in Western countries. A semi-permeable wall surrounded Soviet biology. Anti-Lysenko views could not reach it,[35] while pro-Lysenko writings were translated and reprinted from sympathetic Western sources.[36] Some of the latter experienced a profound crisis of faith and ended up abandoning their allegiance to the communist doctrine. The former had little choice and voiced little open criticism.

The revolutions that ushered in totalitarian regimes did not have a uniform effect on science in different countries. In spite of the lengthier and more violent transition to Soviet rule, Russian science experienced fewer changes directed at scientists and research institutes than did Germany in 1933. One must remember that while the youthful revolutionary idealism still excited the Nazis in the 1930s, it had been by then largely substituted in the USSR with Stalin's great power nationalism. This idealism led to the systematic expulsion of Jews from Germany's civil service, and therefore from science, beginning in 1933. The Bolsheviks were never nearly as consistent in cleansing Soviet science of "vestiges of bourgeoisie and aristocracy," and devised ways to use their expertise for the benefit of the state. Protective coloration and other means of feigning political conformity were quite effective in the context of Soviet science, even in the worst years of Stalin's repressions. Another explanation of these differences may have to do with a relative dearth of scientists in Russia and the centrality of science for Bolshevik ideology. The effects of the collapse of the USSR on scientific institutions and research were far greater than the turmoil following the revolution of 1917. Similarly, whatever the extent of the changes that took place in German science under Hitler, they pale in comparison with massive firings and closures of research institutes that characterized the transition of German science from Nazism in 1945, and, *a fortiori*, from communism after 1989.[37]

Science and freedom

Freedom of scientific pursuit is a crucial aspect of scientific cultures. A mere comparison of titles of publications devoted to the history of science in totalitarian societies shows the unusual frequency of the word "under." It suggests that scientists found themselves subjected to a regime that oppressed them and from which they were somewhat alienated. This is one of the best known aspects of the history of science in totalitarian societies, the history of "science without freedom." It is quite clear that scientists were deprived of human rights and political freedoms. So were most other citizens of those societies. Whatever the virtues of international solidarity of scientists exhibited in this century, the freedom of scientists must be compared to that of their fellow citizens rather than to that of their colleagues in democratic countries.

Once one distinguishes between political and scientific freedoms,[38] a clearer picture emerges. As has already been discussed, scientists were sometimes ideologically restricted in the choice of certain research methods and themes. These restrictions were often initiated by other scientists who adroitly used the totalitarian power structures to concentrate control of entire disciplines in their hands. This concentration of power was easier to attain in totalitarian societies that were by definition quite centralized. However, in other fields of research, totalitarian regimes appear to offer greater research freedoms and better provide for science and the scientists than societies that preceded them. To take an extreme case, German medical researchers enjoyed unprecedented freedom of research on human objects in the concentration camps. Many medical researchers, even those who had misgivings about the scope and nature of repression they found there, agreed to work in the camps because of these unprecedented, virtually unlimited research opportunities. They could practice vivisection, conduct hypothermia and other lethal experiments on real humans, something that could only be done in the concentration camps.[39]

In a less dramatic vein, many Soviet researchers enjoyed greater freedom of research because their work was usually shielded from peer review and other forms of collegial input from abroad that was common prior to the "triumph of socialism." One may even argue that Lysenko and his allies benefited from international isolation and enjoyed unprecedented scientific freedoms within their discipline. It is another question whether Michurinist biology, racial theory or Aryan physics were "real" sciences. Once one uses the standards of contemporary science practiced elsewhere in the world, some of these practices might indeed be disqualified. Others that followed in the mainstream of science would be accepted as bona fide science even though it may have been practiced in morally repugnant conditions.

Totalitarian regimes were often quite generous in supporting science. Investment made under the direct influence of scientists, often could not have been made under the economic circumstances by a regime other than a totalitarian one. Gigantic projects were approved and financed, and the scientists who initiated them played a risky game. In the case of success, they would be elevated to the highest economic and social standard attainable in the Soviet Union or Nazi Germany. However, failure frequently signified imprisonment or execution in the "the world's first socialist country" and less lethal, but still unpleasant punishment in the Third Reich. Easier access to the highest echelons of power enabled some Soviet scientists to exercise greater influence on science policies of the government than would be possible under a democratic regime, as the history of post-Soviet science amply illustrates. They had built up an "empire of science" that found it hard to adjust to post-communism. In contrast, scientists in Nazi Germany, even the most devoted to the Nazi cause, did not enjoy the same access to, and did not enjoy comparable influence on, political leaders in matters of science policy. They could do so only in a narrow range of disciplines related to racial science.

The incarceration of scientists and the spread of research institutes operating behind prison bars characterized both Nazi Germany and the Soviet Union. Since the 1930s scientists and engineers were rounded up, threatened with regular concentration camps, and then, as a token of consideration, offered the privilege to continue their professional work in a research establishment behind iron bars. In the Soviet Union this system of organization of research took the name *sharaga*,[40] while the best description of life in the *sharaga* was produced by the Russian novelist Alexander Solzhenitsyn in his novel *The First Circle*. Important discoveries and innovations came from these institutes. Rocket science, for example, was largely developed by imprisoned researchers in both countries. These programs were predicated on the absence of personal freedom but offered reasonable, often superior conditions for research. In all of these circumstances scientists continued their work, provided they had access to scientific equipment and literature. Golden cages turned out to offer a propitious environment for scientific research. Whether scientists themselves or their research objects were actual inmates, there is an important lesson to be learnt from these experiences with respect to the relationship between science and freedom. Science can and has blossomed in politically oppressive milieux. While political freedom may be attractive to many scientists qua citizens, there seems to be little substance in the oft repeated belief that science and scientists require political freedom to conduct their work. The contrast between the largely weakened and pauperized post-Soviet science and the robust and lavishly supported science under Stalin, makes this point particularly poignant. Nor can be it demonstrated that science is inherently democratic or that it foments political freedom like American scientists were fond of arguing, particularly after 1945. It would appear that the association between science and political freedom was a temporary phenomenon in the intellectual history of the twentieth century, largely magnified in the context of confrontation between democratic and dictatorial regimes.

Conclusions

A comparative history of science in totalitarian societies offers insights into the nature of science and into its integration in totalitarian ideologies and practices. Neither science nor totalitarianism still enjoy the enthusiasm they did throughout most of this century. Symbiosis between the two is a page of history that may never be repeated. Yet the intellectual and moral questions they raised are likely to remain important for scholars, scientists and lay people well beyond the end of the last century.

Science, because of its enhanced prestige in the context of contemporary secularism, has been used to support very different ideas about freedom, morality and society. It was used as an important part of both Communist and National Socialist ideologies. This use of science helped numb the moral sensitivity of those who were put in charge of carrying out the grand designs aiming at "scientifically" transforming society and creating a new man to inhabit them. Moral detachment, common in scientific research, was successfully disseminated as a positive value in societies that were forcefully directed towards respective radiant futures.

31

"Proletarian science," "Aryan physics," and other varieties of the politically correct science formulated by scientists prior to the emergence of these strong regimes, flourished in conditions of totalitarianism. They were often used by scientists in competition for power and influence, sometimes drawing top leaders, such as Stalin and Himmler, into the fray. Those who defended the more transnational varieties of science and those who promoted the "appropriate" sciences, resorted to political authority to the same degree in order to win the cause. Arbitration by political authority was common in all spheres of life in totalitarian societies.

Many scientists have shown a capacity to adjust to totalitarian regimes and to draw benefits from a centralized pattern of organization privileged by these regimes. Science can develop in conditions of totalitarianism, and, in certain cases, totalitarianism accounts for investment in science which would otherwise have been impossible in these countries in conditions of democracy. The nightmarish vision of Alexander Hertzen, a Russian dissident of the nineteenth century, of "a Genghis Khan with a telegraph," has become a reality. *Technodélice*, the delight of research, brought scientists to relish opportunities to conduct experiments on humans, to engage in most barbaric experiments the history of science has known, to develop powerful weapons of mass destruction.

The history of science in totalitarian societies makes associations between science and freedom appear tenuous at best. Science cannot be credibly used as a shield against political oppression since science can be made to support any social or political claim. Moral, ethical, and religious convictions must once again come to the fore as bulwarks against human inhumanity.

Notes

1 "Riadom s nami," *Mesto vstrechi Monreal*, 25 (9 April 1999): 7.
2 See, for example, Isaiah Berlin, *The Sense of Reality* (New York: Farrar, Straus & Giroux, 1997), and a documentary film by Peter Cohen, *The Architecture of Doom* (Stockholm: POJ Film Produktion AB, 1990).
3 Boris Paramonov, "Kommunizm kak proizvedenie iskusstva," *Za Rubezhom* (Novosti Nedeli, Tel Aviv), 41 (9 December 1999): 14–15.
4 Abbott Gleason, *Totalitarianism* (Oxford: Oxford University Press, 1995).
5 Michael T. Florinsky, *Fascism and National Socialism* (New York: Macmillan, 1936).
6 M. Ardemagni, "Deviazione russe verso il fascismo," *Gerarchia*, XIV (July 1934): 571, cited in A. James Gregor, *The Ideology of Fascism* (New York: Free Press, 1969), 449.
7 Leon Trotsky, *The Revolution Betrayed* (Garden City, NY: (s.n.), 1937), 278.
8 See, for example, Walter Laqueur, "Is There Now, or Has There Ever Been, Such a Thing as Totalitarianism?," *Commentary*, 80, 4 (1985): 29–35.
9 J.D. Bernal, *The Social Function of Science* (Cambridge, MA: MIT Press, 1967), 210–31.
10 For example, Mark A. Popovsky, *Manipulated Science* (New York: Doubleday, 1979); Alan Beyerchen, *Scientists under Hitler* (New Haven, CT: Yale University Press, 1977).
11 Max Weinreich, *Hitler's Professors* (New York: YIVO, 1946; second edition Yale University Press, 1999); Benno Muller-Hill, *Murderous Science* (New York: Oxford University Press, 1988).

12 Evgenii Zamiatin, *We* (New York: Dutton, 1924).

13 For example, Zhores Medvedev, *The Medvedev Papers* (London: Macmillan, 1971), 170–1.

14 Varlam Shalamov, *Kolyma Tales* (New York: Norton, 1980); Alexander Solzhenitsyn, *The First Circle* (New York: Harper and Row, 1968).

15 Sergo Beriia, *Moi otets–Lavrentii Beriia* (Moskva: Sovremennik, 1994).

16 For example, Langdon Winner, *Autonomous Technology* (Cambridge, MA: MIT Press, 1977).

17 Mikhail Romm's film, *Deviat' dnei odnogo goda* (Moskva: Mosfilm, 1962).

18 Douglas Weiner, *Models of Nature: Conservation and Ecology in the Soviet Union 1917–1935* (Bloomington, IN: Indiana University Press, 1988).

19 Robert N Proctor, *Racial Hygiene: Medicine under the Nazis* (Cambridge, MA: Harvard University Press, 1988).

20 This is as true of pro-socialist scholars such as Eric Hobsbawm, *Age of Extremes: the Short Twentieth Century, 1914–1991* (London: Abacus, 1994), as of the more conservative theorists of totalitarianism such as Leonard Schapiro, *Totalitarianism* (New York: Praeger, 1972).

21 Karl Marx, *Capital* (Moscow: IML, 1950), 763; Friedrich Engels, *The Condition of the Working-Class in England in 1844* (London: Allen & Unwin, 1952), 18.

22 Evgenii Rashkovsky, "Opyt totalitarnoi modernizatsii Rossii (1917–1991) v svete sotsiologii razvitia," *Voprosy filosofii*, 141 (1992).

23 Karl Jaspers and Rudolf Bultmann, *Myth and Christianity: An Inquiry into the Possibility of Religion without Myth* (New York: Noonday Press, 1958), 6.

24 Michael Kauffman, *Feminism and Judaism* (Jerusalem: Heritage Press, 1995), 145–58.

25 Zhores Medvedev, *The Rise and Fall of T.D. Lysenko* (New York: Columbia University Press, 1969); David Joravsky, *The Lysenko Affair* (Cambridge, MA: Harvard University Press, 1970); Dominique Lecourt, *Proletarian Science? The Case of Lysenko* (Norfolk, VA: NLB, 1977); Mark Popovsky, *Delo akademika Vavilova* (Ann Arbor, MI: Hermitage, 1984); Valerii Soifer, *Lysenko and the Tragedy of Soviet Science* (New Brunswick, NJ: Rutgers University Press, 1994); Nikolai Krementsov, *Stalinist Science* (Princeton, NJ: Princeton University Press, 1997).

26 Kirill Rossianov, "Editing Nature: Joseph Stalin and the 'New' Soviet Biology," *Isis*, 84 (1993), 728–45.

27 Simon E. Shnol', *Geroi i zlodei rossiiskoi nauki* (Moskva: Kron Press, 1997), 77–102.

28 A.S. Sonin, *Fizicheskii idealizm. Istoriia odnoi ideologicheskoi kampanii* (Moskva: Fiziko-matematicheskaia literatura, 1994).

29 For examples, see M.G. Yaroshevsky (ed.), *Repressirovannaya nauka*, 2 vols, (Leningrad: Nauka, 1991 and 1994).

30 Yakov M. Rabkin, "On the Origins of Political Control over the Content of Science," *Canadian Slavonic Papers*, 21, 2 (1979): 225–37.

31 Beyerchen, *Scientists under Hitler*.

32 Jeffrey Herf, *Reactionary Modernism: Technology, Culture and Politics in Weimar and the Third Reich* (Cambridge: Cambridge University Press, 1984); Paul Forman, "Scientific Internationalism and the Weimar Physicists," *Isis*, 64 (1973): 151–80.

33 Alexander Vucinich, *Science in Russian Culture*, vol. 2, (Stanford: Stanford University Press, 1970), 77, 98–9, 200–02; see also Murad Akhundov, "Soviet Science and the Pressure of Ideology," *The Soviet and Post-Soviet Review*, 20, 2–3 (1993): 185, and Boris G Kuznetsov, *Patriotizm rossiiskikh estestvoispytatelei i ikh vklad v nauku* (Moskva: MOIP, 1949).

34 Johannes Stark, "The Pragmatic and Dogmatic Spirit in Physics," *Nature*, 141 (30 April 1938): 770–2; see also his articles "The Attitude of the German Government toward Science," *Nature*, 133 (21 April 1934): 614; and "International Status and Obligations of Science," *Nature*, 133 (24 February 1934): 290.

35 M.V. Konashov, "Lysenkoizm pod zaschitoi spetskhrana, in Yaroshevsky," *Repressirovannaya nauka*, part II, 97–111.

36 For example, Dzheims Faif (James Fyfe), *Lysenko prav* (Moskva: Inostrannaya literatura, 1952).

37 Mitchell Ash, "Scientific Changes in Germany: 1933, 1945 and 1990: Towards a Comparison," *Minerva*, 37 (1999): 329–54.

38 Yakov M Rabkin, "Scientific and Political Freedoms," *Technology in Society*, 13 (1991): 53–68.

39 Alexander Mitscherlich and Fred Mielke, *Doctors of Infamy: The Story of the Nazi Medical Crimes* (New York: Henry Schuman, 1949).

40 L.L. Kerber and Von Hardesty (eds), *Stalin's Aviation Gulag* (Washington, DC: Smithsonian Institution Press, 1996).

3

"IDEOLOGICALLY CORRECT" SCIENCE

Michael Gordin, Walter Grunden, Mark Walker, and Zuoyue Wang[1]

Introduction

The historical study of science and ideology is really a twentieth-century phenomenon, for it is only after the First World War that stark differences appeared in political ideology and regime: liberal capitalist democracy, Marxist-Leninist communism, fascism and National Socialism. Furthermore, as the introduction to this volume argues, this historiography has been profoundly influenced by the Cold War, with the result that certain themes have been dominant. This essay will investigate perhaps the most striking examples of science being influenced by ideology, which here will be called "ideologically-correct-science" (ICS). The French Revolution will be included because, as the above-mentioned introduction also describes, it was both the first case study for the interaction of science and ideology, and a surrogate for other Cold War case studies. Not all relevant types of ICS will be, or even could be covered here. Indeed, this article will sacrifice depth in favor of breadth and use the comparative approach in order to provide a suggestive analysis of science under different ideological regimes.

Scholars have generally assumed that a political environment can influence science, but relatively little is known as to how this functioned in particular circumstances or across national boundaries. ICS refers to attempts by the state (or at least some representatives of, or forces within the state) to not only use science, but also to transform it into a more ideologically acceptable form, both with regard to scientific content and institutions. These efforts were often inconsistent, and not always entirely rational, but they existed all the same. Jacobins called for a "democratic," not "aristocratic" science in the French Revolution. Bolsheviks called for a "Marxist," not a "bourgeois" science in the Soviet Union. National Socialists in Germany called for an "Aryan," not a "Jewish" science. Ideologues in Second World War Japan demanded a nationalistic, "Japanese" science and technology. During the McCarthy era in the U.S.A., politicians and some scientists tried to reshape science to help win the Cold War, sometimes calling for what Jessica Wang

35

has described as an "anti-communist" science.[2] Finally, the Red Guards demanded a "people's science" during Mao Zedong's Cultural Revolution in China.

ICS often followed the same pattern: (1) purge of unacceptable scientists and purge or transformation of unacceptable scientific institutions; (2) the enlistment/recruitment of acceptable scientists; (3) the training of new scientists and creation of new institutions; and (4) the production of ICS. This essay will examine this admittedly ideal pattern in order to shed light on ICS in particular and the interaction of science and ideology in general.

In cases of ICS, the state often rewarded scientists and sciences that were, or appeared to be, politically and ideologically correct, while those who deviated from the prescribed path might receive punishment. This pressure was sometimes overt, as during the Chinese Cultural Revolution, or subtle, as in the McCarthy era of the American national security state. ICS could also be self-imposed. Some scientists, voluntarily or otherwise, sought to apply what they viewed as the official ideology in their scientific work. Finally, ICS could also provide "protective coloration,"[3] whereby a straightforward piece of research was wrapped in the official ideology for self-protection or self-promotion. These attempts to make a science that had merely the forms and trappings of ideology should be distinguished from efforts which were made to influence the content, but which fell short and only had an external effect on the social position of the sciences.

Undoubtedly the classic examples of ICS are the "Aryan Physics" (*Deutsche Physik*) movement during the Third Reich and the "Lysenko Affair" in the Soviet Union. Indeed, these are two of the most-studied examples of ideology influencing science. But Aryan Physics, like its counterparts in other sciences under National Socialism, was neither typical of science in the Third Reich, nor very successful. A search for ICS under Hitler would certainly include "Aryan Science," but must also go beyond it. Similarly, Lysenkoism was not typical of science under Stalin. Although it is debatable whether or not it was "successful," it did not spread from plant breeding and genetics to other sciences, and like Aryan science, it did not go unchallenged.

Furthermore, the history of "ideologically-correct-science" is not merely the story of the perversion or destruction of "good" science. Although this was sometimes the case, it is equally true that there are many examples in which ICS either failed to have a particularly harmful effect, or even produced benefits and positive scientific results. This latter case can be made both for a direct effect, as Loren Graham has argued with regard to dialectical materialism sometimes facilitating scientific progress, and an indirect effect, such as Mark Walker's thesis that the fight against Aryan Physics actually strengthened the hands of some in the German physics community.

ICS is a useful concept, and was very real, but it is also something to be used carefully, for it can also be seen as a straw man. If by ICS one means the total, coordinated, systematic, and intentional implementation of an ideologically determined program in science by the state, then ICS never happened, indeed never came close. But interesting and important things did happen. This article will therefore examine several case studies, comparing and contrasting them, with an eye towards gauging the limits and usefulness of this concept.

France

The varied fortunes of science in the French Revolution, like the other case studies, is a story of complexity and nuance, which can hardly be given adequate justice here. The two most celebrated incidents of the "attack" on science during the Jacobin Terror – the closing of the Academy of Sciences (*Académie des Sciences*), and the execution of the chemist Antoine Lavoisier – both fall short of being conscious attempts to impose an ideology on science.

Jacobin ideology was only dominant for about a year, and was not synonymous with the French Revolution. Moreover, all sciences were not treated equally: important Jacobins were hostile to abstract, theoretical, and mathematical science, but were favorable to natural history. After much debate about how to reform education and how to structure expertise in the Republic of Virtue, in 1793 the Academy was closed as an institution and its resources dispersed. But the issues here were not just science, and even not centrally science. Resentment of royal privilege and corporate prestige were mainly responsible for the closing of the Academy.[4] While the Academy was eliminated as a vestige of royalist corporatist elitism, the Jardin du Roi was kept intact.[5] Moreover, Academy scientists (for the most part) were subsequently employed by the government of Revolutionary France.

The case of Lavoisier, on the surface the purge of an unacceptable scientist who was hardly a strong advocate of Robespierre's government, upon inspection had even less to do with science than the closure of the Academy. Lavoisier was executed by the machinery of the Terror, but not for any reason connected to his positions in science. His association with the Tax Farm sealed his fate. Similarly, Condorcet died at the hands of the Revolution, not because of his mathematics, rather because of hostility to his rationalist Enlightenment views.[6]

The Terror prematurely ended the lives of several prominent scientists, and interrupted the careers of many others, forcing twenty academicians out of forty-eight into "exile" (most went to the provinces, and only about four actually emigrated during the Jacobin Republic). But other natural scientists in educational institutions, government branches, and other venues were actively recruited by the regime. In the case of the Jardin des Plantes an old royal corporate research institution was maintained, albeit with profound alterations (see below). During the period of revolutionary wars, members of the Academy of Sciences received government contracts to fulfill old academic projects like the metric system, gunpowder production, and military engineering. Most aided the Revolution with little grumbling, or even with little attention to other political events once the guillotine lost its central prominence.[7] In contrast, some scientists, like the chemist A. Fourcroy (1755–1809), not only did not resent the Jacobins, they were active members of the state administration.

During Napoleon Bonaparte's Empire, the state recruited scientists even more avidly than before, and many scientists were quite eager to serve Napoleon's "technocratic" regime and appreciated the return to stability he represented in their eyes. The state now poured an enormous amount of money into military institutions, and scientific and technical expertise were richly rewarded in new institutions like the École Polytechnique, which was founded under the Directory and blossomed under Napoleon.[8] Given the prominence of scientists in the advisory apparatus of the Old Regime, especially in the form of the Academy of Sciences, this development under the Directory and the Empire can be seen as a continuation of previously established practices. The persecution of "unacceptable" scientists seems more the aberration from French practice, and the recruitment of those old specialists who were acceptable continued as before, albeit in new institutions.

Since it is difficult to discern a clear ideology in the French case governing the selection of acceptable as opposed to unacceptable scientists, it is also difficult to find an ideological criterion for training new scientists. After the ideological excesses that led to the closure of the Academy had passed and some of the dust of the Terror had settled, France still faced a series of foreign wars, and sorely needed technical expertise and noticeably lacked qualified young practitioners.

The solution was the creation of a new system of Grandes Écoles, headed by the École Polytechnique, designed to inculcate military discipline into young minds drawn from all corners of France with the necessary technical and mathematical skills required by the emerging modern bureaucratic state and a conscripted army.[9] The plans for such a system of education originated in the Old Regime, whereby places in military schools would be attained through nationwide competition and nominations by local authorities. The subject matter of all schools, especially the Polytechnique, was heavily imbued with mathematics applied to concrete problems of military necessity.

The results were impressive: a new generation was trained, filling the officer ranks of the army, the upper levels of the bureaucracy, and other positions at the top of the modern nation-state. The ideology of discipline through mathematical scholarship and service to the State has been so well inculcated into French society that it is scarcely noticed how large an influence this, the most ideological of Napoleonic institutions, still plays in modern France.

The attack on Newtonianism and the retention of the Jardin du Roi as the Jardin des Plantes after the dissolution of the Academy of Sciences are the two most prominent candidates for ideology affecting science. Charles Gillispie and L. Pearce Williams have made the most forceful case for an anti-Newtonianism, arguing that an ideology of anti-elitist Romanticism, deriving from Rousseau and Diderot, motivated the Jacobins' hostility to an atomized, mathematized, Newtonian universe.[10] Many of the individuals in question expressed hostility to some aspects of the Newtonian worldview, and specific examples of ideology did seep into the way debates about the content of science were conducted.[11] Similarly, because of the French Revolution, an exact, scientific biology was displaced by a discursive descriptive natural history within the Jardin des Plantes.[12]

But these examples of ideology and science do not cut to the actual core of the politics of science during the French Revolution. It is unclear how much of a case can be made for the application of specific Rousseauvist anti-Newtonian ideals during the height of the Terror. Even the leading anti-Newtonian ideologue of the time, Marat, was motivated less by an ideological stance towards Newton than by revenge against the Académie des Sciences for its treatment of him during an earlier scientific dispute.[13] Perhaps more startling is the fact that these trends were not followed up once the Academy was closed and Lavoisier had been executed. Laplace wrote his great Newtonian masterpiece after the debates recorded by Williams and Gillispie, for example. But the fact that the attempt to reject or replace Newtonianism did not succeed does not mean that it should be dismissed.

The reformation of the Jardin du Roi into the Jardin des Plantes was more successful during this period. It is noteworthy that, despite the rhetoric of anti-corporatism that led to the demise of the Academy, the Jardin remained well financed and continued to perform the same sort of botanical research as it had under the Old Regime.[14] The secret of this persistence was the articulation by the botanists of a natural historical program they favored that also seemed to accord with the hostility to abstraction noted by Williams and Gillispie. As a result, this kind of empirical botany was encouraged. A particular type of ideo-logically-correct science was indeed adopted as a form of basic research, but basic research led the way here in defining what it meant to be ideologically correct.

ICS during the French Revolution was fleeting and superficial. At the same time that the Reign of Terror reached its height in Paris and de-Christianization was ravaging the French countryside, most scientists, who were products of the Old Regime, were put to work for the Revolution; others needed only to repackage their research in an ideologically congenial wrapper. The scientific community did have to make concessions to the new political order, and a few individual scientists (for example, Condorcet and Lavoisier) suffered and died, but in general French science and scientists benefited from the French Revolution and Napoleon. The push for ICS did not produce a significant change in the content of research. However, ICS and the greater political and ideological currents of the French Revolution did create new scientific institu-tions and influence research programs and thereby led the way in terms of political control of scientists and scientists' accommodation.

The Soviet Union

The reconstruction of Russian science was well on its way to a socially more useful institition before 1917, but the chaos of the First World War and the transformation of Russian society by the Bolshevik Revolution also brought profound change to Russian science. Perhaps surprisingly in a nation where science had never penetrated much farther than a vanishingly small percentage of the literate intelligentsia, the Bolsheviks and their allied parties saw reform of higher education, and therefore indirectly, science as one of the first tasks on their agenda. At almost the first opportunity they approached the Imperial Academy of Sciences in Petrograd with both carrot and stick in hand.[15] On the one hand, the largely bourgeois membership of the Academy was reluctant to extend a badly needed hand to the new regime's quest for rapid economic improvement and technical expertise. On the other hand, compelling reasons kept the vast majority of Russian scientists from emigrating and induced them to cooperate, however reluctantly at first, with the regime.

It was by no means certain that the regime would last, and there seemed to be every hope of the regime moderating its rhetoric against bourgeois specialists (as indeed it did). Such hope gave those with a sincere patriotic bent an opportunity to actually put some of their practical suggestions to work. The thrust of the early period of science policy in the Soviet Union was more the establishment of new scientific institutions rather than the destruction of older ones. Moreover, the Bolsheviks offered unheard-of blandishments in the form of prestige, equipment, and funding in an effort to persuade the academicians to lend some assistance.[16] Finally, the Bolshevik regime seemed the lesser of several evils, as rabidly "anti-specialist" movements like *Proletkul't* hovered on the horizon.

During the period of War Communism (1918–21), *Proletkul't* was a vibrant cultural movement, one that marked the first effort to establish ideologically-correct-science in the Soviet context. *Proletkul't* was by no means directed exclusively, or even principally, towards science. It argued that a proletarian state required a proletarian culture, not the realism of capitalist art, the individualism of capitalist literature, or the technocracy of capitalist science.[17] Technocratic specialists and other remnants of tsarist capitalist culture would have to go, and a more democratic and proletarian science would be imposed in its stead.[18]

Thus for quite some time, bourgeois specialists did not know how they would be treated by the Bolsheviks, especially since there was now a "Communist Academy" alongside the traditional Academy of Sciences. But Lenin had little patience for such efforts to alienate much needed specialists, and his eventual suppression of *Proletkul't* and the closing of the Communist Academy served as a signal to bourgeois scientists that their kind would be tolerated as long as they were amenable to the new regime's demands.[19] In fact, except for an exile of some 200 dissident intellectuals, there was not much of a purge in science until the period 1928–31.

What it meant to be an "acceptable" scientist in the early Soviet Union fluctuated widely with the attitudes and needs of the fledgling regime. During the period of War Communism, bourgeois specialists were (officially) "unacceptable," but were nevertheless used. As the Soviet economy began to falter, the regime began to accommodate those in possession of needed technical skills. The heyday of the bourgeois specialist – the period of the New Economic Policy (NEP, 1921–27) – induced many scientists who were ideologically opposed to Bolshevism to make a temporary peace, while giving the Soviet regime time to lick its fiscal wounds and gear up for socialist industrialization.

The essence of the NEP attitude towards bourgeois scientists and engineers had already been expressed by Lenin in his opposition to the iconoclastic fury of *Proletkul't*. Technical expertise would always be necessary, and as long as those who had the knowledge would only share it if given sufficiently high salaries and ideological breathing room, then they should be afforded those luxuries.[20] This did not mean, however, that ideological constraints were put on hold. Ideologically "acceptable" scientists, like the young Lev D. Landau, for example, were actively encouraged by the regime and promoted over old bourgeois specialists who staffed the old universities.[21] Indeed, the communists made a fundamental distinction between research institutes where scientific ability was most important, even if not accompanied by appropriate political conviction, and the universities, where only politically reliable scientists would be used to train young scientists. It is interesting to note that National Socialists in Germany and to a certain degree the Communist Party in China made the same distinction.

For the time being, the Soviet state only mildly harassed those who chose to hold to their old views – provided their skills were truly indispensable to the industrialization of the new regime. The Shakhty trial of 1928 changed all this, however, when bourgeois specialists were accused of "wrecking" and industrial sabotage designed to cripple Soviet power.[22] The honeymoon had ended, and under the rising power of Joseph Stalin, bourgeois specialists were not tolerated during the years 1928–31. Thereafter the term "bourgeois specialist" was no longer used, and many former "specialists" quietly returned to the positions of prominence in science they had enjoyed before. The Soviet nuclear weapons project, which used both former specialists and younger scientists trained under the Soviet educational system, was typical in this regard.[23]

The generation of new cadres of ideologically suitable scientists and technicians constituted one of the most important aspects of early Soviet science policy. The splitting of research and education was the first stage in this development. Education was placed entirely in the hands of ideologically sanitized pedagogues within the People's Commissariat of Popular Enlightenment (*Narkompros*). Research institutes were left under various economic Commissariats, and permitted a more eclectic personnel. At the same time, the State transformed the research institutes, borrowing some aspects of Western organization for individual labs, but placing them all into a Soviet framework.[24]

While some bourgeois specialists were prominent in this framework, by the early 1930s and the conclusion of the first of Stalin's Five-Year Plans, most had either blended into the woodwork and adopted appropriate ideological colors or had been executed or exiled during the purges.[25] However, as late as the 1950s, non-Party scientists, former "bourgeois specialists," occupied the majority of high-level administrative positions in Soviet scientific research.

The Soviet Union grudgingly used its "bourgeois specialists" while simultaneously training new cadres of "red" scientists. But these new ideologically correct scientists had the same professional aspirations as their "bourgeois" mentors – concern for international scientific standards, the need for international contacts – and therefore sometimes clashed just as forcefully with the Stalinist regime's desire for ideologically fidelity.

The growth of cadres of communist researchers in the various fields of science was really quite extraordinary.[26] As more and more competent (and ideologically "clean") individuals were generated, they began to take over positions held by disgraced members of the older generation.[27] By the onset of the Second World War, Stalinists had essentially completed the ideological purification and installation of "red" specialists in almost all levels of the Soviet research empire.

Dialectical materialism, the official philosophy of science of the Soviet Union, complemented historical materialism – the Marxist theory of historical and economic development – to compose the complete orthodox set of beliefs about the social and natural world. The nature of dialectical materialism and the extent of its epistemological and ontological grasp had been a source of debate since the early interpretations of Engels' and Lenin's writings on the natural world during the 1920s.[28] But unlike the meaning of dialectical materialism, the unwritten "requirement" that scientists hold to some of its tenets (or at least not openly contradict them) was more or less constant through Soviet history – and reached some disastrous consequences during the Lysenko years. Yet the historian Loren Graham has pointed to another side of dialectical materialism, which he calls the "authentic phase."[29] Graham argues that dialectical materialism was sometimes used by scientists freely as a positive force for scientific reasoning.[30] Graham's persuasive argument thus makes the intriguing point that scientists often adopted ICS as basic research in the Soviet Union voluntarily (or semi-voluntarily) and occasionally used it to produce significant results.

Whereas dialectical materialism provides the most important example of ICS as basic research in the Soviet Union, the most famous instance of ICS as applied research is the well-known case of Lysenkoism.[31] Scion of a Ukrainian peasant family, Trofim Denisovich Lysenko began his work as an agronomist in the 1920s in an agricultural station near Baku. While there, he claimed to have discovered a biological process he dubbed "vernalization" (*iarovizatsiia*): the treatment of germinated seeds of various plants with abnormal conditions of heat, cold, and other forms of environmental exposure, in order to make plants develop in a more appropriate way – essentially a neo-Lamarckian biological program. Lysenko's attempts to present his results to the Soviet agronomic and genetics community were rebuffed as contrary to all known facts about genetics. The famed geneticist N.I. Vavilov initially supported Lysenko's research as potentially producing innovations in agronomic practice, but broke with Lysenko when he started to push his neo-Lamarckian views on plant breeding and genetics in the mid-1930s.

In the early 1930s, Lysenko teamed up with ideologist I.I. Prezent, who convinced Lysenko to link his neo-Lamarckian views of inheritance with Darwinism, and to couch both in a Marxist framework. This marriage of dialectical materialism and agronomic practice in opposition to genetic theory caught the attention of Stalin in the late 1930s, who praised Lysenko openly in various contexts as a means of supporting the regime's disastrous and bloody collectivization campaign in the countryside. Lysenko grew in power, becoming president of the Lenin All-Union Academy of Agricultural Sciences (VASKhNIL) while its former president, Vavilov, was arrested on grounds of counter-revolutionary and anti-Soviet activity, and died of malnutrition shortly before being released from prison in the early 1940s. The actual banning of genetics did not happen until after the Second World War, when Lysenko's star actually seemed to be waning and the Cold War got underway. But in 1948, Lysenko read a speech (toned down but supported by the personal editing of Stalin)[32] condemning genetics as a "bourgeois" science and banning almost all research on it in the Soviet Union.

While Nikita Khrushchev liberalized much of the terror apparatus of Stalin's state after the latter's death in 1953 and especially after the Twentieth Party Congress in 1956, he liked Lysenko personally and continued to fund him lavishly and support a series of disastrous agricultural programs which his favorite proposed. After Khrushchev's fall in October 1964, however, Lysenko's days in power were numbered. Genetics was restored in 1965, but the recovery process was painful and the loss of Vavilov hard to forgive. The scars caused by Lysenkoism remain to this day.

Lysenkoism, which influenced Soviet science for decades, was obviously ideologically-correct-science. But it was eventually overthrown by forces within Soviet science and society, and the science it had discredited and dismantled, genetics, was reinstated and rebuilt. Lysenko and his followers also failed to extend their influence to other Soviet sciences. In particular, physics was able to rebuff Lysenko-inspired attacks on certain aspects of modern physics, both because of the relevance and irrelevance of physics to Stalin's foreign and domestic policies. At first physics was not very important to the Soviet leadership, for in contrast to Lysenko's theories, it promised neither to solve the country's problems, nor fit particularly well into Soviet ideology. When the Second World War began and the potential of nuclear weapons was clear, physics became far too important to purge or distort. Thus Lysenkoism was arguably exceptional, and reveals little about the other major purges of Soviet intellectuals that were attempted in the period after the Second World War.[33]

Germany

During the first year of National Socialist rule in Germany, a significant percentage of scientists (perhaps as much as 15 percent) were forced out on racial and political grounds.[34] This purge was not aimed particularly at scientists or science – the campaign against Albert Einstein is the exception that proves the rule – but rather was a consequence of the National Socialist "cleansing" of the entire civil service. This larger purge was itself apparently a largely unplanned, if not spontaneous reaction to the failure of the nationwide boycott of Jewish businesses in April 1933. The effect on science was tremendous, but this purge does not demonstrate any plan or intention on the part of the National Socialist leadership to create an Aryan science.

There is evidence for more direct interest by Hitler's government in the transformation of scientific institutions, but again a close look reveals different priorities. The universities were purged and transformed right away because they were educational institutions charged with the training of German youth.[35] Their transformation was profound, but also hidden. The structure of the university remained largely intact, at least on paper, but most autonomy was robbed by the introduction of the "leadership principle": a strict hierarchy, whereby one had to obey everyone above, but could order about everyone below. In principle, the faculty still met and voted, prepared lists of candidates for positions, and so on, but in practice the deans and rector – political appointees, of course – often had almost dictatorial power. Other scientific institutions, including the Kaiser Wilhelm Society and its research institutes, as well as the various Academies of Science, were transformed in a similar way, although significantly later in the Third Reich.[36]

The dates by which all Jewish members had been purged provides one of the most telling indications of how relatively unimportant these research institutes and academies were for the National Socialist leadership. Whereas most Jewish scientists in the universities had lost their positions in 1933, other Jewish scientists managed to remain at research institutions for many years. The Prussian Academy of Sciences, for example, under pressure from the Ministry of Education, finally asked the last Jewish member to resign in 1938, shortly before *Reichskrystallnacht*, the nationwide orgy of violence against German Jews. In 1933 the Education Ministry had been keen to publicize its treatment of Einstein. After he had already resigned from the Prussian Academy, the Ministry pressured its leadership to issue a press release, essentially saying good riddance. But in 1938 the Ministry wanted to keep the final purge of Jewish scientists quiet, so as not to publicize the fact that there were still Jews in the Prussian Academy.[37]

The purge of German science by the National Socialists makes clear how little interest Hitler and his followers had in scientific research. Their interest was aroused only when scientists demonstrated that modern science could serve National Socialism. It is striking and depressing to see how quickly the vacancies caused by the purge were filled by generally competent, racially and politically acceptable scientists eager to serve the new regime. For those who were already in place, it usually sufficed to demonstrate Aryan status and an apolitical attitude to keep their jobs, although they were pressured to yield greater political and ideological cooperation with the National Socialist movement.

Those moving up – i.e., who did not already have permanent positions – had to be both Aryan and willing participants in the political and ideological rituals introduced by the National Socialists into the universities. Such rituals included or encompassed attendance at "political" indoctrination camps, membership in National Socialist organizations, including, of course, the National Socialist German Workers Party (NSDAP), and other forms of participation in National Socialism. The authorities paid little attention to their research when it came to judging political acceptability, although this might well enter into whether or not they were hired.

Rearmament, especially beginning in 1936, offered great opportunities to scientists who had something to offer the regime; similarly, the new racial hygiene policies (sterilization, "euthanasia," restrictions on marriage) provided great opportunities for physicians, biologists, anthropologists, psychologists, and psychiatrists.[38] In general, most scientists did not make the transition to the racist, Aryan science, but did adapt themselves and their research in order to work under National Socialism.

Perhaps the greatest failure of National Socialist science policy concerned the training of the next generation of scientists. The politicization of education and emergence of National Socialist youth organizations like the Hitler Youth eroded both the quality and the quantity of scientific education. The creation of new institutions was not much more successful. The National Socialists did not really try very hard to create new scientific institutions. Even the *Ahnenerbe* ("Ancestral Heritage"), the scientific research arm of the SS, relied mainly on research contracts in order to encourage certain types of scientific work.[39] Most truly new research institutions created by the National Socialists had little to do with science, and were so ideological and politicized that they were really incapable of producing significant basic or applied research. Scientific work was done for the National Socialists, but it usually took place in institutions Hitler's movement had inherited. The rocket project is the exception that proves the rule: Army officers and engineers, not National Socialists, created and developed it during the Third Reich.

The politicization of the universities was compounded by the carnage of war, as very many students were offered up as cannon fodder during the Second World War. The result was a generation lost to German science: with few exceptions, only scientists who had entered the university system during the Weimar Republic survived the Third Reich. Despite the regime's attitudes towards women, by the middle of the war women made up a large percentage of university students because so many of their male counterparts were fighting and dying on the front.[40]

The regime wanted the help of scientists and physicians in order to provide a scientific basis for their racist, and eventually murderous race hygiene. But despite the active participation of scientists and physicians in this program, and the infamous experiments carried out at Auschwitz and elsewhere on unwilling concentration camp inmates,[41] researchers could not deliver scientific proof of the supremacy of the Aryan race. In retrospect, this was revealed by the infamous Nuremberg Laws in 1935. When the National Socialist State finally issued the binding legal definition of what "non-Aryan" meant, it had to fall back on to a religious, not racial definition of who was a Jew.

However, there were also scientists who claimed to be practicing Aryan science when carrying out their basic research. These were not race hygienists – although a biologist claimed to be fighting for an "Aryan Biology" (*Deutsche Biologie*) – but rather physicists, mathematicians, psychologists, chemists, and engineers. These Aryan movements in German science and technology[42] eventually failed in their efforts to seize control of their disciplines precisely because they were barren in the National Socialist sense. The rulers of Germany wanted science useful to them, and it was the leaders of the established scientific communities, not the Aryan scientists, who were able to gain and retain the backing of influential patrons in the National Socialist state.

Thus the calls by the Nobel laureates and "Aryan Physicists" Philipp Lenard and Johannes Stark to eliminate the influence of Jews in physics, and the "Aryan Mathematician" Ludwig Bieberbach's assertion that Aryans and Jews made different types of mathematics eventually fell on deaf ears, while the applied mathematician Ludwig Prandtl and the theoretical physicist Werner Heisenberg offered their expertise in designing wind tunnels and nuclear weapons, respectively. Established scientists like Heisenberg and Prandtl were thereby able to sideline or neutralize their "Aryan" colleagues by convincing political leaders that their basic and applied research might facilitate both military conquest by creating new and improved weapons[43] as well as the racial engineering of Europe by providing new methods for distinguishing Aryans and non-Aryans.

Ideologically-correct-science had a profound effect on German science, far beyond the well-known case of the ideological attacks by Aryan Physics on modern physics. But the end effect was almost always defeat for the ideologues, as the established scientific communities sought refuge and support from leading National Socialists. The main effect of "Aryan Science" was to drive most other scientists further and faster down into the arms of National Socialism, making themselves more useful and relevant for the often murderous policies of the regime.

The very anti-intellectual climate within the ranks of leading National Socialists worked against the advocates of Aryan Science. A National Socialist ideologue disdainful of science also had little interest in its Aryan variant, while the technocrats scattered throughout the National Socialist hierarchy naturally threw their support behind established scientists who could deliver the goods.

Scientists helped build rockets[44] and jet planes,[45] researched new biological and chemical weapons (which fortunately were not used), and dangled the prospect of "Wonder Weapons" like the atom bomb[46] before leading National Socialists. Researchers ranging from physicians, biologists, psychiatrists, psychologists, anthropologists, economists and geographers[47] helped implement the murderous race hygiene policies of "euthanasia" and "Germanization," and finally murder and genocide by helping to select the victims and create new and more effective ways to torture and kill them.

Japan

During the Second World War in Japan there was no need for a racial or ethnic cleansing in the sciences because, with the exception of a few notable Koreans, all scientists were Japanese and shared essentially the same racial, ethnic, and cultural identity. Moreover, because of the relative paucity of scientists with advanced scientific training in Japan, even Koreans – whose homeland Japan had occupied before the war and who were considered second-class citizens in Japan – were allowed to retain their positions at the university.[48]

There were also few incidents in which scientists were jailed for expressing anti-imperialist views. In one case, several members of an academic research group, including the physicist Taketani Mitsuo, were arrested for advocating resistance to Japanese imperialism through their serial publication, *Sekai bunka*(World Culture). Taketani was detained – allegedly for his research activities on natural dialectics – accused of helping to promote the Communist Party in Japan, and forced to state that he had acted under instructions from the Comintern. The judge who reviewed his case, however, suspended prosecution and released him to the custody of his colleague and close friend, Yukawa Hideki.[49] On the whole, such instances were rare.

Under military influence, the government enacted numerous laws in the 1930s to acquire greater control over the people and the economy. As part of the militarization of the nation, institutions of scientific research, such as the imperial universities and the prestigious Institute for Physical and Chemical Research (also known by its Japanese acronym "Riken"), were also brought under the aegis of the military. By influencing budget allotments for basic and applied research, as well as the production orders for the resulting manufactured items, the military began to have a significant impact on scientific research. As a result, there was no need to transform such institutions; only greater administrative and economic control proved necessary.[50]

Nevertheless, the recruitment of first-rate scientists became a critical problem for the military. Nearly all of the nation's most famous and competent scientists had spent years abroad studying in Western nations, and they were thus considered suspect by the military leadership, which was by and large xenophobic in its worldview. Moreover, most scientists at the university had little interest in suspending their own research for military projects.[51] Out of

47

necessity, however, the military consulted leading scientists at the universities when those in uniform – who were usually little more than higher-school educated technicians – proved incapable of advanced level research. Both the army and navy, for example, turned to physicist Nishina Yoshio, Director of the Physics Department at the Riken, to complete a feasibility study of the possible exploitation of nuclear energy for military purposes.[52]

Like Nazi Germany, Japan proved terribly shortsighted in preparing the nation's scientific infrastructure for a prolonged and total war. One particularly notable area of failure was in the training of new scientists and the creation of new institutions of science and technology. The military continued to draft students from university science programs and departments throughout the war. This was halted only through the concerted efforts of senior scientists like Nishina, who insisted upon military deferments for designated students of exceptional ability in exchange for agreeing to conduct research for the military. Only by such means was the older generation able to preserve the next generation of scientists.[53]

As for new institutions of science, by the early 1940s, the resources for their construction and maintenance were dwindling and the move was toward consolidation and rationalization, not expansion or the creation of new scientific institutions. The capstone of the trend came in early 1942 with the establishment of the Board of Technology. Roughly analogous to the Office of Scientific Research and Development in the United States, the purpose of the Board of Technology was to coordinate scientific research and development of new technologies between civilian and military institutions, as well as between the army and navy.

The organization looked impressive on paper, but in reality, the Board of Technology proved a dismal failure. It never acquired sufficient authority or capability to supersede the numerous administrative boundaries that such a task entailed. It could not overcome the substantial compartmentalization of civilian and military research. Neither could it redirect the complex network of financial arrangements, production contracts, and social ties that each military service had to its preferred *zaibatsu* (industrial combines) and university cliques. The board had little success in overcoming the bitter enmity that existed between the army and navy to convince them to collaborate on key projects until the war was already all but lost.[54]

There was no single ideology in wartime Japan comparable to Aryan Physics in Nazi Germany, nor was there any ideological movement in the sciences that gained such comprehensive state support and promotion as Lysenkoism did in the Soviet Union. Beyond the ubiquitous rhetoric of national militarism that emphasized sacrifice and service to the Emperor and nation, there was no prevailing ideology to impact science as there was in Germany or the Soviet Union. Yet there was a call for a distinctly Japanese form of technological development based on the nation's situational imperatives, that is, the rise in demand for military production in the face of rapidly diminishing raw materials.

Japan's progress in science and technology had been dependent upon the cooperation of Western nations and foreign teachers since the late nineteenth century, but as the military leaders of Japan drove the nation inexorably toward world war in the 1930s, increased hostility toward the West compelled bureaucrats and intellectuals alike to question the West as a model for Japan's technological development. Having no indigenous model as a substitute, a distinction had to be drawn between American and German paths of development. The rationalized "German path," it was argued, was better suited to Japan, as it aimed to limit the use of raw materials and promoted the use of substitutes. When its German ally proved niggardly in technology transfers, however, Japan was forced again to look inward.

Imitation of the West was to be rejected in favor of a uniquely Japanese path of technological development in accord with the nation's paucity of natural resources.[55] As one ideologue stated:

> The resources of the Greater East Asia Co-Prosperity Sphere are awaiting the creation of the new technologies that will make most effective use of them. It is only then that these resources will acquire value. The existence of scientific research, which may give birth to this new technological creativity, will provide a firm basis for the cultivation of the Co-Prosperity Sphere, and for this reason the promotion of such research is currently an urgent necessity.[56]

Despite the flurry of mobilization and rationalization laws and measures that were enacted to realize this vision, no distinctively "Japanese" science emerged, nor was Japan able to free itself of its pattern of technological borrowing and dependence upon the West. Toward the end of the war, when necessity and desperation drove the nation's leadership to extremes, the government and military called upon scientists and engineers to draw inspiration from Japan's traditional past and the unique characteristics of the Japanese people, all in the effort to create an ideological rallying point for Japanese scientists in the development of some new weapons technology that could turn the tide of war in their favor.[57]

By 1943 the National Socialist state was also calling upon its scientists, engineers, and even inventors to create "wonder weapons" which would use qualitative superiority to overcome the quantitative superiority of its opponents. The Japanese military's answer, however, was no miracle weapon of science, such as a rocket or an atomic bomb. Rather, it responded with crude suicide craft, such as the *Ohka* piloted missile and the *Kaiten* midget submarine. No "Japanese" style science emerged from the war, and the ultimate "Japanese path" of technological development resulted in death for many of the nation's youth.

In the case of Japan, there was no readily identifiable ideologically-correct-science. Rather there was only a vague policy objective to guide technological development that was in accord with the nation's situational imperatives. The lack of natural resources, coupled with the nation's Spartan industrial infrastructure and its limited scientific and technological capacity, predestined Japan's fate in a total war against the United States. Japan proved incapable of producing such wonder weapons as the atomic bomb, long-range guided missiles, and advanced radar, but its army did, for a time, strongly support the development of biological weapons and made significant advances in this field.[58]

Biological weapons were easy to mass-produce with few resources, and thus fit within the military's vision of a new weapon derived from a "Japanese technological path." Ultimately, however, the fear of a response in kind from the United States and the ever-present possibility of a boomerang effect appear to have deterred Japan's use of biological weapons on a wider scale beyond the war in China. As a result, biological weapons never became a significant factor in determining the outcome of the Pacific War, and the military never acquired a uniquely Japanese wonder weapon of his own.

United States

If communism served as a powerful ideology of science in the East, especially in the Soviet Union and China, did anti-communism play the same role in the West during the Cold War? Since American McCarthyism in the late 1940s and 1950s represented the peak of this political and cultural phenomenon, we might expect to see signs of the search for an ideologically correct science.[59] However, there were very few unambiguous examples of efforts to influence scientific content that were motivated or constrained by anti-communism as a political ideology. Rather, what does emerge clearly is the pervasive influence of anti-communism on the *political* roles of scientists whose professional identity assumed significant but not overwhelming importance.

Scholars now generally agree that the two characteristics of McCarthyism, domestic anti-communism and the denial of due process to those accused of communism, existed both before and after the period when Senator Joseph McCarthy made the cause his personal crusade in 1950 – 4.[60] The Cold War ideology of anti-communism not only saw a direct threat to American security in potential Soviet expansion on the international front, but also from perceived communist subversion at the domestic front. The national security state organized national life around national defense and fed on the Cold War ideology of anti-communism. It dominated science and technology policy, and thus indirectly but powerfully shaped American scientists' political and scientific activities.

On the one hand, a large number of American scientists engaged directly in the making and testing of nuclear weapons in the national laboratories of the Atomic Energy Commission (AEC). Those in academic and industrial settings also came to depend on the defense establishment for funding. As Paul Forman has argued, this dependence tended to make scientists choose, consciously or unconsciously, research directions that would benefit their patrons.[61] On the other hand, scientists were persecuted during the McCarthy era because their past association, political opinions, or policy advice deviated from the political orthodoxy prevailing at the time.

The security clearance case of J. Robert Oppenheimer was perhaps the best-known example of McCarthyist attacks on scientists. As the famous director of the Los Alamos laboratory that created the atomic bomb during Second World War, Oppenheimer was nevertheless stripped of his security clearance in 1954 by the Atomic Energy Commission (AEC). The AEC decision cited past association with radical causes, opposition to the hydrogen bomb, and "defects in character."[62] Hundreds of other, less-well known scientists suffered similar or worse treatment both before and after the Oppenheimer case.[63] The U.S. State Department denied passports to a number of American scientists with liberal reputations so they could not travel abroad. It also refused to issue visas to some foreign scientists who wanted to visit the U.S.A.[64]

The Oppenheimer case evoked a most vehement protest from the scientific community, which generally blamed the injustice on a paranoid security system.[65] It undoubtedly marked a profound deterioration in the relationship between many American scientists and the national security state. The far-reaching repercussions did not escape the top government officials. Eisenhower, while agreeing with the AEC's decision, nevertheless worried about the case's effect on scientists in various defense projects. Aware of the potentially explosive impact on scientists and dangerous exploitation by McCarthy, Eisenhower told his aides that "we've got to handle this [Oppenheimer case] so that all our scientists are not made to be Reds."[66] Eisenhower wrote to New York writer Robert Sherwood shortly before the AEC decision, stating that because he was "so acutely conscious of the great contributions the scientists of our country have made to our security and welfare," he shared the hope that Oppenheimer could be cleared.[67]

It was to the President's and the AEC's relief that the "mass exodus" from weapons laboratories, as predicted by various scientific groups, failed to materialize in the wake of the Oppenheimer case.[68] While Eisenhower may have feared losing the services of scientists, there was no lasting damage to the Cold War partnership between science and the State. Scientists had warned since the late 1940s that unfair security procedures would lead scientists to desert government positions, but the threat was more rhetoric than reality.[69]

The fact that few scientists left their government research positions in the aftermath of the Oppenheimer case indicates that scientists, especially younger scientists, learned to live with the new Cold War political economy of science, what Eisenhower would call the "military-industrial complex." As physicist Herbert York, first director of the AEC's Livermore Laboratory on nuclear weapons in the 1950s, later reflected, young scientists had the practical needs of finding jobs that matched their training and supported their families.[70] Others simply recognized the need to combat two "Joes," both Joe McCarthy and Joe

Stalin, especially after the outbreak of the Korean War.[71] In any case, since the newly-founded National Science Foundation only slowly gained substantial budgets, the funding structure of American science was skewed toward the military, which left few alternative sources of financial support to scientists, including the training of new ones.[72] Consciously or unconsciously, scientists were integrated into the national security state.

In the United States during the McCarthy era, there was a general acceptance, either tacit or explicit, by both individuals and institutions, of loyalty oaths and security clearances. Most universities, for example, refused to hire known communists. In 1950 the regents of the University of California adopted a requirement that all university employees had to sign a loyalty oath stating that they were not members of the Communist Party. As a result, dozens of faculty members, many of them scientists, left for other institutions in protest while some others, including tenured professors, were fired from the university.[73]

The same year, scientific organizations such as the Federation of American Scientists and the National Academy of Sciences fought successfully to remove amendments in the National Science Foundation Bill which would have required applicants to the foundation for unclassified research to undergo security clearance and an FBI background check. But, with dismay, they felt that they had to accept the requirement of loyalty oath in the bill as a compromise to get it passed.[74] There were a few nuclear scientists who deliberately switched to fields where they did not need security clearance. Leo Szilard, for example, turned to molecular biology.[75] Others, like Philip Morrison, continued in their fields but avoided work that would have required them to apply for a security clearance.[76]

Without belittling the pains and the injustice that the victims of Cold War anti-communism suffered, a distinction should be made between Stalinist/Maoist communism and American McCarthyism. Few American scientists were persecuted through State-sponsored violence for their particular beliefs in science, in contrast to what happened in Stalinist Russia and Maoist China, although they were repressed in other ways. While Vavilov starved to death in Stalin's prison for resisting Lysenkoist theory and while dozens of prominent Chinese scientists were killed during the Cultural Revolution for being bourgeois "reactionary academic authorities" (see below), the worst that happened to American scientist-victims of anti-communism was, with few exceptions, that they lost their jobs, or as in Oppenheimer's case, their security clearances.

The consequences of McCarthyism went beyond the harm done to individuals, however. In fighting dictatorial communism, American anti-communist crusaders adopted the same anti-democratic tactics employed by the enemy. The American left, including that in science, was largely silenced and social reforms aborted. Fearing the charge of "being soft on communism," Cold War liberals energetically led the U.S.A. into a costly and misguided war in

Vietnam.[77] Few of those American scientists who served as major advisors to the government in the 1950s and 1960s, for example, represented the liberal and left-wing positions briefly influential in the immediate post-Second World War years.[78] The significance of McCarthyism "may well have been in what did not happen rather than in what did."[79] In science, for example, we will never know what scientific research could have been pursued during the Cold War had there not been pressure to work in the national security state system. Likewise, we can only speculate on whether scientific research more closely related to civilian technology could have been advanced further and earlier than it did.

Scientists, especially nuclear scientists associated with government labs, were certainly subject to special scrutiny from the government because of their perceived access to "atomic secrets." Yet, the ideological impact of anti-communism on American science was only general and indirect. In contrast to Marxism, which attempted to function in Stalinist Russia and Maoist China as an all-encompassing ideology with specific doctrines governing science and philosophy, American anti-communism was primarily a political ideology. It guided American foreign policy and influenced domestic politics, but never set down a number of doctrines to be followed in science and philosophy.

The excesses of the McCarthy era in the United States had a profound effect on scientists and their relationship with the state, but it did not lead directly to a new type of an ideologically-correct "anti-communist" science like Aryan Physics or Lysenkoist biology. In the United States, ICS manifested itself as the connection between research and the goals of the State, in particular the integration of science into the national security state, rather than the establishment of ideological tests for the content of science. Not even the "Oppenheimer affair" could halt or even slow the flow of scientists into military-related research.

China

As with the Soviets, science occupied a special position in the ideology of the Chinese Communist Party. The founders of the party in the 1920s turned to Marxism as a "scientific" explanation of history and communism as a natural course of social changes. Mao Zedong later developed an essentially instrumentalist ideology of science that might be called "revolutionary utilitarianism." "Revolutionary" referred to the goal of the Communist Party's science policy to ensure the political loyalty of the scientists, while "utilitarian" spoke to the pressure on scientists to produce immediate, practical results.

Even before their take-over of mainland China in 1949, Mao and his followers had launched a notorious "rectification" campaign in their stronghold in Yanan during the early 1940s. It resulted in a complete re-structuring of the scientific establishment under communist control. A number of scientists and science administrators had insisted on the priority of basic science in education and research. Now they were removed from their positions and punished in favor of those who advocated the re-orientation toward meeting immediate needs in production and military technology. Some of those suspected of harboring bourgeois thoughts – that is, thought not sufficiently revolutionary – were "sent down" to the countryside to learn from the peasants and thereby set an ominous precedent. As the historian James Reardon-Anderson points out, the narrow enforcement of revolutionary utilitarianism marginalized fundamental science "with long-term repercussions on the modernization of China."[80]

Ideological purification of the scientists, often Western-trained, started almost immediately in the People's Republic of China (PRC), and intensified during the "thought reform" campaign at the height of the Korean War in 1952–3. In a scene paralleling that in the U.S.A., war hysteria turned into a hunt for internal enemies, resulting in the suicides of a number of scientists in Shanghai.[81] Many scientists were accused of following bourgeois scientific theories, such as Mendelian genetics, cybernetics, resonance chemical theory, and Gestalt psychology, and forced to renounce them. The Anti-Rightist Campaign of 1957, however, outdid all these previous purges. Attacking those scientists who had criticized the party's mishandling and distrust of scientists as part of a conspiracy to overthrow the new government, Mao ordered the purge of hundreds of thousands of intellectuals, including scientists.[82] Many of the brightest scientists were thus taken away from science and education and placed in forced labor for many years. Despite periods of relaxation, the pressure on scientists for ideological purification never completely relented over the next two decades.[83]

Mao's distrust of scientists reached a crescendo during the Cultural Revolution of 1966–76. He and his supporters unleashed a harsh reign of terror by the radical Red Guards against anyone, including scientists, who could be accused of deviating from Mao's correct political line. Along with other intellectuals, scientists were again purged for their bourgeois ideology and their elitism; they had to be cleansed and reformed. Red Guards and other rebels took over scientific and educational institutions and stopped virtually all research. Scientists, especially those formerly in administrative positions, were criticized and persecuted, and sometimes beaten, tortured, and killed. By 1969, many scientists who survived the ordeal were sent to the countryside or factories to perform physical labor and help make a "people's science." Only after Mao's death in 1976 was it possible for a full-scale restoration of utilitarian science policy under the leadership of Deng Xiaoping.[84]

Scientists were usually purged in the Mao era not so much for the ideological content of their scientific theories as for their political opinions and even personal background, such as training in the West and working under the Nationalists before 1949. Yet, in a few cases, notably Lysenkoism, persecution did fall on those with the "wrong" beliefs. There were numerous cases where the pursuit of basic research by itself could bring on the indictment that one was ignoring the practical duties of a scientist and thus deviating from the correct Maoist model of integrating theory with practice. Perhaps the most striking feature of the treatment of Chinese scientists under Mao was the wide swings between liberalization and harsh tightening, which reflected the divisions within the Communist party leadership over the future course for China and the complexities of modernization.

In 1948–9, communist leaders encouraged scientists who had worked under the rival Nationalists to stay where they were, instead of following the fleeing Nationalist forces to Taiwan. After the establishment of the PRC in October 1949, many of these carry-over scientists attained important administrative positions in the reconstructed scientific institutions of the new regime, including the Chinese Academy of Sciences. Characteristically, however, the most important policies on personnel and research directions were determined by party officials and the few scientists who were also party members.[85] In the early 1950s, the party and government pursued a policy of encouraging scientific research and education, in part to persuade those who had worked under the Nationalists to stay with the new regime, and in part to attract those Chinese students and scientists training or working in the West to return to their homeland. Despite the various political purges in this period, by and large, the latter succeeded as thousands of them overcame obstacles in the West to return to China.[86]

Distrusting these carry-over and returnee scientists, however, the party launched efforts to train its own "red" experts almost immediately following the establishment of the PRC. Typically, the new recruits undertook narrowly focused undergraduate studies in one of the Chinese universities, which were radically restructured in the early 1950s according to the Soviet model to emphasize specialized technical fields, such as metallurgy or geology. The best of these students – in terms of both technical competence and political loyalty – were sent to the Soviet Union for graduate study. Upon their return, they were expected to become leaders in the Chinese scientific enterprise. For example, Zhou Guangzhou, a physics student, followed this path. He would later become a leader in the Chinese nuclear weapons project and president of the Chinese Academy of Sciences in the 1980s.[87]

During the early stage of the Cultural Revolution, there was little effort to train new scientists, except for what existed in the nuclear and military space projects. Most of the universities were shut down from 1966 to about 1971, with no students admitted or graduated. In 1971, universities were re-opened and operated under a radical new direction: freshmen were to come not from high school graduates based on national entrance examinations, as before, but from peasants, workers, and soldiers with practical experience but with junior high school preparation, selected on political criteria. The standards in this new educational regime proved so low that years later, following the end of the Cultural Revolution, the so-called peasant-worker-soldier students had to be re-trained after graduation to reach university level.[88]

The experiment in recruitment and enlistment in science turned out to be a complete failure: it may have produced a "red," but not by any means "expert," generation. When the pragmatist party leader Deng Xiaoping, who was purged by Mao during the Cultural Revolution, returned to power following Mao's death in 1976, he brought the returnee/carryover generation of scientists back into power. The older scientists often bypassed the peasant-worker-soldiers of the Maoist era and began to train a new generation of scientists who came through a restored educational system. Many of the latter also began to pursue studies abroad, especially in the U.S.A. and Western Europe.

Deng's advocacy of utilitarianism continued to dominate Chinese science policy in the 1980s and 1990s as market-driven economic reform brought another wave of structural changes to Chinese science. A number of scientists did run afoul of the regime in this period, but, again, because of their political beliefs and activities, not their scientific theories. Fang Lizhi, the prominent astrophysicist and political dissident, lost his position as vice president of the University of Science and Technology of China, Hefei, and was expelled from the party when he was blamed for student unrest. However, in a move that echoed that of the Soviet Communist Party, Fang was allowed to work at an observatory in isolation from students. In fact, the regime intentionally publicized his research to indicate that it continued to value science although it discouraged political dissent.[89]

Despite the claims by the Red Guards and other Maoists at the time, the Cultural Revolution produced few, if any, ideologically correct scientific theories. Much of the energy of the radical Maoist theorists was focused on attacking what they viewed as bourgeois scientific beliefs within and outside of China, rather than constructing new ones. These included Albert Einstein's relativity theory, which the Maoists denounced as politically capitalistic and philosophically idealistic.[90] Perhaps the only plausible case of ideologically-correct-science in China took place on the eve of the Cultural Revolution.

On August 18, 1964, Mao Zedong invited several Chinese philosophers of science to his residence in Beijing's Forbidden City to chat about the philosophical implications of new theories of elementary particles. "The world is infinite," he asserted, "time and space are infinite. Space is infinite at both the macro and micro levels. Matter is infinitely divisible." Mao's comments, which were prompted by a recent article on elementary particles written by the Japanese Marxist and physicist Shoichi Sakata, led to widespread, officially sponsored discussions on philosophy of science. A number of physicists participated in the discussion, at the end of which they concluded that, based on Mao's version of dialectical materialism, elementary particles were not "elementary," but could be further divided into constituent parts – stratons – to signify the infinite stratification of matter.[91] Starting in 1965, these physicists, many of whom had participated in the making of the Chinese hydrogen bomb, began to construct a theoretical model of how stratons made up hadrons (protons and neutrons).[92]

56

At the time, there were already models of hadrons in the West, which assumed that they were composed of quarks, but it was not clear whether quarks were merely mathematical devices or real particles. The Beijing group claimed that its model differed from the quark model in that it was relativistic and assumed that there were real sub-particles that were made up of hadrons. However, the model was essentially compatible with, and later subsumed under the quark model, which eventually did assume that quarks represented not merely mathematical constructs but real physical entities in what is now called the Standard Model.[93]

There seems little doubt that the initiation and the philosophical interpretation of the straton theory had much to do with Mao's pronouncements on the infinite divisibility of matter in the Chinese context. Maoism provided both the political and ideological justification of a fairly esoteric branch of science that could easily have been branded a bourgeois exercise in an ivory tower. One might argue that these scientists would probably have received protection due to their contributions to the Chinese bomb project, as in the case of Soviet nuclear physicists described by David Holloway. But the fact that some of these scientists were criticized and not allowed to present the theory to an international gathering of scientists in Beijing after the start of the Cultural Revolution, despite the apparent ideological correctness of their work, seems to indicate that the "nuclear umbrella" worked less in Maoist China than in Stalinist Russia.[94] The straton model was thus more a case of protective coloration than a genuine scientific creation of dialectical materialism. Yet, perhaps more than any other scientific endeavors in China in the Maoist years, the straton model represented an attempt at ideologically-correct-science.

Not surprisingly, in comparison with the often critical reception of theoretical research by Maoist "revolutionary utilitarianism," applied research fared much better. In addition to the nuclear weapons project, which was both politically correct and utilitarian to the extreme, a number of other fields also received ample moral and material support in the Mao era. The pivotal Twelve-Year Plan in Science and Technology, formulated by hundreds of scientists under the leadership of Zhou Enlai in 1956, for example, identified about fifty-six applied research areas for heavy investment in resources and personnel. Only as an afterthought was "major theoretical problems in natural sciences" tagged on as the last item in the plan. Even in these basic research fields, applications were the major motivation.[95]

The Communist Party provided lavish rewards for achievements in applied research. Since the early 1990s, the faith in science and technology as the basis of China's modernization has flourished in China under Party leader Jiang Zemin, who had trained as an engineer. By applying modern science and technology to the economy and management, Jiang and his supporters hoped for both a robust economy and a stable social order under Communist rule.

Perhaps no modern state has purged and persecuted its scientists as ruthlessly and repeatedly as the People's Republic of China, but these waves of purges, as well as the intervals of relative tranquility which followed them, had little to do with ideologically-correct-science. Mao and his successors have wanted first and foremost obedience from their scientists. Ideologically correct conduct is part of this, but in this regard scientists have been no different than other Chinese.

Conclusion

If one assumes that the state (or forces or individuals within it) was imposing ideology on science and scientists, then this implies that science and ideology are separable. In fact, as the above examples demonstrate, even in the case of ICS science is not being determined by ideology, but also is not free of it. No political regime has ever tried consistently and comprehensively to impose ICS on its scientists. There have been individuals or portions of a regime that have tried to impose some ideologies on some aspects of science and on some scientists. However, such individual cases are often "overdetermined": there are alternative explanations other than ICS, both for the specific ideological attacks and the efficacy of the victims' resistance. On the other hand, ICS did happen. Just because the entire state was not behind it for the entire period does not contradict that. Although ICS in the ideal sense always failed, it also always had a significant effect, whether an intended one, or not.

No single ideology, including liberal democracy, has historically proven more effective than another in driving science or leading to intended results. Communist regimes appear to have been more likely to try and impose ideological standards on their science, perhaps because Marxism is so comprehensive a political philosophy. In other regimes, ideological attacks on science remained more crude and overt, like National Socialist calls to eliminate "Jewish Science" or anti-communists in the United States during the 1950s denouncing "internationalism" in science.

Although communist regimes were sometimes the most ruthless oppressors of their own scientists, they were also sometimes more flexible and pragmatic.[96] Thus in the Soviet Union and the Peoples' Republic of China, scientific representatives of the ideological enemy, the "bourgeois specialists," were kept on for a while, sometimes even pampered, in order to bridge the gap of time between the beginning of the revolution and the point at which the new regime had trained its own, ideologically acceptable (or more ideologically acceptable) scientists. In some cases, these holdovers from the previous regime held on long enough to shed their label and remain productive and sometimes integral parts of their national scientific efforts.

This would have been inconceivable in Hitler's Germany. The few exceptions of Jewish or part-Jewish scientists who managed to survive the Third Reich working for the regime are the exceptions that prove the rule. The overwhelming majority of "non-Aryan" scientists were purged because of their race and irregardless of their expertise or the scientific needs of the regime. Similarly, scientists suspected or accused of communist sympathies during the McCarthy period were damaged goods and not to be used by the government. It should also be noted that, whatever the actual loss of scientific manpower to Hitler's Germany and Eisenhower's America, in both cases there were sufficient numbers of ambitious, competent, and politically and ideologically acceptable scientists to fill the vacancies.

It is probably no coincidence that all of these examples also deal with military conflict or preparations for it. The French Revolution and National Socialism unleashed war, the Soviet Union and the People's Republic of China were born of war, Japan was fighting the Second World War, and in the 1950s America and the Soviet Union were battling in the Cold War. The militarization of society and of science bring the two closer together: the former's chances for success may hinge on the military potential of its science and technology; while the latter's service to its nation at war may entangle and immerse it more deeply in the regime's political and ideological goals. One pattern thus emerges quite often in the examples studied by this essay: the military potential of science and scientists outweigh and overrule attempts to purify science ideologically.

It was never easy to introduce ideologically driven and compliant science and technology. Purges of scientists or attacks on scientific theories on ideological grounds were often self-defeating and short-lived. Scientists sometimes suffered, but not because they were scientists, rather because they were part of a greater ideological or racial group perceived by the regime to be a potential threat. When challenged, scientists did not defend themselves by claiming the ideological neutrality of their work and demanding intellectual freedom. Instead they strove to blunt attacks by either winning over their critics or enlisting other patrons. Of course, this meant that they had to work closely, or at least more closely with at least some forces in the state and demonstrate their usefulness to them.

The end result was usually partial success for the established (and now sometimes embattled) scientific community, for their critics were eventually silenced and they did safeguard some of their professional prerogatives, but also partially entailed an accommodation to, and collaboration with ideological aspects of the regime, sometimes directly related to their scientific work, other times in a more general form

This is not what the sociologist Robert Merton or physicist Samuel Goudsmit predicted in the aftermath of the Second World War and the beginning of the Cold War.[97] Science is not especially suited to democracy, as Goudsmit claimed, and is able to compromise some of Merton's norms for scientific work. Indeed it is striking, that, despite the great differences between these often very different examples, most of the scientists responded in a similar way when their apolitical science was threatened. ICS tells us that (1) no matter how ruthless, totalitarian, racist, or intolerant a regime might be, when it needs its scientists, it will do what it has to in order to harness them; and (2) whether they support the regime whole-heartedly or not, most scientists, or perhaps better put, scientific communities, will do what they have to in order to be able to do science. Thus science is independent of particular political and ideological regimes: just not in the way most people believe it is.

Notes

1 Society of Fellows, Harvard University and History Department, Princeton University; Department of History, Bowling Green State University; Union College; California State Polytechnic University, Pomona. We would like to thank the colleagues who read and commented on a preliminary version of our manuscript: Mitchell Ash, Charles C. Gillispie, Loren Graham, Alexei Kojevnikov, and Jessica Wang.

2 Jessica Wang, *American Science in an Age of Anxiety: Scientists, Anti-Communism, and the Cold War* (Chapel Hill, NC: University of North Carolina Press, 1998).

3 Doug Weiner, *Models of Nature* (Bloomington, IN: Indiana University Press, 1988).

4 Harold T. Parker, "French Administrators and French Scientists during the Old Regime and the Early Years of the Revolution," in Richard Herr and Harold T. Parker (eds), *Ideas in History* (Durham, NC: Duke University Press, 1965), 85–109.

5 Roger Hahn, *The Anatomy of a Scientific Institution: The Paris Academy of Sciences, 1666–1803* (Berkeley, CA: University of California Press, 1971); Charles C. Gillispie, *Science and Polity in France at the End of the Old Regime* (Princeton, NJ: Princeton University Press, 1980), 81–99.

6 Keith Michael Baker, *Condorcet* (Chicago: University of Chicago Press, 1975), 350–2.

7 Henry Moss, "Scientists and Sans-culottes: The Spread of Scientific Literacy in the Revolutionary Year II," *Fundamenta Scientiae*, 4 (1983): 101–15; Gillispie, *Science and Polity in France at the End of the Old Regime*, 143–84.

8 Ken Alder, *Engineering the Revolution: Arms and Enlightenment in France, 1763–1815* (Princeton, NJ: Princeton University Press, 1997), ch. 8; Terry Shinn, *L'École Polytechnique, 1794–1914* (Paris: Presses de la fondation nationale des sciences politiques, 1980); Ambroise Fourcy, *Histoire de l'Ecole Polytechnique* (Paris: Belin, 1987); Nicole et Jean Dhombres, *Naissance d'un nouveau pouvoir: Sciences et Savants en France, 1793–1824* (Paris: Payot, 1989); Janis Langins, "The Ecole Polytechnique and the French Revolution: Merit, Militarization, and Mathematics," *LLULL*, 13 (1990): 91–105; and Janis Langins, *La République avait besoin de savants* (Paris: Belin, 1987).

9 For a short account, see Langins, "The Ecole Polytechnique". More detailed treatments are provided by the Alder and Langins works cited above.

10 See Charles C. Gillispie, "The *Encyclopédie* and the Jacobin Philosophy of Science: A Study in Ideas and Consequences," in Marshall Clagett (ed.), *Critical Problems in the History of Science* (Madison, WI: University of Wisconsin Press, 1959), 255–89; and L. Pearce Williams, "The Politics of Science in the French Revolution," in Clagett, 291–320.

11 Jessica Riskin, "Rival Idioms for a Revolutionized Science and a Republican Citizenry," *Isis*, 89 (1998): 203–32.

12 Charles C. Gillispie, "De l'Histoire naturelle à la biologie: Relations entre les programmes des recherche de Cuvier, Lamarck, et Geoffrey Saint-Hilaire," in *Collecter, observer, classer* (forthcoming).

13 J.W. Dauben, "Marat: His Science and the French Revolution," *Archives Internationales de l'Histoire des Sciences*, 22 (1969): 235–61; Gillispie, *Science and Polity in France at the End of the Old Regime*, 290–330.

14 See Hahn, *passim*, for references to primary literature on the Jardin.

15 Loren R. Graham, *The Soviet Academy of Sciences and the Communist Party, 1927–1932* (Princeton, NJ: Princeton University Press, 1967); and Alexander

Vucinich, *Empire of Knowledge: The Academy of Sciences of the USSR (1917–1970)* (Berkeley, CA: University of California Press, 1984).

16 Graham, *The Soviet Academy of Sciences*, 24–79; and Kendall E. Bailes, *Technology and Society under Lenin and Stalin: Origins of the Soviet Technical Intelligentsia, 1917–1941* (Princeton, NJ: Princeton University Press, 1978).

17 Lynn Mally, *Culture of the Future: The Proletkult Movement in Revolutionary Russia* (Berkeley, CA: University of California Press, 1990). Katerina Clark, "The Changing Image of Science and Technology in Soviet Literature," in Loren R. Graham (ed.), *Science and the Soviet Social Order* (Cambridge, MA: Harvard University Press, 1990), 266–67.

18 Mally, chapter 6.

19 On specialists during this period, see Bailes.

20 Jeremy R. Azrael, *Managerial Power and Soviet Politics* (Cambridge, MA: Harvard University Press, 1966). On the economics and politics of the NEP period, see Stephen F. Cohen, *Bukharin and the Bolshevik Revolution: A Political Biography, 1888–1938* (Oxford: Oxford University Press, 1980), and Robert V. Daniels, *Conscience of the Revolution: Communist Opposition in Soviet Russia* (Boulder, CO: Westview Press, 1988).

21 On the tense compromise between ideological promotion of young Soviet acolytes like Landau and the surviving tsarist professorate, see Karl Hall, "Purely Practical Revolutionaries," (Ph.D., Harvard University, 1999).

22 For a detailed account of the Shakhty Trial and the subsequent Industrial Party Affair, see Bailes, chapter 3.

23 David Holloway, *Stalin and the Bomb* (New Haven: Yale University Press, 1994).

24 On the education system generally in this period, see Sheila Fitzpatrick, *The Commissariat of Enlightenment: Soviet Organization of Education and the Arts under Lunacharsky, October 1917–1921* (Cambridge: Cambridge University Press, 1970). On the reform of institutes, see Loren R. Graham, "The Formation of Soviet Research Institutes: A Combination of Revolutionary Innovation and International Borrowing," *Social Studies of Science*, 5 (1975): 303–29. On Leningrad Physico-Technical Institute in this period, see Paul Josephson, *Physics and Politics in Revolutionary Russia* (Berkeley, CA: University of California Press, 1991).

25 See Graham, *The Soviet Academy of Sciences*, 24–79; Vucinich, 72–129; and Azrael, chapters 3–4.

26 Bailes.

27 Sheila Fitzpatrick, *The Cultural Front: Power and Culture in Revolutionary Russia* (Ithaca, NY: Cornell University Press, 1992).

28 David Joravsky, *Soviet Marxism and Natural Science, 1917–1932* (London: Routledge and Kegan Paul, 1961).

29 Loren R. Graham, *Science in Russia and the Soviet Union: A Short History* (Cambridge: Cambridge University Press, 1993), ch. 5.

30 Loren R. Graham, *Science, Philosophy, and Human Behavior in the Soviet Union* (New York: Columbia University Press, 1987), chaps 1–2.

31 David Joravsky, *The Lysenko Affair* (Cambridge, MA: Harvard University Press, 1970); Krementsov; Valery N. Soyfer, *Lysenko and the Tragedy of Soviet Science* (New Brunswick, NJ: Rutgers University Press, 1994); Zhores A. Medvedev, *The Rise and Fall of T.D. Lysenko*, (New York: Columbia University Press, 1969); David Joravsky, "Soviet Marxism and Biology Before Lysenko," *Journal of the History of Ideas*, 20 (1959): 85–104; Dominique Lecourt, *Proletarian Science? The Case of Lysenko* (Norfolk, VA: NLB, 1977); Nikolai L. Krementsov, *Stalinist Science* (Princeton, NJ: Princeton University Press, 1997).

32 Kirill O. Rossianov, "Stalin as Lysenko's Editor: Reshaping Political Discourse in Soviet Science," *Configurations*, 3 (1993): 439–56; and Kirill O. Rossianov, "Editing Nature: Joseph Stalin and the 'New' Soviet Biology," *Isis*, 84 (1993): 728–45.

33 Alexei Kojevnikov, "Rituals of Stalinist Culture at Work: Science and the Games of Intraparty Democracy circa 1948," *Russian Review*, 57, 1 (1998): 25–52.

34 Mitchell Ash and Alfons Söllner, "Forced Migration and Scientific Change after 1933," in Mitchell Ash and Alfons Söllner (eds), *Forced Migration and Scientific Change: Emigré German-Speaking Scientists and Scholars after 1933* (Cambridge: Cambridge University Press, 1996), 1–19.

35 Alan Beyerchen, *Scientists under Hitler: Politics and the Physics Community in the Third Reich* (New Haven, CT: Yale University Press, 1977).

36 Rudolf Vierhaus and Bernhard vom Brocke (eds), *Forschung im Spannungsfeld von Politik und Gesellschaft–Geschichte und Struktur der Kaiser-Wilhelm/Max-Planck-Gesellschaft* (Stuttgart: DVA, 1990); Kristie Macrakis, *Surviving the Swastika: Scientific Research in Nazi Germany* (Cambridge, MA: Harvard University Press, 1993).

37 Mark Walker, *Nazi Science* (New York: Plenum, 1995).

38 Robert Proctor, *Racial Hygiene: Medicine under the Nazis* (Cambridge, MA: Harvard University Press, 1988); Paul Weindling, *Health, Race, and German Politics between National Unification and Nazism, 1870–1945* (Cambridge: Cambridge University Press, 1989); Uwe Hoßfeld, "Staatsbiologie, Rassenkunde und Moderne Sythese in Deutschland während der NS-Zeit," in Rainer Brömer, Uwe Hoßfeld, and Nicolaas Rupke (eds), *Evolutionsbiologie von Darwin bis heute* (Berlin: VWB, 2000), 249–305; Uwe Hoßfeld and Thomas Junker, "Synthetische Theorie und 'Deutsche Biologie': Einführender Essay," in Brömer, Hoßfeld, and Rupke, 231–48; Thomas Junker, "Synthetische Theorie, Eugenik und NS-Biologie," in Brömer, Hoßfeld, and Rupke, 307–60.

39 Walker, *Nazi Science*.

40 Jacques Pauwels, *Women, Nazis, and Universities: Female University Students in the Third Reich* (New Haven, CT: Greenwood Press, 184).

41 Alexander Mitscherlich and Fred Mielke, *Doctors of Infamy: The Story of the Nazi Medical Crimes* (New York: Henry Schuman, 1949); Robert J. Lifton, *The Nazi Doctors: Medical Killing and the Psychology of Genocide* (New York: Basic Books, 1986); Benno Müller-Hill, *Murderous Science: Elimination by Scientific Selection of Jews, Gypsies, and Others, Germany 1933–1945* (Oxford: Oxford University Press, 1985).

42 For biology, Ute Deichmann, *Biologists under Hitler* (Cambridge, MA: Harvard University Press, 1996); for physics, Beyerchen, David Cassidy, *Uncertainty: The Life and Science of Werner Heisenberg* (New York: Freeman, 1991), Mark Walker, *German National Socialism and the Quest for Nuclear Power, 1939–1949* (Cambridge: Cambridge University Press, 1989), and Walker, *Nazi Science*; for mathematics, Herbert Mehrtens, "Ludwig Bieberbach and 'Deutsche Mathematik,'" in E. Phillips (ed.), *Studies in the History of Mathematics* (Washington, DC: American Mathematical Association, 1987), 195–247; for psychology see Ulfried Geuter, *The Professionalization of Psychology in Nazi Germany* (Cambridge: Cambridge University Press, 1992), Mitchell Ash, "From 'Positive Eugenics' to Behavioral Genetics: Psychological Twin Research under Nazism and Since," *Historia Pedagogica – International Journal of the History of Education, Supplementary Series*, 3 (1998): 335–58, and Mitchell Ash, "Constructing Continuities: Kurt Gottschaldt und Psychological Research in Nazi and Socialist Germany," in Kristie Macrakis and Dieter Hoffmann (eds), *Science and Socialism in the G.D.R. in Comparative Perspective* (Cambridge, MA.: Harvard University Press, 1999), 286–301, 360–65; for chemistry Ute Deichmann, *Flüchten, Mitmachen, Vergessen* (Weinheim: Wiley-VCH, 2001); for technology, Karl-

Heinz Ludwig, *Technik und Ingenieure im Dritten Reich* (Düsseldorf: Droste Verlag, 1974) and Helmut Maier, "National sozialistische Technikideologie und die Politisierung des 'Technikerstandes': Fritz Todt und die Zeitschrift 'Deutsche Technik,'" in Burkhard Dietz, Michael Fessner, and Helmut Maier (eds), *Technische Intelligenz und "Kulturfaktor Technik"* (Münster: Waxmann, 1996), 253–68.

43 Cassidy; Walker, *Nazi Science*.

44 Michael Neufeld, *The Rocket and the Reich: Peenemünde and the Coming of the Ballistic Missile Era* (New York: Free Press, 1995).

45 Helmuth Trischler, *Luft- und Raumfahrtforschung in Deutschland 1900–1970* (Frankfurt am Main: Campus Verlag, 1992); Ulrich Albrecht, "Military Technology and National Socialist Ideology," in Renneberg and Walker (eds), 88–125, 358–63.

46 Walker, *German National Socialism and the Quest for Nuclear Power*.

47 For physicians, Lifton, Weindling, Proctor; for biologists, Deichmann, Junker; for psychologists, Geuter, Mitchell Ash, "Denazifying Scientists and Science," in Matthias Judt and Burghard Ciesla (eds), *Technology Transfer out of Germany* (Amsterdam: Harwood, 1996), 61–80; for anthropologists, Hossfeld; for geographers, Mechtild Rössler, "'Area Research' and 'Spatial Planning' from the Weimar Republic to the German Federal Republic: Creating a Society with a Spatial Order under National Socialism," in Renneberg and Walker (eds), 126–38, 363–6.

48 Two examples are Dr. Pak Ch'ul Jai, an X-ray physicist, and Dr. Lee Tai Kyu, a professor of chemistry, both of whom held positions at Kyoto Imperial University during the war. See Walter E. Grunden, "Hungnam and the Japanese Atomic Bomb: Recent Historiography of a Postwar Myth," *Intelligence and National Security*, 13 (Summer 1998): 32–60.

49 Yukawa was awarded the Nobel Prize in physics in 1949. Taketani Mitsuo, "Methodological Approaches in the Development of the Meson Theory of Yukawa in Japan," in Nakayama Shigeru, David L. Swain, and Yagi Eri (eds), *Science and Society in Modern Japan: Selected Historical Sources* (Tokyo: University of Tokyo Press, 1973), 24–38.

50 Hiroshige Tetu, "The Role of the Government in the Development of Science," *Journal of World History*, 9 (1965): 320–39. Itakura Kiyonobu and Yagi Eri, "The Japanese Research System and the Establishment of the Institute of Physical and Chemical Research," in Nakayama, Swain, and Yagi (eds), *Science and Society in Modern Japan*, 158–201. Kamatani Chikayoshi, "The History of Research Organization in Japan, *Japanese Studies in the History of Science*, 2 (1963): 1–79.

51 S. Watanabe, "How Japan Has Lost a Scientific War" (September 1945), File: "Historical Information," RG 38, Box 111, U.S. National Archives, Washington, DC.

52 Yomiuri Shimbunsha (ed.), "Nihon no genbaku" (Japan's Atomic Bomb), in *Shôwa shi no Tennô* (The Emperor in Shôwa History), vol. 4 (Tokyo: Yomiuri Shimbunsha, 1968), 78–229.

53 John W. Dower, "'NI' and 'F': Japan's Wartime Atomic Bomb Research," *Japan in War and Peace: Selected Essays* (New York: New Press, 1993), 55–100. Morris Fraser Low, "Japan's Secret War? 'Instant' Scientific Manpower and Japan's World War II Atomic Bomb Project," *Annals of Science*, 47 (1990): 347–60.

54 Walter E. Grunden, "Science under the Rising Sun: Weapons Development and the Organization of Scientific Research in World War II Japan" (Ph.D., University of California at Santa Barbara, 1998).

55 Tessa Morris-Suzuki, *The Technological Transformation of Japan: From the Seventeenth to the Twenty-First Century* (Cambridge: Cambridge University Press, 1994), 144–5.

56 Morris-Suzuki, 145.

57 Excerpts of Lieutenant-General Tada Reikichi, President of the Board of Technology, from Captain George B. Davis to Major Francis J. Smith, "Transmittal of Items from the 'Daily Digest of World Broadcasts'" (27 July 1945), "Japan, Misc" RG 77, Entry 22, Box 173, Folder #44.70, U.S. National Archives, College Park, MD; See also, Major H.K. Calvert to Major F.J. Smith, "Japanese Militarists Want Miracle Weapon" (29 May 1945) "Japan, Misc" RG 77, Entry 22, Box 173, Folder #44.70, U.S. National Archives, College Park, MD.

58 Peter Williams and David Wallace, Unit 731: Japan's Secret Biological Warfare in World War II (New York: The Free Press, 1989).

59 For a concise treatment of McCarthyism in general, see Ellen Schrecker, No Ivory Tower: McCarthyism and the Universities (New York: Oxford University Press, 1986) and Ellen Schrecker, The Age of McCarthyism: A Brief History with Documents (Boston: Bedford Books, 1994).

60 Lawrence Badash, Scientists and the Development of Nuclear Weapons: From Fission to the Limited Test Ban Treaty, 1939–1963 (Atlantic Highlands, NJ: Humanities Press, 1995), 102.

61 Paul Forman, "Behind Quantum Electronics: National Security as Basis for Physical Research in the United States, 1940–1960," Historical Studies in the Physical and Biological Sciences, 18 (1987): 149–229.

62 Richard G. Hewlett and Jack M. Holl, Atoms for Peace and War, 1953–1961: Eisenhower and the Atomic Energy Commission (Berkeley, CA: University of California Press, 1989), ch. 4, "The Oppenheimer Case," 73–112.

63 Wang, Anti-Communism, 289–95.

64 Wang, Anti-Communism, 274–9.

65 "Scientists Affirm Faith in Oppenheimer," Bulletin of the Atomic Scientists, 10 (May 1954): 188–91. "Scientists Express Confidence in Oppenheimer," Bulletin of the Atomic Scientists, 10 (September 1954): 283–6.

66 Steven E. Ambrose, Eisenhower: The President (New York: Simon and Schuster, 1984), 166.

67 Eisenhower to Robert E. Sherwood, April 21, 1954, in The Diaries of Dwight D. Eisenhower, reel 4, 193. See also Dwight D. Eisenhower, Mandate for Change, 1953–1956 (Garden City, NY: Doubleday and Company, Inc., 1963), 314.

68 Hewlett and Holl, 111–12.

69 Wang, Anti-Communism.

70 Interview with Herbert York by Zuoyue Wang, La Jolla, CA, July 18, 1992.

71 Daniel Kevles, "Cold War and Hot Physics: Science, Security, and the American State, 1945–56," Historical Studies in the Physical and Biological Sciences, 20 (1990): 239–64.

72 Wang, Anti-Communism, 261–2, 281.

73 Schrecker, No Ivory Tower, and Schrecker, The Age of McCarthyism; interview with Wolfgang Panofsky by Zuoyue Wang, Stanford, CA, March 5, 1992.

74 Schrecker, No Ivory Tower; Wang, Anti-Communism, 254–261.

75 William Lanouette, Genius in the Shadows: A Biography of Leo Szilard, the Man Behind the Bomb (Chicago: University of Chicago Press, 1994).

76 Wang, Anti-Communism, 277.

77 Wang, Anti-Communism, 289–295, Schrecker, The Age of McCarthyism, 92–94.

78 Zuoyue Wang, In the Shadow of Sputnik: American Scientists and Cold War Science Policy (forthcoming). Wang, Anti-Communism.

79 Schrecker, No Ivory Tower; Schrecker, The Age of McCarthyism, 92.

80 James Reardon-Anderson, The Study of Change: Chemistry in China, 1840–1949 (Cambridge: Cambridge University Press, 1991), 339–64, on 359.

81 Yao Shuping et al., Zhongguo kexueyuan (Chinese Academy of Sciences), 3 volumes (Beijing: Contemporary China Press, 1994), vol. 1, 26–7.

82 Jonathan Spence, *The Search for Modern China* (New York: W.W. Norton, 1990), 572.

83 Yao *et al.*, 87–9.

84 See, for example, Peter Neushul and Zuoyue Wang, "Between the Devil and the Deep Sea: C.K. Tseng, Ocean Farming, and the Politics of Science in Modern China," *Isis*, 91, 1 (2000): 59–88.

85 See, for example, several articles in the Chinese official journal *Bai nian chao* (Hundred Year Tide), June 1999, especially Li Zhenzhen, "Interview with Yu Guangyuan and Li Peishan," 23–30, and "Interview with Gong Yuzhi," 31–27.

86 See Jin Chongji, *Zhou Enlai zhuan*(A Biography of Zhou Enlai), 2 volumes (Beijing: Central Documentation Press, 1998), vol. 1, 234.

87 Dai Minghua, *et al.*, "Zhou Guangzhao" in Lu Jiaxi (ed. in chief), *Zhongguo xiandai kexuejia zhuanji* (Biographies of Modern Chinese Scientists), 6 volumes (Beijing: Science Press, 1991–4), vol. 6, 187–96.

88 Wu Heng, *Keji zhanxian wushinian* (Fifty Years on the Scientific and Technological Front) (Beijing: Science and Technology Documentation Press, 1992), 348–9.

89 See H. Lyman Miller, *Science and Dissent in Post-Mao China: The Politics of Knowledge* (Seattle: University of Washington Press, 1996).

90 Xu Liangying and Qiu Jingcheng, "Guanyu wuoguo 'wenhua dageming' shiqi pipan ai-in-si-tan he xiangduilun de chubu kaocha" (A Preliminary Study of the Movement to Denounce Einstein and Relativity During the Period of the Great Cultural Revolution in Our Country), in Xu (ed. in chief), *Ai-in-si-tan yanjiu* (Einstein Studies), 1 (1989): 212–50.

91 Gong Yuzhi, "Mao Zedong yu zhiran kexue" (Mao Zedong and the Natural Sciences), in Gong, *Zhiran bianzhengfa zai zhongguo* (Dialectics of Nature in China) (Beijing: Peking University Press, 1996), 87–112.

92 He Zuoxiu, "A History of the Establishment of the Straton Model," *Yuanshi zhiliao yu yanjiu* (Documentation and Research in the History of the Chinese Academy of Sciences), 1994, issue no. 6, 16–30.

93 He.

94 Holloway.

95 See "1956–1967 nian kexue jishu fazhan guihua gangyao" (Outline of a Long-Term Plan for the Development of Science and Technology, 1956–1967), in Chinese Communist Party Central Documentation Institute (ed.), *Jianguo yilai zhongyao wenxian xuanbian* (Selected Key Documents Since the Founding of the Country) (Beijing: Central Documentation Press, 1992–), vol. 9, 436–535.

96 Also see for the example of the German Democratic Republic the following articles: Mitchell Ash, "Wissenschaft, Politik und Modernität in der DDR: Ansätze zu einer Neubetrachtung," in K. Weisemann, P. Kroener, and R. Toellner (eds), *Wissenschaft und Politik: Genetik und Humangenetik in der DDR (1949–1989)* (Münster: LIT-Verlag, 1997), 1–26; Mitchell Ash, "1933, 1945 und 1990 – drei Bruchstellen in der Geschichte der deutschen Universität," in A. Söllner (ed.) *Ostblicke – Perspektiven der Hochschulen in den neuen Bundesländern* (Opladen: Westdeutscher Verlag, 1998), 212–37; Mitchell Ash, "Scientific Changes in Germany 1933, 1945 and 1990: Towards a Comparison," *Minerva*, 37 (1999): 329–54.

97 See the introduction to this volume, as well as Jessica Wang, "Merton's Shadow: Perspectives on Science and Democracy since 1940," *Historical Studies in the Physical and Biological Sciences*, 30, No. 1 (1999): 279–306, and David Hollinger, "The Defense of Democracy and Robert K. Merton's Formulation of the Scientific Ethos," in David Hollinger, *Science, Jews, and Secular Culture: Studies in Mid-Twentieth Century American Intellectual History* (Princeton, NJ: Princeton University Press, 1996), ch. 5.

4

FROM COMMUNICATIONS ENGINEERING TO COMMUNICATIONS SCIENCE

Cybernetics and information theory in the United States, France, and the Soviet Union

David Mindell[1], *Jérôme Segal*[2], *and Slava Gerovitch*[3]

Technical content and national context

When it appeared in 1953, the Russian translation of the "Mathematical Theory of Communication" hardly read like the same text written in English by Claude Shannon a few years before.[4] Purged of any trace of man-machine analogies, the translation portrayed communications engineering as an ideologically neutral technical field. The Russian editor replaced Shannon's original title with the Russian, "The Statistical Theory of Electrical Signal Transmission," and he rid the work of the words "information," "communication," and "mathematical" entirely, put "entropy" in quotation marks, and substituted "data" for "information" throughout the text. The editor assured the reader (and the censor) that Shannon's concept of "entropy" had nothing to do with physical entropy and was called such only on the basis of "purely superficial similarity of mathematical formulae."[5] Thus the editor carefully avoided the anthropomorphic connotations of the words "information" and "communication" and at the same time distanced the use of the term "entropy" in the text from its controversial discussions in physics and biology. Trying to avoid any reference to the links between information theory and linguistics, the cautious editor even removed the entire third section of Shannon's paper, the one dealing with the statistical analysis of natural language. The editor drew a sharp line between what he called ideologically deficient, pseudo-scientific attempts to "transfer the rules of radio communication to biological and psychological phenomena" and the practically useful, firmly scientific statistical theory of electrical signal transmission.[6] His discursive strategy was simple: to portray information theory as a purely technical tool with no connection to the ideology-laden biological and social sciences. The translation typified Soviet communications engineers' attempts to remove ideology from their work to place emphasis on technical applications of information theory rather than its potential conceptual innovations.

This translation of Shannon's work occurred at the time of the Soviet anti-cybernetics campaign, about which more will be said later in this paper. Taken on its own, however, the episode points to the charged relationship of information theory, and its Cold War cousin, cybernetics, to ideology and the social sciences in the cultural and political worlds of the 1940s and 1950s. In both cases, highly technical theories put forward by mathematicians acquired ideological baggage which some took up with enthusiasm and others vehemently rejected. In both cases, claims were made for the broader significance of the work outside the technical realms in which they originated, although the authors participated in this process to differing degrees. Norbert Wiener explicitly expanded his theory of prediction and smoothing to a universal science, an extrapolation Claude Shannon always resisted for his information theory. French scientists took up cybernetics and information theory, at least in part, because of their perceived ideological implications. In contrast, Shannon's Russian editors clearly thought they could distinguish between the political and cultural implications of information theory and the raw, technical content which concerned transmission of signals through noisy channels.

A cross-national comparison of the generation and reception of the two new sciences reveals how they acquired, shaped, and were formed by the cultures in which they were embedded. In the United States, information theory and cybernetics emerged out of highly technical, but also irreducibly social, military problems presented by the Second World War. In his post-war *Cybernetics*, Norbert Wiener attempted to abstract his mathematics out of the technical culture which gave birth to it and simultaneously to extend its reach beyond technology into biology, economics, and social systems. Claude Shannon, with a less ambitious but more analytically specific theory, made more modest claims but with equally broad implications: his theory of entropy and channel capacity could model not only technical communications but also human language, and hence a broad array of human activities. In France, commentators and scientists variously saw these new American sciences as bourgeois conjecture, full of mythology and mystification, or as exciting meta-theories capable of uniting diverse disciplines. Similarly, in Russia, an early anti-cybernetics campaign saw Shannon and Wiener's work as embodiments of idealist, reactionary, American pseudoscience. After Stalin's death, however, Russian scientists made a complete reversal of their attitude toward the two new sciences. They extracted from information theory "natural laws" of information processing and made cybernetic feedback the foundation of a dialectical description of language and society. Following initial skepticism and discussion, cybernetics was institutionalized in Europe in a way it never was in the United States.

In the pages that follow, we examine the conception of information theory and cybernetics in the United States and their relationship to the technical cultures within which Shannon and Wiener worked. Then we move on to their reception in France, and how ensuing debates shaped significant French contributions to the information sciences. Similarly, in Russia we trace how an initial ideological hostility transformed into grudging acceptance and then total embrace. As with any such story, the choice of origins is somewhat arbitrary, as both Shannon and Wiener had important influences from France, Russia, and elsewhere, but the two formed significant nodes in ongoing international networks of mathematics. In the interests of analytical simplicity and brevity, we do allow some slippage between information theory and cybernetics, as did the actors under study, though the relationship between them is worthy of a study in its own right.

In all these debates, and indeed in the subsequent history, the essence of cybernetics and information theory prove hard to pin down. Were they embodiments of an American, overarching military-industrial mindset, or new sciences of everything? The intellectual equivalents of the Marshall Plan, or useful new descriptions of electrical signals? Did Shannon and Wiener represent the antecedents of computer science, or an updated expression of Taylorist industrial rationality? Information theory and cybernetics were, perhaps, all of these things and none of them. Their very malleability makes a cross-national comparison worthwhile, as it highlights the difficulty of culling discrete political messages from mathematics, and also the difficulty of harvesting a pure, apolitical mathematics from its historical soil.

The origins of cybernetics: expanding control

Norbert Wiener, in his 1948 book *Cybernetics: or Control and Communication in the Animal and the Machine*, articulated the marriage of communication and control for a generation of engineers, systems theorists, and technical enthusiasts of varied stripes. Wiener declared the merger occurred instantly, obviously and completely in the course of his work on anti-aircraft prediction devices. "I think that I can claim credit," Wiener wrote in his memoir, "...for transferring the whole theory of the servomechanism bodily to communication engineering."[7] Recently, historians have revisited this account, exploring the genesis of Wiener's project, its roots in his earlier work, and its short-term failure and profound long-term effects. But these views center on Wiener: the academic, the intellectual, and the mathematician; they tend not to address his connection to a broader technical culture.[8]

Indeed we have little historical understanding of cybernetics in relation to engineering practice in control, computing, electronics, and communications. Some things were genuinely new about the human/machine relationship articulated by cybernetics, others were derived from existing ideas in engineering. Wiener's cybernetics emerged from the world of automation, military command, and computing during and after the Second World War. Wiener's own work on control systems during the war existed within a set of projects and a technical agenda which aimed to automate human performance in battle through a tight coupling of people and machines. Indeed before Wiener's cybernetics, American technology was already suffused with what would later be called "cybernetic" ideas. Several strong pre-war traditions of feedback mechanisms – including regulators and governors, industrial process controls, military control systems, feedback electronics, and a nascent academic discipline of control theory – suggests a broader and more gradual convergence of communications and control than the strict "Wienerian" account.[9] Servo engineers turned to techniques common in the telephone network to characterize the behavior of powerful feedback devices. Radar engineers adapted communications theory to deal with noise in tracking. Human operators were always necessary but problematic components of automatic control systems. Military technologists had wrestled with the notion of prediction since at least the turn of the century. These were but a few of the features of the technological terrain onto which Norbert Wiener stepped in 1940 when he began working on control systems. Yet Wiener eventually presented cybernetics as a specifically *scientific* discourse of communication and control, distinct from its practitioners. Like Shannon's Russian translator, Wiener attempted to divorce the content of his work from its social soil, and to embed it in a different tradition.

The NDRC and the fire control problem

In 1940 Vannevar Bush formed the National Defense Research Committee (or NDRC), to bring university and industrial research to bear on military problems. Led by Warren Weaver, then also head of the Natural Sciences Division of the Rockefeller Foundation, the NDRC established a committee, called section D-2, responsible for control systems. Under this committee, control engineers developed the technology, indeed the practical philosophy, that Wiener would articulate so effectively in his postwar writing on cybernetics. During the war, much of that philosophy coalesced around difficult problems of antiaircraft fire control. Using artillery to hit fast-moving airplanes pressed to the limits the engineering knowledge of dynamic performance, mathematical precision, corrupted data, and the human operator. Research in data smoothing and prediction – two key elements of fire control – began to formalize an engineering approach based on abstracting the physical world and manipulating it as electrical signals, the basis of later strategies of computing and information processing. Engineering practice evolved in parallel with this theoretical work, and sometimes preceded it.

Weaver's control systems committee brought institutional pressure to bear on communications and control; it placed dual emphasis on Bell Labs, temple of communications, and MIT, which had a strong program in feedback control and servomechanisms (a servomechanism, or servo, is an electric or hydraulic motor that, with the addition of a feedback loop, is able to precisely hold and control its position). During the war, the NDRC control systems committee funded eighty research contracts in feedback theory, devices, and computing, totaling about $10 million at Bell Labs, MIT, and a number of other academic and industrial laboratories. Nearly every American computer pioneer (Atanasoff, Eckert & Mauchly, Shannon, Stibitz) of the time worked on at least one of these contracts during the war. Two of the eighty projects funded Norbert Wiener at MIT.[10]

Wiener had studied electrical networks during the 1930s, and in 1940 he proposed to apply network theory to servo engineering. This work had already been done by Henrik Bode at Bell Labs, however, so in late 1940 the NDRC asked Wiener to bring his knowledge of networks to prediction in fire control. This tricky problem required anticipating the future position of a target aircraft so an antiaircraft gun could lead the target and hit it with a shell, after some finite delay of the shell's time of flight (as much as one minute for high-flying aircraft). Wiener simulated a prediction network on MIT's calculating machine, the Differential Analyzer, and showed encouraging results. On December 1, 1940 the NDRC let a contract to MIT for "General Mathematical Theory of Prediction and Applications." For the contract, Wiener and his assistant, engineer Julian Bigelow, would devise a theory to follow a given curve, chosen to represent the path of an airplane, and estimate the value of that curve at some time in the future. During early 1941 the two designed and built a machine to simulate their ideas on prediction.[11]

Wiener and Bigelow quickly ran into a stability problem. Wiener's network was highly sensitive, even unstable, in the presence of high frequency noise.[12] This was a cousin of the stability problem facing other engineering disciplines which dealt with feedback loops – transient inputs caused oscillations. Wiener quickly realized the problem was fundamental, "in the order of things," (he compared it to Heisenberg's uncertainty principle) and that he would need a new approach. He and Bigelow now turned to statistics, designing a new predictor based on "a statistical analysis of the correlation between the past performance of a function of time and its present and future performance." The new network continually updated its own prediction as time passed and it compared the target's flight path with previous guesses. A feedback network converged on guesses which minimized this error.[13] In modern terms, this device might be described as a one-dimensional neural network, which learned about the world as it gathered new data.

Through the remainder of 1941, Wiener worked out in detail the theory behind his statistical approach, scribbling on a blackboard as Bigelow took notes. Warren Weaver agreed that Wiener's theory could produce an optimal predictor, and let another NDRC contract for Wiener to write up his theoretical results.[14] The product of that contract, Wiener's report, *Extrapolation, Interpolation, and Smoothing of Stationary Time Series*, was published by the NDRC for restricted circulation in early 1942. Here Wiener explicitly brought together statistics and communications theory with engineering of high-power systems:

In that moment in which circuits of large power are used to transmit a pattern or to control the time behavior of a machine, power engineering differs from communication engineering only in the energy levels involved and in the particular apparatus used suitable for such energy levels, but is not in fact a separate branch of engineering from communications.[15]

Building on his own work in harmonic analysis and operational calculus, Wiener constructed a general theory of smoothing and predicting "time series": any problem (including economic and policy questions) expressed as a discrete series of data. While he gestured at electric power and servo design as well as communications, Wiener did not explicitly address any previous work in feedback theory.

Yet Wiener's work, however theoretically important, did not have immediate applications. In late 1942, Weaver reported that for predicting actual recorded target tracks, Wiener's "optimal" method proved only marginally more effective than a far, far simpler design of Henrik Bode's. At its next meeting, the NDRC decided to terminate Wiener's work; the project ended in January, 1943 (Bigelow left to join a statistical fire control group at Columbia).[16]

Wiener's civilian elaboration

Disappointed by his failure to produce a practical device for the war effort, Wiener plunged into elaborating on his work in a context separate from the NDRC's concrete demands. Wiener had a long time interest in physiology, and the previous spring he and collaborators physician Arturo Rosenblueth and neurologist Walter Cannon had begun addressing physiological and neurological feedback. In the spring of 1942 Wiener's papers first mention the idea of the human operator as a feedback element, an integral part of the system. He discussed the "behaviorist" implications of his work in control, "the problem of examining the behavior of an instrument from this [behaviorist] point of view is fundamental in communication engineering."[17] This period, the last few months of Wiener's NDRC program, marked the conception of his "cybernetic vision," which would make him famous after the war. Wiener placed his understanding of the servomechanical nature of the mechanisms of control and communication in both humans and machines at the core of cybernetics, and his program sought to extend that understanding to biological, physiological, and social systems.

For Norbert Wiener, in the midst of the technological war, cybernetics became a civilian enterprise. The cancellation of his NDRC contracts in 1943 put him outside the massive wartime research effort, with access to only civilian resources. His 1943 paper, "Behavior, Purpose, and Teleology," written with Rosenblueth and Bigelow, allied servomechanisms with the "behavioristic approach" to organisms and classified behavior by level of prediction.[18] The paper's philosophical tone and biological metaphors reflected the strictures of secrecy surrounding Wiener's prior work and his new alliance with the life sciences: topics and researchers which, like Wiener, were comparatively free of the war effort. Later Wiener acknowledged the role fire control and prediction played in his thinking, but beginning with "Behavior, Purpose, and Teleology," cybernetics recast military control in a civilian mold.

Most indicative of this alienation and reconstruction is Wiener's consistent hesitation to acknowledge any of the multiple traditions of feedback in engineering which preceded him. In all his writing on cybernetics, he *never* cited Elmer Sperry, Nicholas Minorsky, Harold Black, Harry Nyquist, Hendrik Bode, or Harold Hazen: all published on the theory of feedback before 1940 (their publications became standard citations); all were recognized as important to the field; all speculated on the human role in automatic control; some even wrote on the merger of communications and control and the epistemology of feedback. But Wiener only rarely cited *any* servo theory later than Maxwell's 1867 paper "On Governors."[19] The omissions are striking. Wiener must have been aware of the work: he was closely involved in Vannevar Bush's research program in the 1930s including Hazen's work on servos; he worked with MIT's Servomechanisms Lab and its Radiation Laboratory during the war, and was in touch with Hendrik Bode during the wartime work on predictors. Still he wrote, "I think that I can claim credit for transferring the whole theory of the servomechanism bodily to communication engineering." Wiener placed cybernetics at the end of an intellectual, scientific trajectory, separate from the traditions of technical practice from which it sprang. Wiener's chapter on "Cybernetics in history," from *The Human Use of Human Beings*, refers only to Leibniz, Pascal, Maxwell, and Gibbs as "ancestors" of the new discipline.

Wiener reacted to and built on an evolving understanding, pervasive among engineers and psychologists involved with fire control in the 1940s, that the boundary between humans and machines affected the performance of dynamic systems and was a fruitful area of research. Unlike Wiener, however, NDRC researchers remained bound by military secrecy at least until 1945 and busy with contractual obligations (many remained so after the war). With no publication restrictions and no time obligations to wartime research contracts, Wiener could do and say as he pleased.

Wiener's reformulation had ideological implications, especially in light of his own estrangement from military research. After Hiroshima and Nagasaki, Wiener became critical of the American military's dominance of the country's engineering efforts. In the early forties, he had been anything but a pacifist: he suggested to the army filling antiaircraft shells with flammable gasses to burn enemy planes from the sky; he pondered what types of forested areas and grain crops were most susceptible to fire bombing.[20] Still, the atomic bombs, and perhaps his disappointing NDRC project, changed Wiener's attitude toward military research. His primary substantive contact with what he later called "the tragic insolence of the military mind," occurred under NDRC auspices and ended in January, 1943.[21] Though the "Interpolation, Extrapolation..." paper had significant military applications, Wiener's Cybernetics sought primarily to elaborate it as civilian philosophy, rather than military engineering.

Wiener's efforts to bring his model to broad communities of physiologists, physicians, and social scientists are well documented.[22] Through the informal "Teleological Society," the series of Macy Conferences, and a growing identity as a public intellectual, Wiener elevated his thinking on control and communication to a moral philosophy of technology, and enjoyed enthusiastic response. Of this elevation of the A.A. Predictor to the "symbol for the new age of man," Galison argues that Wiener enshrined an oppositional military metaphor into the civilian science of cybernetics and its descendents.[23] In light of Wiener's wartime work, however, the survival of the oppositional model was also ironic, as Wiener's wartime experience suggests he formulated cybernetics also as a specifically non-military, scientific endeavor.

Nor was Wiener's formulation the only one to emerge from the War with broad implications. In 1945, as the NDRC closed down, it issued a series of Summary Technical Reports. The volume on fire control contained a special essay, "Data Smoothing and Prediction in Fire-Control Systems," by Richard B. Blackman, Hendrik Bode, and Claude Shannon, which formally integrated communications and control and pointed toward generality in signal processing. The authors treated fire control as "a special case of the transmission, manipulation, and utilization of intelligence." They assessed control as a problem in electrical communications, developing analogs to the prediction problem, "couched entirely in electrical language." The authors, like Wiener, recognized the broad applicability of their study, "The input data...are thought of as constituting a series in time similar to weather records, stock market prices, production statistics, and the like." Acknowledging the importance of Wiener's work, Blackman, Bode, and Shannon devoted significant effort to summarizing his statistical approach. Ultimately they rejected it, however, due to problems applying the RMS error criterion to fire control, as well as its assumptions about statistical behavior of human pilots. Instead, the paper formulated the problem as one of communications engineering, drawing heavily on Bode's work in feedback control: "there is an obvious analogy between the problem of smoothing the data to eliminate or reduce the effect of tracking errors and the problem of separating a signal from interfering noise in communications systems."[24] While noting "this analogy...must of course not be carried too far," the paper considered inputs and disturbances in fire control systems as signals in the frequency domain, just like those in telephone communications.

At the same time that Wiener was working through his ideas on cybernetics, of course, Claude Shannon developed his own theory of communication, and the case forms something of a contrast to Wiener's expansive moves. Shannon built on his own experiences in fire control, computing, and cryptography as well as ideas from twenty years before at Bell Labs. In his famous 1948 paper, "A Mathematical Theory of Communication," Shannon provided a measure of channel capacity, in bits per second, which describes the maximum amount of information possible to send down a given channel. He added a serious consideration of noise and a statistical approach to the problem. "Communication theory is heavily indebted to Wiener for much of its basic philosophy and theory," Shannon wrote, citing Wiener's NDRC report.[25] Shannon's measure leads to a theory of efficient coding, how to optimally translate a series of "primary symbols," such as English text, into "secondary code" to be transmitted."[26] As if to solidify the connection between Shannon's theory and fire control, Warren Weaver wrote a popular introduction and explication of information theory, published with Shannon's paper in a small book, in which Weaver called for an expanded context for information theory in a hierarchy of human activity.[27] Yet while others built on Shannon's work and applied it to numerous other problems, including everything from biology to psychology and art, Shannon himself did not make the expansive leaps that Wiener did. In fact, Shannon mocked the "scientific bandwagon" which had grown up around information theory, and warned that "the basic results of the subject are aimed in a very specific direction, a direction that is not necessarily relevant to such fields as psychology, economics, and other social sciences."[28]

In light of the NDRC's research program in fire control, and, for that matter, of decades of pre-war control engineering, Wiener's syntheses of communications and control, human and machine, articulated broad converging patterns as much as created new ones. Cybernetic ideas had as much to do with established and evolving engineering traditions as with any radically new military mindset. *Cybernetics*, the book as well as the movement, articulated a vision of changing human/machine analogies which resonated with a broad audience. Its ramifications in the United States and abroad were significant, if as much for the overarching vision as for any concrete results. Its very malleability, however, of the human/machine analogy and its underlying mathematics, would both undermine cybernetics and be a source of its power, especially as it moved into international environments. Wiener's own postwar politics would not be enough to stabilize the ideology of his formulation, for others, as they took up cybernetics, had politics of their own.

Cybernetics and information theory in France

How, then, was cybernetics received outside the United States, as a military tool, an analytical technique, and a philosophical program? While cybernetics is generally thought to have American origins, the book itself was actually published in France. In that country, information theory was hailed as a new general discipline which included cybernetics. Through the first two congresses dealing with cybernetics, held in Paris in 1950 and 1951, the French adopted and modified Wiener's work. Debates became acrimonious as the French Communist Party strongly engaged itself against cybernetics, which it saw as a "bourgeois" science. From the late 1950s onwards, a kind of normalization of the field took place, which correlated both with the promotion of cybernetics in popular science articles and books and with the institutionalization of cybernetics research in Western Europe. This development also relied on the contributions of a few scientists who took advantage of information theory to fill the existing gap between physics, mathematics and engineering science.

In the spring of 1947, Wiener was invited to a congress on harmonic analysis, held in Nancy, France and organized by the bourbakist mathematician, Szolem Mandelbrojt (1899–1983). During this stay in France, Wiener received the offer to write a manuscript on the unifying character of this part of applied mathematics, which is found in the study of Brownian motion and in telecommunication engineering. The following summer, back in the United States, Wiener decided to introduce the neologism "cybernetics" (from the Greek, meaning the man at the wheel or rudder) into his scientific theory. Though the word is found in the *Gorgias* by Plato, it also had a French usage, though Wiener did not know that the French physicist André-Marie Ampère (1775–1836) had already used it for his classification of the sciences to define "how the citizens can enjoy a peaceful time".[29]

Wiener's book was published in English by Hermann Editions in Paris and by M.I.T. Press, in collaboration with John Wiley & Sons in New York. In an introductory chapter about this "explosive science," Pierre De Latil reminds us that M.I.T. Press tried their best to prevent the publication of the book in France, since Wiener, then professor at M.I.T., was bound to them by contract. As a representative of Hermann Editions, M. Freymann managed to find a compromise and the French publisher won the rights to the book.[30] This became financially significant since after three reprints in six months, the book had sold 21,000 copies. A journalist at *Business Week* compared it with the Kinsey Report, also published in 1948, about the sexual behavior of American people: "In one respect Wiener's book resembles the Kinsey report; the publication reaction to it is at least as significant as the content of the book itself."[31]

The French press reacted enthusiastically. On December 28, 1948 in the newspaper *Le Monde*, a whole page was dedicated to "A new science: Cybernetics" with the subtitle "Towards a governing machine..." The author, Dominique Dubarle, stuck close to the myth of the robot, predicting that man would be replaced by machine even for the functions which require man's intelligence. Far from the technical questions linked to servomechanisms, this perspective was clearly driven by a kind of techno-optimism. New kinds of machines were mentioned: "prediction machines" (like air defense systems), "sensitive machines" (so the blind people could "see" again), and "sorting machines". It is noteworthy that Dubarle identified the key common point of these machines, the capacity to treat information, newly defined according to the scientific context introduced by Wiener and especially developed by Claude E. Shannon in his mathematical theory of communication. "Let's say that those machines are designed to collect and elaborate information in order to produce results which can lead to decisions as well as to knowledge." This was how

Dubarle ended his review, reflecting on "a unique government for the planet" which could as a new "political Leviathan," "supply the present obvious inadequacy of the brain when the latter is concerned with the customary machinery of politics."

This article in *Le Monde* was the impetus for a series of articles in the main intellectual journals like *Esprit* and *La Nouvelle revue française*. We again find Dubarle, in 1950, defending an indeterminist conception of science, necessary for him in order to introduce the scientific notion of information.[32] This philosophical debate turned out to be crucial in 1953, when the famous French physicist, Louis de Broglie, commenting on the theories of quantum physics adopted this same position (see below). During this lapse of two years, between 1948 and 1950, Louis de Broglie had already been confronted with cybernetics. A "Circle of Cybernetical Studies" (*Cercle d'Etudes Cybernétiques*) had been created and de Broglie was the Honorary President of this first association with the word "cybernetics" in its name.[33] Vallée, Scotto di Vettimo and Talbotier decided as early as 1949 to gather interested readers of Wiener's book.[34] Whereas many people read and discussed Wiener's book, few scientists were in touch with Americans involved in this field. For instance, the physicist Leon Brillouin (1889–1969), who had lived in New York since 1941, tried to organize visits from French officials to take advantage of the latest developments related to computers.[35] J. Pérès, director of the Institut Blaise Pascal in Paris (created in 1946), went with L. Couffignal to the United States, but Couffignal, then in charge of the construction of the first French computer, preferred advocating a different "French" conception and decided to ignore the American accomplishments related to the construction of the first computers. This turned out to be an important error and the notorious French delay in computing finds an explanation in the fact that so much credit had been accorded to Couffignal.

French political context

Still, the fact that Couffignal decided to ignore U.S. research has to be understood in the French context of this time. In 1947, France was marked by political instability. In the November 1946 legislative election, the French Communist Party came first, with nearly a third of the votes, but in May 1947, the Communist ministers were dismissed by the President Ramadier who followed Truman's appeal from March 15th to all Western countries to exclude all communist forces from governments. French people still had to live with rationing, and by the end of August 1947, the daily bread value per inhabitant went under the 200g level. Strikes in October and November led to the resignation of the Ramadier cabinet.

These strikes affected the reception of cybernetics in France. At the Conservatoire Nationale des Arts et Métiers (C.N.A.M.), a series of five public lectures had been announced, dealing with servomechanisms. In his introductory remarks, Albert Métral (1902–62) advocated the "French technology" which could in his view could easily compete with the "science and techniques from abroad." The participants were mostly scientists who also worked with the military, especially from the engineering sciences or telecommunication research. Trained in mechanics, Métral praised the French "grey matter potential." This kind of scientific nationalism was indeed associated with a vague anti-Americanism, as in some of the public declarations made at this time by General de Gaulle.[36] One of the lectures organized in Paris in the last week of October 1947 had to be cancelled because of the strikes.[37] The general climate of opinion at this time was somewhat hostile to American culture and science, and this was only the beginning of the Cold War era. France was really "between the East and the West," which was made clear whenever the Marshall Plan or the status of Germany were discussed.[38]

This political context, then, set the stage for cybernetics, viewed as an American theory, to be introduced in France. In 1950, the French mathematician G.-Th. Guilbaud in his article entitled, "Cybernetical Divagations," criticized the use of a "fashionable name" and wondered if, in the development of cybernetics, there were not some "improper associations," "fuzzy meaning" and constitution of "myths." Nevertheless, he recalled that cybernetics was born out of a desire for unification and that, as such, it was worthy of consideration.[39] This idea of unification provided the impetus for the first scientific congresses on cybernetics.

Discussions at the first French congresses

The first two congresses dealing with cybernetics or information theory gathered scientists from different backgrounds with different goals.[40] Mathematicians, for instance, were not much interested in the very general considerations contained in Cybernetics, while some engineers, who in France were somewhat despised by the intellectual elite, were intrigued by this book, in which they saw as a possibility of gaining social recognition. Generally, these congresses allowed a first timid institutionalization of cybernetics.

Instead of a congress, between April and May 1950 Louis de Broglie organized a series of lectures. The general title was "Cybernetics," with the subtitle, "Signal and Information Theory."[41] Dennis Gabor was the only scientist from abroad who insisted, like Louis de Broglie, on bringing cybernetics into the physical sciences to avoid it becoming a part of mathematics. Studies on Brownian motion, for instance, were considered helpful for telecommunication engineers. Engineers involved in this field generally accepted this suggestion, and Julien Loeb, from the National Center for the Study of Telecommunications (C.N.E.T.), who also had presented a paper at the C.N.A.M. in 1947, recalled that "If sciences like biology, sociology etc. should benefit from the theoretical works exposed in these series of lectures, the telecommunication techniques themselves should also profit."[42]

It was only after these lectures that information theory was progressively recognized as an autonomous scientific discipline. André Blanc-Lapierre, a trained physicist who decided to work on noise effects, remembers that prior to this lecture series, his colleagues found his work too impregnated with mathematics and that in the mathematics community, he was criticized for not having thoroughly studied probability theory.[43]

Cybernetics appeared again one year later in a congress titled "computing machines and human thought," held in Paris in January 1951. This congress was aimed at a larger public; as we can read in a report written by Paul Chauchard, it was "the first manifestation in France of the young cybernetics, with the participation of N. Wiener, the father of this science."[44] The anti-Americanism expressed at the end of the 1940s had almost vanished. The Marshall Plan had been accepted, two countries had been created in Germany and France was now clearly on the Western side.[45] For this congress, sponsored by the Rockefeller Foundation, a number of foreigners were invited, including Howard Aiken, Warren McCulloch, Maurice Wilkes, Grey Walter, Donald MacKay and Ross Ashby, along with Wiener who was staying in Paris for a couple of months at the Collège de France. It is no surprise that the two French scientists who organized the conference were the two who had visited the U.S. laboratories, Couffignal and Pérès.

Three hundred people attended the congress where fourteen machines from six different countries were demonstrated, including a mechanical chess player by Torres y Quevedo and the famous "Turtle" conceived by Grey Walter, two machines specially designed to imitate human behavior. Studying the thirty-eight presented papers and the script of the reported discussions, one can make two points. First, whereas in France the mathematicians seemed to dominate research related to computing machines, one finds physicists in the same position in the U.K. Secondly, information theory already played an important role in the development of the analogy between the human brain and computing machines. McCulloch for instance suggested that the nervous system makes use of "logarithmic processes," which are also utilized by telecommunications engineers. So, at the end of these two conferences, there was already a kind of "French cybernetics," which had been recognized in the scientific establishment. In 1952, a first assessment of American cybernetics was made by Louis de Broglie, who also attended this congress. He estimated that overall, cybernetics had not been as innovative as it could have been and that in fact, servomechanism theory had already been established as an independent discipline without it.[46] This was the time when cybernetics became the focal point of an important ideological debate.

Ideological attack from the French Communist Party

Since the outbreak of the Cold War, philosophy of science had developed into a major ideological battlefield, and cybernetics quickly became the subject of several lively discussions. Already in May 1948, before the publication of *Cybernetics*, Jacques Bergier had written an article on "an ongoing new revolution even more important than that of the atomic age," the general theory of automata. His position was ambiguous: on the one hand he expressed his enthusiasm, and on the other, he feared that "robots will take the place of workers."[47] He referred mostly to Soviet science, and, as in the American case, his two reference fields which led to a general theory were automatic exchange systems and anti-aircraft technology.

One year later, in the same weekly, Jean Cabrerets attacked cybernetics.[48] Referring to the project of a French computing machine that Couffignal had begun to conceive, Cabrerets proudly announced that "the universal French machines have chosen intelligence" and will soon "eclipse those [American] electronic brains like the Eniac." His comments on the visit that Brillouin organized for Couffignal were typical of the period: "It is to us significant that these new 'electronic brains' were born after and not before the visit that Brillouin and Couffignal made last year to the universities of Harvard and Philadelphia."

To understand these vehement attacks against American science, one has to consider the action of the Kominform, created in September 1947, and the evolution of the Soviet positions (see the section on Soviet cybernetics below). The anti-cybernetics campaign in France culminated with an article by André Lentin, in the official monthly journal of the communist party, *La Pensée* (The Thought). Lentin wrote on "Cybernetics: Real Problems and Mystifications".[49]

In an interview, Lentin remembers that he was annoyed at the beginning of the 1950s because "bad scientists" used cybernetics' popularity to publish and Wiener's book was used in almost all disciplines as a kind of panacea.[50] In his article, he directed the reader to the proceedings of the 1947 conferences held at the C.N.A.M. to understand a real general theory of servomechanisms. He tried to show that what Wiener did was more or less merely commentary on James Watt's work on the governor of a steam engine (which is, of course, a ridiculous claim). Cybernetics was simply described as a "gigantic enterprise of mystification." The only interest Lentin saw in Wiener's theory was the description of negative feedback which showed for him a "clear and conscious expression of the dialectic laws." Apart from this point, he believed that cybernetics should be rejected because it is a legitimatization of three dangerous bourgeois ideologies: Taylorism, robots without class consciousness instead of workers; idealism, interpreting a formal analogy between information and entropy as an identity; and, above all, capitalist economy, if one thinks of such feedback laws as "offer and demand determine the market."[51]

French contributions to information theory

Apart from these ideological critics, the mid-1950s were also marked by French contributions to the development of information theory. Brillouin, a central figure for the exchanges he organized between France and the United States, was also one of the first promoters of information theory in physics. His first publication on this theme, which emphasized the analogy between entropy and information as defined by Shannon in order to explain paradox like that of the Maxwell's Demon, dates from 1949.[52] A few years later he managed to rewrite most of the chapters of physics using information theory. The corresponding book became known worldwide as a milestone in the development of information theory.[53] With his involvement in information theory, Brillouin managed to fill the gap between his interests for engineering science (for instance, he contributed during the war to the development of magnetron) and his general conceptions of physics.

In mathematics, two names are particularly significant and information theory was again at the crossing between different interests. Benoît Mandelbrot (born in 1924), who proofread with Walter Pitts Wiener's manuscript for *Cybernetics*, made for his Ph.D. in mathematics a clear connection between game theory and information theory.[54] He showed, for instance, that both thermodynamics and statistical structures of language can be explained as results of minimax games between "nature" and "emitter." He also made the connection between the definitions of information given by the British statistician Ronald A. Fisher in the 1920s, by the physicist Dennis Gabor in 1946 (he was born in 1900 in Budapest but exiled in Great Britain since 1934) and the already well-known definition proposed by Shannon.

Beyond a mathematical generalization of all these definitions, one finds also an important development of the unifying character of information theory. This is a noteworthy aspect of another Ph.D. written by Marcel-Paul Schützenberger (1920–96) and also published in 1953.[55] As early as 1951, preparing this work, Schützenberger had showed that a generalized information theory could be used as well for the analysis of electrical circuits as for the determination of liminal sensibility values in drug design or for botanic taxonomy.[56]

These French contributions to information theory quickly found recognition in the United States more than in their own country. From 1952 to 1954, for instance, Mandelbrot was at M.I.T. and then at the Institute of Advanced Studies in Princeton, and it is emblematic that when a new journal was founded on information theory, *Information and Control*, in 1958, Brillouin was one the three editors (with the British Colin Cherry and the American Peter Elias), and among the scientists of editorial board, one finds L. Couffignal and B. Mandelbrot next to Shannon and Wiener.[57]

A French consensus regarding the place of cybernetics

After Stalin's death in May 1953, the general strike that took place in France in August and the election of René Coty in December, a period of normalization set in. This was profoundly marked for the cybernetics field by the popularization of Wiener's theory with a European reappropriation and the beginning, even in the 1950s, of the institutionalization of cybernetics research.

A few journalists decided to write books recasting the work of Wiener and all the American scientists who had participated in the birth of cybernetics in a wider (French!) tradition beginning with Ampère and including for instance, the work of Lafitte presented in his *Thoughts on the Science of Machines*, published in 1932.[58] The main books, assuring a large audience for cybernetics, were written by the journalists Pierre de Latil, Albert Ducrocq in France and also by the mathematician Vitold Belevitch in Belgium.[59] With this new rebuilt history of cybernetics, the aim was to present an alternative to the Anglo-Saxon empiricism which gave significant importance to simulations. For the French, a few realizations are indeed shown as examples, but the "as if" does not become an "is" like in America. Presenting the latest work on machine languages, Belevitch showed the underestimated difficulties related to the understanding of language. General enthusiasm had given place to moderate assessment.

At the same time, the institutionalization of cybernetics took place along a French – Belgian axis.[60] The International Association for cybernetics was created in Namur, and it is in this Belgian city that international conferences are regularly held, the first one in 1956 and the most recent in 1998. In France, cybernetics helped to reshape the boundaries between mathematics, physics and engineering, making them more permeable. After identifying the ideological load of cybernetics and using it in the debates of the national context, cybernetics was regarded as a pool of new ideas to promote interdisciplinarity.

The Second International Congress on Cybernetics in Namur in September 1958 was attended by a small "reconnaissance mission" (three delegates) from the Soviet Academy of Sciences. Upon their return, the delegates complained in their report that "the small size of the Academy of Sciences delegation does not correspond to the scale of our country and to the tasks put before Soviet science in the field of cybernetics."[61] In subsequent years, however, Soviet delegations remained small and low-profile. In 1960, Communist Party bureaucrats rejected a proposal for the Soviet Academy Council on Cybernetics to join the International Association for Cybernetics.[62] As in the French case, the history of Soviet cybernetics was marred by a controversy.

Cybernetics and information theory in the Soviet Union

The evolution of Soviet attitudes toward cybernetics and information theory in many ways paralleled the French story, but with stronger accents: the initial Soviet rejection of cybernetics was more decisive, while the later embrace of this field proved more wide-spread and profound. In 1954, the *Short Philosophical Dictionary* – a standard ideological reference – defined cybernetics as a "reactionary pseudo-science," "an ideological weapon of imperialist reaction."[63] By the mid-1950s, cybernetics was portrayed as an innocent victim of political oppression and "reha-bilitated," along with political prisoners of the Stalinist regime. In the early 1960s, cybernetics was canonized in a new Party Program and hailed as a "science of communism." Soviet intellectuals of the Khrushchev era put cybernetics forward as a project for radical reform, challenging the Stalinist legacy in science and society. By the early 1970s, however, former cybernetics enthusiasts were left disillusioned, while cybernetic discourse was appropriated by the political establishment and cybernetics turned into a tool for maintaining the existing order rather than changing it.[64]

The anti-cybernetics campaign in the USSR

The early 1950s – the time when cybernetics and information theory became known to the Soviet reader – was the wrong time to propagate in the Soviet Union ideas originated in the West. That applied not only to political doctrines, but to scientific theories and engineering approaches as well. In the Cold War wave of anti-American propaganda in the early 1950s, nearly a dozen sharply critical articles appeared in Soviet academic journals and popular periodicals, attacking cybernetics and information theory as products of American imperialist ideology and totally ignoring Russian traditions in these fields.[65]

Soviet critics charged that Shannon's theory of communication reduced the human being to a "talking machine"[66] and equated human speech with "just a 'flow' of purely conditional, symbolic 'information,' which does not differ in principle from digital data fed into a calculating machine."[67] Wiener's formula, "information is information, not matter or energy," provoked a philosophical critique of the concept of information as a non-material entity.[68] Repeating Lenin's criticism of some philosophical interpretations of relativity physics in the early twentieth century, Soviet authors castigated cyberneticians for replacing material processes with "pure" mathematical formulae and equations, in which "matter itself disappears."[69] Cybernetics was labeled a "pseudo-science produced by science reactionaries and philosophizing ignoramuses, the prisoners of idealism and metaphysics."[70]

The anti-cybernetics campaign turned into a relentless war on words like "information" and "entropy," which crossed the boundaries between the animate and the inanimate. In his Cybernetics, Wiener attempted to bring together Shannon's concept of entropy as a measure of uncertainty in communication and Erwin Schrödinger's concept of negative entropy as a source of order in living organisms. Wiener identified information with negative entropy and aspired to create a common language for describing living organisms, self-regulating machines (for example, servomechanisms and computers), and human society. The critics argued that such crossing of disciplinary boundaries was illegitimate and accused cyberneticians of philosophical, ideological, and eventually political errors. As in the French case, the Soviet critique of cybernetics served a particular political agenda; it was inspired if not directly commissioned by the Communist Party and became part of a general wave of anti-American propaganda in the context of the Cold War.[71]

Schrödinger's analysis of life within the framework of the chromosome theory became a prominent target in Trofim Lysenko's crusade against genetics.

Lysenko, backed by high-ranking Soviet officials, attempted to discredit the use of physical methods and conceptual apparatus in biology.[72] In his 1940 dispute with the leading Soviet specialist on probability theory, Andrei Kolmogorov, Lysenko argued that "biological regularities do not resemble mathematical laws" and equated the use of statistics in support of genetics with the submission to "blind chance."[73] Trying to protect their political and institutional domination in Soviet biology, the Lysenkoites erected a philosophical Chinese wall between biology, on the one side, and physics and mathematics, on the other. This posed serious obstacles before information theory and cybernetics, which attempted to breach that wall.

Seeking to avoid political complications, Soviet mathematicians and engineers working in the field of control and communications kept their studies strictly technical and eschewed man/machine analogies. In the late 1930s and early 1940s, Kolmogorov developed a prediction theory of stationary processes similar to Wiener's, but did not make any attempt to extend its applications to the life sciences or social sciences.[74] Kolmogorov was also among the first mathematicians to appreciate the significance of Shannon's "Mathematical Theory of Communication."[75] Kolmogorov and his students developed a rigorous mathematical foundation of information theory, providing precise definitions and meticulous proofs of major theorems. The 1953 Russian translation of Shannon's work unfortunately transformed the original nearly beyond recognition; working under Soviet ideological censorship and self-censorship, a cautious editor removed not only ideologically suspicious passages, but also Appendix 7, which seemed too abstract for a technical paper. As Kolmogorov later discovered with great disappointment, some of his important theoretical results had already been published by Shannon in the cut-out fragments.[76]

The rehabilitation of cybernetics and the new era

In March 1953, with the death of Stalin, the Soviet Union entered a new era. The political "thaw" brought significant changes to all spheres of Soviet life, including science and technology. The period of forced isolation of Soviet science and technology from its Western counterparts came to an end. The division into "socialist" and "capitalist" science no longer held; claims were made for the universality of science across political borders. The Soviet leadership embarked on a course of rapid assimilation of modern Western scientific and technological advances. In March 1955, a special governmental committee prepared a classified report, "On the State of Radioelectronics in the USSR and Abroad and Measures Necessary for Its Further Development in the USSR." This report emphasized the Soviet lag in communications engineering, control engineering, and computing and blamed it on the anti-cybernetics campaign: "As a result of irresponsible allegations by incompetent journalists, the word 'cybernetics' became odious and cybernetic literature was banned, even for specialists, and this has undoubtedly damaged the development of information theory, electronic calculating machines, and systems of automatic control."[77] In October 1955, the Academy of Sciences, the State Committee on New Technology, and the Ministry of Higher Education submitted to the Party Central Committee a top secret report, "The Most Important Tasks in the Development of Science in the Sixth Five-Year Plan," which, in particular, called for a significant expansion of studies in the theory of probabilities, including information theory. "It is imperative," the report stressed, "to achieve a radical improvement in the application of probability theory and mathematical statistics to various problems of biology, technology, and economics. The void existing here must be filled."[78]

As a sign of recognition of the importance of information theory for the national defense, the Soviet authorities became concerned with potential leaks of Soviet results in this field to the West. In August 1955, when Kolmogorov was invited to Stockholm to give a series of lectures on the theory of probabilities, the Party Central Committee allowed him to go only under the condition that he would not lecture on information theory. The head of the Science Department of the Central Committee argued that "certain aspects of information theory, if developed further, may become very important for secret work."[79] Ironically, as soon as ideological obstacles to the development of information theory were removed, the policy of military secrecy imposed new, even more severe restrictions on this field.

In August 1955, in a drastic reversal of the earlier philosophical critique, the journal *Problems of Philosophy* published the first Soviet article speaking positively about cybernetics and non-technical applications of information theory, authored by three specialists in military computing: Aleksei Liapunov, a noted mathematician and the creator of the first Soviet programming language; Anatolii Kitov, an organizer of the first military computing centers; and Sergei Sobolev, the deputy head of the Soviet nuclear weapons program in charge of the mathematical support. They presented cybernetics as a general "doctrine of information," of which Shannon's theory of communication was but one part. "Cybernetics," they wrote, "combines common elements from diverse fields of science: the theory of communication, the theory of filters and anticipation, the theory of tracking systems, the theory of automatic regulation with feedback, the theory of electronic calculating machines, physiology, and so on. Cybernetics treats various subjects of these sciences from a single point of view – as systems that are processing and transmitting information."[80] The three authors interpreted the notion of information very broadly, defining it as "all sorts of external data, which can be received and transmitted by a system, as well as the data that can be produced within the system."[81] Under the rubric of "information" fell any environmental influence on living organisms, any knowledge acquired by man in the process of learning, any signals received by a control device via feedback, and any data processed by a computer.

Treating information theory as an "exact science," Soviet specialists saw its mission in bringing rigor into disciplines deeply corrupted by ideological and political pressures. Kolmogorov insisted that now, with the advent of cybernetics and information theory, "it is impossible to use vague phrases and present them as being 'laws,' something that unfortunately people working in the humanities tend to do."[82] "The laws of existence and transformation of information are objective and accessible for study," wrote the mathematician Igor' Poletaev, the author of *Signal*, the first Soviet book on cybernetics. "The determination of these laws, their precise description, and the use of information-processing algorithms, especially control algorithms, together constitute the content of cybernetics."[83] Soviet cybernetics transcended the domain of engineering and fashioned itself as a *science*, a systematic study of the laws of nature. The "nature" that cybernetics studied, however, was of a special kind: it was an "objective" world constituted by information exchanges and control processes.

Liapunov and his colleagues soon put forward an ambitious project for the comprehensive "cybernetization" of Soviet science. Lecturing in diverse scientific, engineering, and public audiences, Liapunov carried with him a huge human-size table, whose rows represented twelve methods of cybernetic analysis (determining information exchanges, deciphering information code, determining the functions and elements of the control system, etc.) each of which was applied to eight fields of study (economics, computer science, hardware design, production control, linguistics, genetics, evolutionary biology, and neurophysiology), represented by columns.[84] Biologists and linguists, physiologists and economists, computer programmers and engineers all found a place for themselves in this grand design. In 1956–7, Liapunov and his associates delivered over one hundred lectures on cybernetics in various academic institutions. Soviet cybernetics spread over a wide range of disciplines and became a large-scale social movement among Soviet scientists and engineers.

In April 1959, the Academy of Sciences created the Council on Cybernetics to coordinate all Soviet cybernetic research, including mathematical and engineering aspects of information theory. The Academy also established the Laboratory for the Systems of Information Transmission, later transformed into the Institute for the Problems of Information Transmission, which became the leading Soviet research center in communications engineering.[85] Institutionally and conceptually, Soviet communications engineering was brought under the roof of cybernetics; the Laboratory director Aleksandr Kharkevich became deputy chairman of the Cybernetics Council.

Soviet cybernetics as a trading zone

Soviet cybernetics served as a "trading zone," in which information theory concepts could transcend the boundaries of communications engineering and spread into the life sciences and the social sciences.[86] Bringing genetics under the cybernetic umbrella, in particular, served an important purpose: to protect Soviet geneticists from Lysenkoites' attacks. "A 'unit of hereditary information' sounded less anti-Lysenkoist than a 'gene,'" recalled geneticist Raisa Berg.[87]

Soviet genetics found an institutional niche among the communication sciences, the domain of mathematicians and engineers, where the Lysenkoites could not reach. Mathematicians Liapunov and Sobolev declared: "A living organism develops out of certain embryonic cells in which somewhere lies information received from the parental organisms. This is not physics; this is not physiology; this is the science of the transmission of information."[88] They argued that since Lysenko could not prove the flow of hereditary information from an organism as a whole to its embryonic cells, his claim of the inheritance of acquired traits must be false. On the other hand, they asserted the validity of classical genetics on the basis of its "full agreement with the ideas advanced in cybernetics."[89] The prominent evolutionary biologist Ivan Shmal'gausen, one of the main targets of Lysenko's 1948 speech, defended his theory of stabilizing selection by "translating Darwin's theory into the language of cybernetics."[90] The Council on Cybernetics provided support for persecuted biologists; the series *Problems of Cybernetics*, edited by Liapunov, regularly published articles on genetics, which could not appear in the biological journals, controlled by the Lysenkoites.

In the field of linguistics, a crucial mediating role was played by the prominent Russian émigré linguist Roman Jakobson, who since 1949 taught at Harvard University. Jakobson was fascinated by Shannon's work and applied Shannon's method of calculating the entropy of printed English in his analysis of spoken Russian.[91] Jakobson saw a deep similarity between Shannon's choice of binary digits (bits) as minimal units of information and his own earlier idea of using binary oppositions as the structural basis for organizing phonemic distinctive features into a phonological system. In Jakobson's view, Shannon's theory helped generalize Jakobson's insight about the underlying binary structure of spoken language to human communication in general.[92] In 1957, Jakobson became an Institute Professor at MIT, where he helped establish the Center for Communications Sciences; in 1958 he joined the editorial board of the journal *Information and Control*. Starting from the mid-1950s, Jakobson regularly visited the Soviet Union and actively propagated the innovations brought into linguistics by information theory.

Models of communication as exchange of information

The model of human communication as information exchange became very popular among young linguists who challenged traditional Soviet linguistics, which relied on intuitive concepts and ideological declarations. Ironically, they elaborated a new concept of meaning based on Shannon's notion of information, even though Shannon himself had intentionally excluded any consideration of meaning from his communication theory. Linguists Igor' Melčuk and Alexander Zholkovsky developed a formal model of natural language, in which they turned Shannon's definition of information as "that which is invariant under all reversible encoding or translating operations"[93] into a definition of meaning as "what is common in all texts that are intuitively perceived as equivalent to the original text."[94] In a Soviet context, Shannon's model of communication crossed the boundaries between engineering and science to serve as a basis for an alternative to the dominant linguistic discourse.

Searching for rigorous laws in linguistics, Kolmogorov and his students conducted a series of experiments on measuring the entropy of printed texts, using a modified version of Shannon's letter-guessing method.[95] Kolmogorov was particularly pleased to remark (in private) that from the viewpoint of information theory (Soviet) newspapers were less informative than poetry since political discourse employed a large number of stock phrases and was highly predictable in its content.[96] On the other hand, brilliant poetry, despite the strict limitations imposed by the poetic form, carried more information, for original poetic expressions were much more difficult to guess.

Kolmogorov's poetic studies had a surprising outcome, leading to the elaboration of an original mathematical theory of complexity related to the concepts of information and entropy. While Shannon interpreted entropy as a measure of uncertainty, and Wiener as a measure of disorder, Kolmogorov viewed it as a measure of complexity. Kolmogorov put forward an algorithmic approach to the definition of information as an alternative to Shannon's probabilistic approach. In his view, the main problem with the probabilistic approach was that it precluded the possibility of calculating the amount of information in the case of a unique message, for example, Tolstoy's novel *War and Peace*. "Is it possible to include this novel in a reasonable way into the set of 'all possible novels,'" Kolmogorov asked sarcastically, "and further to postulate the existence of a certain probability distribution in this set?"[97] He proposed to measure the amount of information in an individual object with relation to another individual object, based on the notion of "relative complexity," or entropy, of those objects. He defined the relative complexity of an object (depending on the "method of programming") as the minimal length of a "program" that can produce that object.[98] "If some object has a 'simple' structure," he explained, "then for its description it suffices to have a small amount of information; but if it is 'complex,' then its description must contain a lot of information."[99] Kolmogorov argued that, within his algorithmic approach, the complexity of the novel *War and Peace* could be "uniquely determined," given certain *a priori* information about the language, style, and content of the text.[100] His reformulation of both information theory and probability theory in terms of complexity was perceived in the mathematics community as "almost a cultural revolution, turning both subjects inside out, and reversing the order in which they are normally considered."[101]

Paradoxically, cybernetics, which was supposed to bring formal rigor and exact reasoning to all disciplines, was itself conspicuously lacking a formal definition. Soviet cyberneticians often had very different notions about the content and boundaries of cybernetics. In his 1958 article in *The Great Soviet*

Encyclopedia, Kolmogorov defined cybernetics as a discipline studying "the methods of receiving, storing, processing, and using information in machines, living organisms, and their associations."[102] In the same volume, Kolmogorov also published an entry on information, which he introduced as the "main concept of cybernetics."[103] Mathematician Andrei Markov, Jr., ridiculed Kolmogorov's definitions, arguing that they produced a vicious circle. Kolmogorov responded by defining information as an "operator which changes the distribution of probabilities in a given set of events." Markov dismissed that definition too, mockingly describing how "a given computer would receive a given operator, which changes the distribution of its probabilities, and store this operator on its magnetic drum."[104] In cybernetic discourse, the word "information" had two very different meanings: in information theory, the "amount of information" characterized the uncertainty removed by the "information source"; in computing, on the other hand, the term "information" stood informally for any kind of data processed by a computer. The mechanical unification of information theory and computing in the Soviet Union under the rubric of cybernetics mixed the two uses of the term "information" together and produced the confusion pointed out by Markov. The insurmountable difficulty of forging a common language for all members of the diverse cybernetic community to a large extent undermined the entire project for the "cybernetization" of science. Soviet cybernetics, which at first had emerged as an alternative to official philosophy and a movement for radical reform, eventually lost its rebellious spirit and turned into a pliable philosophical doctrine of the "dialectical rotation of information and noise."[105]

Conclusion

The main difference between Soviet cybernetics and its American and French counterparts is not to be found in the range of cybernetic applications or the types of mathematical models used. In this sense, there was a great similarity across the borders, due to the systematic Soviet efforts to appropriate the latest American and Western European techniques and technologies. The main difference lies in the political and cultural meanings attached to cybernetic ideas.

The history of cybernetics and information theory is one of crossing cultural, political, and disciplinary boundaries. Wiener abstracted a general scientific theory out of technical culture, and his theory was broadly interpreted by American political scientists, anthropologists, economists, and social scientists. Through Wiener's supple hands, what started out as an applied method of military computing transformed into a vision of the new bio-machine age. In the West, cybernetics contributed to the already strong contemporary traditions of mathematical reasoning in biology, physiology, linguistics, and economics by expanding the arsenal of mathematical and engineering tools used in those disciplines for modeling and implementation of control and communication mechanisms. In the United States, Wiener's formulation of cybernetics as civilian science of technology and society helped to legitimize ideas originally developed and continually applied to warlike purposes. Ironically, that vision's military roots, and many of its Cold War military applications, were at odds with Wiener's personal pacifist stand.

Crossing international borders placed cybernetics and information theory in completely different cultural contexts, in which the question of national origins of scientific ideas suddenly acquired great political significance. The Soviet ideological campaigns against Western influences condemned information theory and cybernetics as reactionary and idealistic. The Soviet position had great impact on French Communists and the subsequent controversy over these theories in France. In both France and the Soviet Union, cybernetics and information theory could be adopted only after their "domestication," i.e. adaptation to the specific cultural situations in the two countries.

In France, reactions of many scientists towards cybernetics were, from the beginning, marked by a kind of diffuse nationalism. The French attempted to appropriate cybernetics as their own by claiming Ampère's priority. Even if Wiener's work had to be mentioned, it was only to add immediately that the book was published in Paris and that Ampère had used the word 'cybernétique' as early as the 19th century. The communist party supported this campaign until it reversed its position following changes coming from Moscow.

Once cybernetics was sufficiently reinterpreted, France and the Soviet Union deployed its ideas differently. In France, from the mid-1950s onwards, cybernetics was used to promote interdisciplinary fields in which engineers as a group found public recognition. Soviet cybernetics, on the other hand, emerged during the post-Stalin era as a cross-disciplinary project and a social movement with a distinct mission – to reform Soviet science, both politically and intellectually – after the years of Stalinism. Western scientists viewed cybernetics as a useful method for solving a wide range of theoretical and practical problems. For Soviet scientists, cybernetics served a higher goal, breaking administrative and disciplinary barriers and liberating Soviet science from ideological and political pressures; they spoke the cybernetic language as a language of objectivity and truth.

Different national versions of cybernetics and information theory did not differ much in the range of cybernetic applications or the types of mathematical models used, considering the active exchange of latest techniques and technologies among the industrialized countries. The main difference lay in the political and cultural meanings attached to cybernetic ideas. Crossing boundaries often provoked attempts to separate the content of information theory and cybernetics from their initial ideological assumptions. Each time a significant cultural/political/disciplinary boundary is crossed, old ideological connotations are questioned and new ones attached. Trying to avoid political complications, Soviet scientists in the early 1950s tried hard to present the two new sciences as politically neutral, value-free technical tools for solving problems. Having failed to de-ideologize cybernetics and information theory, however, they instead re-ideologized these two sciences – but with different ideology. A cross-cultural analysis illuminates both the ideological malleability of cybernetics and information theory and the role of cultural context in shaping the fate of these ideas.

Notes

1 MIT.
2 Institut Universitaire de Formation des Maîtres de Paris.
3 Dibner Institute for the History of Science and Technology and the Russian Academy of Sciences.
4 See Claude E. Shannon and Warren Weaver, *The Mathematical Theory of Communication* (Urbana, IL: The University of Illinois Press, 1949), 29–125, translated as Klod Shennon [Claude Shannon], "Statisticheskaia teoriia peredachi elektricheskikh signalov," in Nikolai A. Zheleznov (ed.), *Teoriia peredachi elektricheskikh signalov pri nalichii pomekh* (Moscow: Izdatel'stvo inostrannoi literatury, 1953).
5 Nikolai A. Zheleznov, "Predislovie," in Zheleznov, *Teoriia peredachi elektricheskikh signalov*, 5.
6 Zheleznov, *Teoriia peredachi elektricheskikh signalov*, 6.
7 Norbert Wiener, *I Am a Mathematician: The Later Life of a Prodigy* (Cambridge, MA: MIT Press, 1956), 265; Also see *Cybernetics: Or Control and Communication in the Animal and the Machine* (Cambridge, MA: MIT Press, 1948), 8 for a similar account and a similar claim.
8 Steve Joshua Heims, *John von Neumann and Norbert Wiener: From Mathematics to the Technologies of Life and Death* (Cambridge, MA: MIT Press, 1980); Peter Galison, "The Ontology of the Enemy: Norbert Wiener and the Cybernetic Vision," *Critical Inquiry*, 21 (Autumn, 1994): 228–66; Paul Edwards, *The Closed World: Computers and the Politics of Discourse in Cold War America* (Cambridge, MA: MIT Press, 1996); Lily Kay, "Cybernetics, Information, Life: The Emergence of Scriptural Representations of Heredity," *Configurations*, 5 (1997): 23–91.
9 David A. Mindell, "Beasts and Systems: Taming and Stability in the History of Control," in Miriam Levin (ed.), *Cultures of Control in the Machine Age* (London: Harwood Academic Publishers, 2000).
10 David A. Mindell, *Between Human and Machine: Feedback, Control, and Computing before Cybernetics* (Baltimore: Johns Hopkins, 2000)
11 Several published accounts narrate of Wiener's work in prediction: Wiener, *I Am a Mathematician*, 242–56; Stuart Bennett, *A History of Control Engineering, 1930–1960* (London: The Institution of Electrical Engineers, 1993), 170–9; Stuart Bennett, "Norbert Wiener and Control of Anti-Aircraft Guns," *IEEE Control Systems*, December (1994): 58–62; Galison, "The Ontology of the Enemy"; Pesi Rustom Masani and R.S. Phillips, "Antiaircraft Fire Control and the Emergence of Cybernetics," in Pesi Rustom Masani (ed.), *Norbert Wiener: Collected Works with Commentaries*, vol. 4 (Cambridge, MA: MIT Press, 1985), 141–79.
12 Norbert Wiener, Final Report on Section D2, Project #6, December 1, 1942, quoted in Masani and R. Phillips, "Antiaircraft Fire Control and the Emergence of Cybernetics," 152.
13 See Bennett, *A History of Control Engineering*, 174, and "Norbert Wiener and Control of Anti-Aircraft Guns," for a technical explanation of this approach; see also Thomas Kailath, "Norbert Wiener and the Development of Mathematical Engineering," (unpublished manuscript, Stanford University, 1996).

14 "Summary of Project #6: Section D-2, NDRC," October 1, 1941. OSRD E-151 Applied Mathematics Panel General Records, Box 24.

15 Norbert Wiener, *The Extrapolation, Interpolation, and Smoothing of Stationary Time Series*(Cambridge, MA: MIT Press, 1949), 3; This is the published version of Wiener's original "Yellow Peril," report (so named because of its yellow cover and difficult mathematics) "Extrapolation, Interpolation, and Smoothing of Stationary Time Series with Engineering Applications," NDRC Report to the Services 370, February 1, 1942.

16 Division 7 Meeting Minutes, January 7–8, 1943 and February 3, 1943. OSRD7 GP Box 72 Division 7 Meetings folder. See also Galison, "The Ontology of the Enemy," 244–5 and Bigelow interview, 8. NW to WW, January 15, 1943 and January 28, 1943 are Wiener's last words on the project to the NDRC. Wiener recognized his predictor barely exceeded the performance of competing smoothers, but he believed there was too little data (only two courses for comparison) and that further work should continue to compare ten or a hundred courses.

17 See, for example, Wiener to Haldane, June 22, 1942. Wiener Papers, Box 2 Folder 64. This letter is marked "NOT SENT"; That May, Rosenblueth mentioned his conversations with Wiener and Bigelow in a presentation at a meeting on the physiology of the conditioned reflex, sponsored by the Macy Foundation; see Steve J. Heims, *Constructing a Social Science for Postwar America: The Cybernetics Group: 1946–1953* (Cambridge, MA: MIT Press, 1993), 14–15.

18 Arturo Rosenblueth, Norbert Wiener, and Julian Bigelow, "Behavior, Purpose, and Teleology," *Philos. Sci.*, 10 (1943): 18–24, reprinted in Masani (ed.), *Collected Works*, vol. 4, 180–6.

19 In one of Wiener's rare references to servo theory, on page 7, *Cybernetics* cites Leroy A. MacColl, *Fundamental Theory of Servomechanisms* (New York: Van Nostrand, 1946). This book synthesizes the Bell Labs approach to servos as developed for the electrical gun director computer, T-10. While Wiener cited Maxwell's paper as fundamental, Otto Mayr has persuasively argued that it was incoherent in terminology and definition and lacked the idea of a closed feedback loop so central to later conceptions of control; Otto Mayr, "Maxwell and the Origins of Cybernetics" in *Philosophers and Machines* (New York: Science History Publications, 1976), 168–88.

20 Lt. Col. C. Thomas Sthole to NW, July 23, 1943. Wiener Papers, Box 1 Folder 57; NW to Bush, September 21, 1940. Box 2, Folder 58, Wiener Papers.

21 Norbert Wiener, "A Scientist Rebels," *Atlantic Monthly* (January, 1947), reprinted in Masani (ed.), *Collected Works* vol. 4, 748; note that in Masani, *Norbert Wiener*, the bibliography of Wiener's military work (391) lists no contributions after January 15, 1943.

22 Steve Joshua Heims, *The Cybernetics Group: Constructing a Social Science for Postwar America* (Cambridge, MA: MIT Press, 1993); Edwards, *The Closed World*, ch. 6, Kay, "Cybernetics, Information, Life," 47.

23 Galison, "The Ontology of the Enemy," 253.

24 R.B. Blackman, H.W. Bode, and C.E. Shannon, "Data Smoothing and Prediction in Fire-Control Systems," in Harold Hazen, *Summary Technical Report of Division 7, NDRC Volume I: Gunfire Control* (Washington, DC: Office of Scientific Research and Development, National Defense Research Committee, 1946); Also see H.W. Bode and C.E. Shannon, "A Simplified Derivation of Linear Least Square Smoothing and Prediction Theory," *Proc. I.R.E.*, 38, April (1950): 425, which addresses Wiener's prediction in more detail; Also see R.B. Blackman, *Linear Data-Smoothing and Prediction in Theory and Practice* (Reading, MA: Addison-Wesley, 1965), an extension of the 1948 work.

25 Claude Shannon, "A Mathematical Theory of Communication," *BSTJ*, 27, July–October (1948): 379–423, 623–56, reprinted in N.J.A. Sloane and Aaron D. Wyner (eds), *Claude Elwood Shannon: Collected Papers* (New York: IEEE Press, 1993), 5–83; Claude Shannon and Warren Weaver, *The Mathematical Theory of Communication* (Urbana, IL: University of Illinois Press, 1949). The relationship between Shannon and Wiener's work is more complex than outlined here. In a later interview, Shannon related "I don't think Wiener had much to do with information theory. He wasn't a big influence on my ideas there [at MIT], though I once took a course from him." Shannon, *Collected Papers*, xix. Semantic confusion sometimes exists over the "Weaver-Shannon" or the "Wiener-Shannon," theory of communication. The former derives from the book listed in the previous note, and is inaccurate because Weaver served only to translate Shannon's work to make it more accessible (Weaver claimed no more).

26 Shannon, "A Mathematical Theory of Communication," 36.

27 Shannon and Weaver.

28 Claude Shannon, "The Bandwagon," *IEEE Transactions on Information Theory*, vol. 2, March, 1956; reprinted in Sloane and Wyner (eds), *Collected Works*, 462.

29 André-Marie Ampère, *Essai sur la philosophie des sciences ou exposition naturelle de toutes les connaissances humaines* (Paris: Mallet-Bachelier: 1834).

30 Pierre de Latil, *La Pensée artificielle* (Paris: Gallimard, 1953); Having lived together in Mexico, Freymann and Wiener were friends and it is Freymann who is supposed to have suggested that Wiener write this book. Wiener thought at first that he would need at least twenty years to produce something on this subject and finally went back to Mexico to write the manuscript.

31 (Anonymous article), "Machines that think," *Business Week* (19 February 1949), 38–42.

32 Dominique Dubarle, "Idées scientifiques actuelles et domination des faits humains," *Esprit*, 18, 9 (1950): 296–317; Emphasis is laid on Shannon's work and information is defined as "what the signal brings to the receptor, not in terms of supports necessary to exist physically, but regards to the different configuration which are going to be identified."

33 Robert Vallée, "The 'Cercle d'Etudes Cybernétiques,'" *Systems Research*, 7 (1990): 205. More directly on the role of de Broglie, see Robert Vallée, "Louis de Broglie and Cybernetics," *Kybernetes*, 19, 2, (1990): 32–3.

34 Among the forty members, one finds Couffignal, Dubarle, Ducrocq, Latil, Lafitte and Mandelbrot, all scientists who played an important role in the introduction of cybernetics in France.

35 Léon Brillouin, "Les machines américaines," *Annales des Télécommunications*, 2 (1947): 331–46, and "Les grandes machines mathématiques américaines," *Atomes*, 2, 21 (1947): 400–04.

36 Métral does not mention the contacts that some of the participants had with American researchers, nor the existence of a similar congress, "on automatic regulator and servo-mechanisms," held in London in May 1947.

37 Archives of the C.N.A.M., folder "Conférences d'actualité scientifique," 1947.

38 Pierre Grosser, *Les temps de la guerre froide* (Bruxelles: Complexe, 1995) or for a rapid overview Pierre Grosser, "Entre l'Est et l'Ouest, la France," *L'Histoire*, 209 (1997): 28. As far as instability is concerned, it is worth mentioning that on November 30th, 1947, 80,000 reservists had to be called up to face the crisis. For the period from the mid-1950s to the beginning of the 1960s, see Kristin Ross, *Fast Cars, Clean Bodies: Decolonization and the Reordering of French Culture* (Cambridge, MA: MIT Press, 1995).

39 G.-T. Guilbaud, "Divagations cybernétiques," *Esprit*, 18, 9 (1950): 281–95.

40 In Europe, information theory usually means cybernetics and communication theory (see F. Stumpers and H.M. Louis "A Bibliography on Information Theory (Communication Theory – Cybernetics)," *I.R.E. Transactions on Information Theory*, 1 (1955): 31–47.

41 Louis de Broglie, *La Cybernétique: Théorie du Signal et de l'Information* (Paris: Edition de la Revue d'Optique Théorique et Instrumentale, 1951). The choice of this subtitle has been very controversial, depending on the definition given to cybernetics (this became evident during interviews with four of the participants).

42 Broglie, *La Cybernétique*, 4.

43 Interview conducted on March 19th, 1997. For a general background, see Michel Atten (ed.), *Histoire, Recherche, Télcommunication, des Recherches au CNET 1940–1965* (Paris: Dif'pop, 1996).

44 Paul Chauchard, "Les machines à calculer et la pensée humaine," *Revue Générale des Sciences Pures et Appliquées*, 58 (1951): 5–7; The proceedings of the congress appeared in 1953 as number 47 of the series *Colloques Internationaux du Centre National de la Recherche Scientifique* (no editors named).

45 The Communist Party was still the first party at the legislative elections from June 17th 1951 (with 26.5 percent) but its influence in the intellectual world was not so decisive.

46 Louis de Broglie, "Sens philosophique et portée pratique de la cybernétique," *Atomes*, 7 (1952): 3–9.

47 Jacques Bergier, "Un plan général d'automatisation des industries," *Les Lettres françaises* (15 April 1948), 7.

48 Jean Cabrerets, "Intelligence et mémoire des Cerveaux électroniques," *Les Lettres françaises*, (13, 20, 27 October 1949).

49 André Lentin, "La cybernétique: problèmes réels et mystifications," *La Pensée*, 47, March–April (1953), 47–61.

50 Interview with André Lentin conducted on June 20th, 1997.

51 Incidentally, on the subject of idealism, one should keep in mind this sentence written by Wiener: "Information is information, not matter or energy. No materialism which does not admit this can survive at the present day." Norbert Wiener, *Cybernetics, or Control and Communication in the Animal and the Machine* (Paris: Hermann et Cie/The Technology Press, 1948).

52 Léon Brillouin, "Life, Thermodynamics, and Cybernetics," *American Scientist*, 37 (1949), 554–68. On the history of the Maxwell's Demon with a particular attention to the use of information theory, see H.S. Leff and A. F. Rex (eds), *Maxwell's Demon: Entropy, Information, Computing* (Bristol: Adam Hilger, 1990).

53 Léon Brillouin, *Science and Information Theory* (New York: Academic Press. Inc., 1956); the book was edited again in 1962 and translated into Russian and French.

54 Benoît Mandelbrot, "Contributions à la théorie mathématique des jeux de communications," *Publication de l'Institut de Statistiques de l'Université de Paris*, 2, fasc. 1 and 2 (1953): 3–124.

55 Marcel-Paul Schützenberger, "Contributions aux applications statistiques de la théorie de l'information," *Publications de l'Institut de Statistique de l'Université de Paris*, 3 (1953), 3–117.

56 Jean Ville and Marcel-Paul Schützenberger, "Les opérations des mathématiques pures sont toutes des fonctions logiques," *Comptes rendus de l'Académie des Sciences*, 232 (1951): 206–07 and Marcel-Paul Schützenberger, "Sur les rapports entre la quantité d'information au sens de Fisher et au sens de Wiener," *Comptes-Rendus de l'Académie des Sciences*, 232 (1951): 925–7.

57 See the first volume of *Information and Control*, published 1958.

58 Jacques Lafitte, *Réflexions sur la Science des Machines* (Paris: Librairie Bloud & Gay, 1932).

59 Pierre de Latil, *La Pensée artificielle* (Paris: Gallimard, 1953); Albert Ducrocq, *Découverte de la cybernétique* (Paris: Julliard, 1955); and Vitold Belevitch, *Langage des machines et langage humain* (Bruxelles: Office de publicité, 1956).

60 For a brief account of this French-Belgian connection, see Brigitte Chamak, *Le groupe des dix ou les avatars des rapports entre science et politique* (Paris: Editions du Rocher, 1997), 104–22.

61 Ivan Ia. Aksenov, Iurii Ia. Bazilevskii, and R.R. Vasil'ev, "Otchet ob itogakh II Mezhdunarodnogo kongressa po kibernetike," Russian Academy of Sciences Archive, Moscow (Arkhiv Rossiiskoi Akademii Nauk), f. 395, op. 17, d. 47, l. 43.

62 Kirillin and Monin to the Central Committee, July 13, 1960; Russian State Archive of Contemporary History, Moscow (Rossiiskii gosudarstvennyi arkhiv noveishei istorii [hereafter RGANI]), f. 5, op. 35, d. 134, ll. 55–56.

63 "Kibernetika," in Mark Rozental' and Pavel Iudin (eds), *Kratkii filosofskii slovar'* (Moscow: Gospolitizdat, 1954), 236–37.

64 On the history of Soviet cybernetics, see Boris V. Biriukov (ed.), *Kibernetika: proshloe dlia budushchego. Etiudy po istorii otechestvennoi kibernetiki* (Moscow: Nauka, 1989); Slava Gerovitch, *From Newspeak to Cyberspeak: A History of Soviet Cybernetics* (Cambridge, MA: MIT Press, 2002); Richard D. Gillespie, "The Politics of Cybernetics in the Soviet Union," in Albert H. Teich (ed.), *Scientists and Public Affairs* (Cambridge, MA: MIT Press, 1974), 239–98; Loren R. Graham, *Science, Philosophy, and Human Behavior in the Soviet Union* (New York: Columbia University Press, 1987), ch. 8; David Holloway, "Innovation in Science: the Case of Cybernetics in the Soviet Union," *Science Studies* 4 (1974): 299–337; Alexander Y. Nitussor, Wolfgang Ernst and Georg Trogeman (eds.), *Computing in Russia: The History of Computer Devices and Information Technology Revealed* (Braunschweig: Vieweg, 2001); Dmitrii A. Pospelov and Iakov I. Fet (ed. and comp.), *Ocherki istorii informatiki v Rossii* (Novosibirsk: OIGGM SO RAN, 1998).

65 Both Wiener and Shannon were indebted to Russian scientists for important insights. Wiener was influenced by the works of the Russian mathematicians Nikolai Bogoliubov, Andrei Kolmogorov, and Nikolai Krylov, physiologist Ivan Pavlov, and linguist Roman Jakobson; see Wiener's acknowledgement of Russian contributions in Wiener, *Cybernetics*, 11, 59, 127, and Wiener, *The Human Use of Human Beings: Cybernetics and Society* (New York: Avon Books, 1954), 255. Shannon employed the apparatus of "Markov processes," developed by the Russian mathematician Andrei Markov, Sr., in the early twentieth century for the same problem of stochastic description of natural language texts; see Shannon and Weaver, 45; Andrei A. Markov, "Primer statisticheskogo issledovaniia nad tekstom 'Evgeniia Onegina', illiustriruiushchii sviaz' ispytanii v tsep'," *Izvestiia Imperatorskoi Akademii Nauk* (1913): 153–62.

66 Bernard E. Bykhovskii, "Nauka sovremennykh rabovladel'tsev," *Nauka i zhizn'*, 6 (1953): 44.

67 Teodor K. Gladkov, "Kibernetika–psevdonauka o mashinax, zhivotnykh, cheloveke i obshchestve," *Vestnik Moskovskogo universiteta*, 1 (1955): 61.

68 Wiener, *Cybernetics*, 132.

69 Bykhovskii, "Nauka sovremennykh rabovladel'tsev," 44.

70 Zheleznov, "Predislovie," 6.

71 One Soviet author cited André Lentin's critique of cybernetics with strong approval; see Materialist [pseudonym], "Whom Does Cybernetics Serve?" (1953), trans. Alexander D. Paul, *Soviet Cybernetics Review*, 4, 2 (1974): 41.

72 In his anti-genetics speech at the July–August 1948 session of the Lenin All-Union Academy of Agricultural Sciences, personally edited and approved by Stalin, Lysenko attacked Schrödinger's book, *What Is Life?*, for bringing physical methods

into biology. See Kirill Rossianov, "Editing Nature: Joseph Stalin and the 'New' Soviet Biology," *Isis*, 84, 4 (December 1993): 728–45.

73 Trofim D. Lysenko, "In Response to an Article by A.N. Kolmogoroff," *Comptes Rendus (Doklady) de l'Académie des Sciences de l'URSS*, 28, 9 (1940): 833.

74 See André N. Kolmogoroff, "Sur l'interpolation et extrapolation des suites stationnaires," *Comptes Rendus de l'Académie des Sciences de Paris*, 208 (1939): 2043–5; Andrei N. Kolmogorov, "Statsionarnye posledovatel'nosti v gil'bertovom prostranstve," *Biulleten' MGU. Matematika*, 2, 6 (1941): 1–40; Andrei N. Kolmogorov, "Interpolirovanie i ekstrapolirovanie statsionarnykh sluchainykh posledovatel'nostei," *Izvestiia AN SSSR. Matematika*, 5 (1941): 3–14. Referring to the 1949 publication of Wiener's book, *Extrapolation, Interpolation, and Smoothing of Stationary Time Series*, Peter Whittle has concluded: "Kolmogorov and Wiener are generally given joint credit for the development of the prediction theory of stationary processes. This surely constitutes insufficient recognition of Kolmogorov's clear ten-year priority," (Peter Whittle, "Kolmogorov's Contributions to the Theory of Stationary Processes," *The Bulletin of the London Mathematical Society*, 22 (1990): 84; cf. Pesi R. Masani, *Norbert Wiener, 1894–1964* (Basel: Birkhäuser Verlag, 1990), 193–95. As stated above, Wiener's report was already in restricted circulation in 1942, which still gives Kolmogorov a lead.

75 Kolmogorov recalled that at the International Mathematical Congress in Amsterdam in 1954 he, a Russian, had to argue the importance of Shannon's information theory before American mathematicians, who were skeptical about the mathematical value of this engineer's work; see Andrei N. Kolmogorov, "Predislovie," in Klod Shennon [Claude Shannon], *Raboty po teorii informatsii i kibernetike*, trans. from English (Moscow: Izdatel'stvo inostrannoi literatury, 1963), 5.

76 See Andrei N. Kolmogorov, Izrail' M. Gelfand, and Akiva M. Yaglom, "Amount of Information and Entropy for Continuous Distributions," 1958, in Al'bert N. Shiryayev (ed.), *Selected Works of A.N. Kolmogorov, vol. III: Information Theory and the Theory of Algorithms* (Dordrecht: Kluwer Academic Publishers, 1993), 33, fn. 1.

77 Nikolai S. Simonov, *Voenno-promyshlennyi kompleks SSSR v 1920–1950-e gody: tempy ekonomicheskogo rosta, struktura, organizatsiia proizvodstva i upravleniie* (Moscow: ROSSPEN, 1996), 259–60.

78 "Vazhneishie zadachi razvitiia nauki v shestoi piatiletke," October 1955; RGANI, f. 5, op. 35, d. 3, l. 6.

79 Rumiantsev to the Central Committee, August 9, 1955; RGANI, f. 5, op. 17, d. 509, l. 214.

80 Sergei L. Sobolev, Anatolii I. Kitov, and Aleksei A. Liapunov, "Osnovnye cherty kibernetiki," *Voprosy filosofii*, 4 (1955): 140.

81 Sobolev, Kitov, and Liapunov, "Osnovnye," 136.

82 Andrei N. Kolmogorov, "Intervention at the Session," 1956, in Shiryayev (ed.), *Selected Works of A.N. Kolmogorov*, vol. III, 32.

83 Igor' A. Poletaev, *Signal: O nekotorykh poniatiiakh kibernetiki* (Moscow: Sovetskoe radio, 1958), 23.

84 See Aleksei A. Liapunov and Sergei V. Iablonskii, "Teoreticheskie problemy kibernetiki," 1963, in Aleksei A. Liapunov, *Problemy teoreticheskoi i prikladnoi kibernetiki* (Moscow: Nauka, 1980), 71–88.

85 In 1959, the Laboratory began publishing the journal *Problemy peredachi informatsii*, which since 1965 has been also published in the Engish translation as *Problems of Information Transmission*.

86 In his recent study of the subcultures of instrumentation, experiment, and theory within the larger culture of microphysics, Peter Galison calls a "trading zone" the area where, despite the vast differences in their symbolic and cultural systems,

different groups can collaborate. See Peter L. Galison, *Image and Logic: A Material Culture of Microphysics*(Chicago: University of Chicago Press, 1997), especially ch. 9.

87 Raisa L. Berg, *Acquired Traits: Memoirs of a Geneticist from the Soviet Union* (New York: Viking, 1988), 220.

88 Sergei L. Sobolev, "Vystuplenie na soveshchanii," in Petr N. Fedoseev (ed.), *Filosofskie problemy sovremennogo estestvoznaniia* (Moscow: AN SSSR, 1959), 266.

89 Sergei L. Sobolev and Aleksei A. Liapunov, "Kibernetika i estestvoznanie," in Fedoseev (ed.), *Filosofskie problemy*, 252.

90 Raisa L. Berg and Aleksei A. Liapunov, "Predislovie," in Ivan I. Shmal'gausen, *Kiberneticheskie voprosy biologii* (Novosibirsk: Nauka, 1968), 13.

91 Jakobson to Shannon (24 April 1951) MC 72, box 45.21, Jakobson papers, MIT.

92 See Roman Jakobson, "Linguistics and Communication Theory," 1960, in Roman Jakobson, *Selected Writings, vol. II: Word and Language* (The Hague and Paris: Mouton, 1971), 570–9.

93 Claude E. Shannon, "The Redundancy of English," in *Cybernetics: Transactions of the Seventh Macy Conference* (New York: Josiah Macy, Jr. Foundation, 1951), 157. See also Roman Jakobson, "Linguistics and Communication Theory," 578.

94 Igor' A. Melčuk, *Cybernetics and Linguistics: Some Reasons for as Well as Some Consequences of Bringing Them Together* (Vienna: Osterr. Studienges. f. Kybernetik, 1977), 15.

95 Akiva M. Iaglom and Isaak M. Iaglom, *Probability and Information*, trans. V.K. Jain (Dordrecht: Reidel, 1983), 198–201.

96 Andrei S. Monin, "Dorogi v Komarovku," in Al'bert N. Shiriaev (ed.), *Kolmogorov v vospominaniiakh* (Moscow: Nauka, 1993), 484.

97 Andrei N. Kolmogorov, "Three Approaches to the Definition of the Notion of Amount of Information," 1965, in Shiryayev (ed.), *Selected Works of A.N. Kolmogorov*, vol. III, 188.

98 Kolmogorov, "Three Approaches," 190.

99 Andrei N. Kolmogorov, "The Combinatorial Foundations of Information Theory and the Probability Calculus," 1983, in Shiryayev (ed.), *Selected Works of A.N. Kolmogorov*, vol. III, 210.

100 Kolmogorov, "Three Approaches," 192.

101 David G. Kendall, "Kolmogorov: The Man and His Work," *The Bulletin of the London Mathematical Society*, 22 (1990): 40.

102 Andrei N. Kolmogorov, "Kibernetika," in *Bol'shaia Sovetskaia entsiklopediia*, 2nd ed., vol. 51 (1958): 149.

103 Andrei N. Kolmogorov, "Informatsiia," in *Bol'shaia Sovetskaia entsiklopediia*, 2nd ed., vol. 51 (1958): 129.

104 Andrei A. Markov, "Chto takoe kibernetika?" in Aksel' I. Berg *et al.* (eds), *Kibernetika. Myshlenie. Zhizn'* (Moscow: Mysl', 1964), 41. Andrei Markov, Jr., was the son of the author of "Markov processes."

105 Il'ia B. Novik, *Kibernetika: filosofskie i sotsiologicheskie problemy* (Moscow: Gospolitizdat, 1963), 80. On Soviet philosophical discussions of the concept of information, see Graham, *Science, Philosophy, and Human Behavior in the Soviet Union*, 281–93.

5

SCIENCE POLICY IN POST-1945 WEST GERMANY AND JAPAN

Between ideology and economics

Richard H. Beyler and Morris F. Low

Introduction

Both Japan and Germany after 1945, under the aegis of victorious occupying powers, faced the task of a dual reconstruction including both material or economic and political or ideological aspects. Both Germany and Japan had undergone major material damage and structural dislocation which needed repair. From the point of view of many inhabitants, mere survival was a pressing question in the short term, and the economic viability of their countries a real concern for the long term. In this climate, pragmatism held a clear priority over introspection. But both Germany and Japan also needed political reconstruction, especially in the view of the victorious Allies. Both countries had in their recent past undergone a political or ideological derangement, the vestiges of which needed to be eradicated and the future recurrence of which ought to be prevented. Allied policy – and in this article we focus primarily on that of the Americans – in both countries was at first, therefore, based on the idea of controlling a dangerous foe.[1] Before long, however, the keywords of occupation policy changed to integration into a new international strategic structure and positive encouragement of desirable political forms, rather than merely prevention of undesirable ones. The context of this change was the beginning of the Cold War. In that context, maintaining stability and relative good will in West Germany and Japan became crucial elements of America's global strategy. Stability required economic prosperity; good will required the settling down of internal political upsets. In both regards, controls which might be perceived as overly stringent became less desirable from the American perspective.

In the complex process of post-war reconstruction, debates over science policy had a significant and revealing place in regard to both economics and ideology. The economic and strategic relevance of scientific research and development was obvious to everyone, particularly in fields such as atomic energy, chemicals, and electronics. But whether the power of German or Japanese science was seen as a danger or a benefit depended on the changing context.

97

Likewise, science could appear as a symbol of ideological purity or, alternatively, danger. Discussions of the political significance of science, and the social role of scientists, thus served in both Japan and Germany as a forum for the advocacy of alternative visions for ideological reconstruction.

Post-1945 West Germany

The material reconstruction of Germany constituted a daunting challenge, to put it mildly. For many Germans, the years immediately after 1945 were the *Trümmerzeit* – the era of rubble – a designation which could be taken literally. In the city of Berlin, for example, there were 55 million cubic meters of rubble; in Cologne, the per capita quantity of debris was 31.2 cubic meters.[2] Ninety percent of Germany's railroads were unusable at the end of the war, and up through the winter of 1946–7, official food rations only covered about half of the number of calories regarded as normal.[3]

Rubble (*Trümmer*) also had a metaphorical meaning for some Germans, and even more for the Allies: the ideological detritus left by National Socialism. German political reconstruction entailed answering the question of how Nazism came to be, with an eye toward ensuring that the future did not resemble the immediate past. How one accounted for Nazism correlated with a vision for Germany's future. Was a complete break with the past needed, or were there ideas, values, and institutions from the past which ought to be salvaged? Both the Germans and the Americans (and the other Allies) debated among themselves the proper course for German reconstruction. Above and beyond this, the relationship between the German and the American visions of reconstruction had to be negotiated: a negotiation, to be sure, in which the parties came to the table with quite different quantities and kinds of bargaining power.

These problems of reconstruction applied to science just as to any other aspect of German life. What kinds of scientific work should be done, and how could this work be supported materially and economically? As in other segments of society, facilities and equipment were an urgent priority for many scientists, though the degree of urgency was widely different in different locations. The physical condition of universities in the U.S. occupation zone – taking this as representative of the problem to be faced – varied from one extreme at Erlangen, which was largely unscathed, to Würzburg, where only about one-tenth of the classrooms and one-fifth of the laboratories were usable in 1945.[4] From the viewpoint of those within the scientific professions and the academic world, the rebuilding of laboratories and the repair and replacement of damaged and lost equipment were urgent needs, but from the broader perspective of the both Allied and German administrations, these tasks had to take a place among many other urgent ones. The necessity or desirability of allocating material resources to science was, at least at the outset, very much open to debate.

Beyond material reconstruction there was the question of the proper role of the scientific professions in the new German society, either according to members of the scientific community, to its various German constituencies, or to the supervening Allied powers. Much of this question was phrased in terms of the past and potential future involvement of scientists in dangerous ideologies or, conversely, the potential of science and scientists to participate in Allied-initiated efforts to "democratize" German political culture. In short, American authorities and the German leaders who took on political leadership in this new context had to be persuaded that the scientific community could be responsible participants in a new, self-consciously democratized public sphere.

Shifting American attitudes towards science in Germany

American (and other Allied) efforts to control German research after 1945 have received extensive historical treatment.[5] In brief, the story starts from an initial position of attempting to strictly control scientific research, and then follows a contour of increasing relaxation to the eventual active promotion of scientific reconstruction. American science policy for occupied Germany at the end of the war was in line with the so-called Morgenthau plan, named after Franklin Roosevelt's treasury secretary Henry Morgenthau, Jr. He had envisioned a docile post-war Germany in which material reconstruction would not progress beyond a rudimentary, agrarian level. To some extent, this vision was embodied in Joint Chiefs of Staff (JCS) directive 1067, issued as the foundation of U.S. military government policy under the Truman administration in April 1945. The directive, *inter alia*, ordered the closure, for the time being, of all research and educational institutions. But JCS 1067 contained a number of moderating provisos. In particular, exceptions could be made to prevent starvation, disease, or unrest. The wording of this provision made it open to flexible interpretation, since almost any material hardship could if desired be termed a potential source of unrest. The high prioritization of pragmatic health concerns meant that medical schools were the first university faculties to re-open; other schools and departments, including the sciences, followed later.[6]

Another manifestation of initially suspicious attitudes was the Allied Control Council (ACC) Law Number 25, promulgated in summer 1946, and drafted primarily by the American chemist and military government advisor Roger Adams. It banned all research that was directly connected to the military, and also banned all applied research in fields with potential strategic value, such as atomic physics, aerodynamics, those areas of chemistry connected to explosives or chemical weapons, and wide areas in electronics. All other research of any variety was to be closely monitored, with scientists obligated to fill out lengthy questionnaires on a regular basis. Furthermore, scientists with any significant prior Nazi or military involvements were to be barred from leadership positions in laboratories and scientific institutions.[7]

In addition to such efforts to defuse any future possible military threat from German research, the Americans (and the other Allies) exploited German research and researchers for their own interests. Vast quantities of documents from research laboratories – for example, those of the chemical combine IG Farben – were copied and appropriated. Many German scientists were brought to the U.S., often with less than careful scrutiny of the very Nazi or military connections that would have barred them from working in Germany. Besides these sometimes covert actions, the Americans and British, through the Field Intelligence Agency – Technical (FIAT) also organized the writing of the so-called FIAT Reviews, comprehensive summaries of German research from the late 1930s and early 1940s in a wide range of disciplines. German scientists were paid for this work, which in many cases was a welcome source of income before the resumption of laboratory and university work; international scientists benefited from the broader disclosure of German research from an era in which it had become increasingly isolated.[8]

Beyond these efforts to defuse any potential remilitarization of German science, and to appropriate research for their own purposes, the American authorities had hopes for a sweeping reform program in German education at all levels from primary school to university, with the goal of making Germany more democratic and less militaristic. These plans had major implications for science policy in the educational sphere. One of the main concerns was that German secondary schools and universities had been too much focused on the classics at the expense of modern perspectives, too over-specialized at the expense of communicating to students a notion of social responsibility, and too much focused on the achievements of an elite at the expense of educating the public as a whole. Perhaps a less elitist, more generalist education system was in order, one with an increased proportion of social sciences and modern humanities, which could acculturate a broad range of German citizens into a new, democratized public sphere. A significantly increased place for natural sciences – provided, however, that they were taught in a "general education" vein, not one of disciplinary specialization – was part of this vision for educational reform. However, most of the Americans' plans for sweeping reforms of schools and universities fizzled out, not least because of staunch German insistence on the validity of much of their own tradition extant before 1933.[9]

The Americans' overall agenda, for their part, was shifting from controlling the Germans towards integrating them into a new international order. Indeed, within only a few months after May 1945, initially stringent policies began to moderate, and these moderating changes gathered momentum throughout 1946 and 1947. Even after the promulgation of the restrictive ACC Law No. 25, enforcement of the provisions soon became relatively lax. Interest in a thorough denazification of the laboratories flagged. The main conflict shifted from the question *whether* scientific research should be supported – in the most punitive interpretations of the foundational JCS 1067, the answer had been "no" – to the question *by whom*. A convoluted struggle began over whether government support for science should administered at the federal or the state level. In this arena, the Americans, suspicious of any central concentration of power, tended to favor the claims of the individual German states. The British, by contrast, encouraged cooperation and even centralization within their occupation zone in the interests of stabilizing German administration. With the formation of the American-British "Bizonia" in January 1947, compromise solutions were in order.[10]

As the divergence in American and British approaches indicated, the modulation in Allied science policy was not so much due to a systematic re-evaluation as to a number of piecemeal, unsystematic reactions to several interrelated economic and strategic developments. Military government officers became alarmed about the possibility of unrest envisioned in the JCS 1067 loopholes, so that meeting material needs became a top priority for the Allies as well as the Germans. Moreover, a stable and prosperous West Germany would be, in American strategic plans, a keystone of an emerging Atlantic alliance. To some degree, the British also led the Americans in this direction, because of their keen sense of the urgency of stabilizing western Europe. For the British, carrots were much better than sticks in this regard.[11] For the purposes of fighting the nascent Cold War and the related task of encouraging a sense of common purpose among the western European countries, a prosperous, satisfied West Germany was much more advantageous than a suffering, discontented West Germany. In turn, science and technology was seen to be a vital part of Germany's development.

The Cold War also led to another worry among the Western powers, though not one that was much publicized officially: preventing scientific experts from falling into the hands of the Soviets and their satellites, ostensibly neutral but potentially unreliable countries, or even members of the emerging Atlantic alliance also deemed to be in some sense unreliable (notably France). This was especially true with regard to atomic physicists.[12] If the remarks of German atomic physicists interned in England at Farm Hall in the closing days of the war are any indication, many German scientists were eager to remain in (or move to) the West, and Western officials were eager to exploit their concerns to ensure cooperation.[13] Conversely, other German scientists, when faced with dim career prospects in the West or enticement from the East bloc, chose to move East or threatened to do so in an effort to gain leverage *vis-à-vis* the Americans, British, and French.[14]

Thus, at the end of the war, the primary concerns of American policy in Germany were denazification and demilitarization. In terms of science policy, this meant controlling or restricting research to meet those goals. As Cold War maneuvering came to the fore of American concerns by 1947 and 1948, productively deploying the power of science became the overriding consideration. German scientists and policy makers readily responded to these changing priorities among the occupation authorities. Simultaneously, however, German scientists had to re-define their role in the light of science and technology coming from other segments of their own society. Traditional anxieties about the scientization and technologization of the modern world, long current among German intellectuals, received new nuances after 1945 as Germany struggled to create a viable democracy.

Post-war German critiques of science and technology

With the increase in freedom of the press after the end of Nazism, German literature and journalism underwent something of a renaissance. In the plethora of cultural-political books and new and revived journals, anxious discussions of the social role of science and technology were a frequent topic.[15] With the beginning of the atomic age it was obvious to everyone that science held unprecedented power to shape human destiny. This power might be used for good ends, but many authors also wondered, in the words of a widespread slogan, whether there was a "Demon of Technology."[16] The "demonic" here signified a powerful force autonomous from human control and probably hostile to human life and interests. In part, these critiques represented an attempt by traditionalistic humanistic intellectuals to assign blame for the rise of Nazism to the social dominance of scientific-technical experts. Critiques of science and technology also represented a defense of the German tradition in a fashion that would not be too conspicuously anti-Western or anti-democratic.[17]

One of the most conspicuous examples of anti-science critique was the book *Die Perfektion der Technik*, by essayist and poet Friedrich Georg Jünger. The book was written in the first half of 1939, but the first edition was not published until 1946.[18] Like his better-known brother Ernst, F.G. Jünger was fascinated with technology, but unlike Ernst, technology had little aesthetic appeal for him. *Die Perfektion der Technik* was an unrelievedly dire and pessimistic account of the effects of scientific thinking and technology. The term "perfection" was meant in an ironic sense; as is made clear in the title of the English translation, it referred to a *Perfection without Purpose*. Jünger's indictments were heavy. Machines had not increased leisure but rather increased the total amount of work. Wealth, which Jünger conceived as a mode of "being" rather than "having" was not increased by technology; rather, there was only a redistribution of poverty. Under the guise of increased productivity, modern technology was in fact "pillaging" the earth. In a Bergsonian gesture, Jünger unfavorably contrasted vitally experienced time with the "dead time" measured out by mechanical devices. He decried the "tyranny of causalism" which was the corollary of "dead time." Examples of the perfidious effects of this way of thinking Jünger saw in, among other things, the "fraud" of "'scientific' nutrition" and in the "mechanical sterility of modern sports."[19] More momentously, Jünger saw technical achievement and warfare to be inextricably linked: "Just as a technically organized economy becomes more and more a war economy, so technology develops more and more into a war technology...In our dynamic age, technology steps up its pillage of world resources; but while it devours material for war preparations, it reduces at the same time our living standards...[T]he total consumption demanded by a total war may well consume whatever gains result even from the winning of the war. What must be anticipated is a condition where there is neither victor nor vanquished, but only general exhaustion."[20] In the post-war era in which Demilitarization was a major concern of the Allies and their German protégés, such charges were serious challenges to the viability of science in the new order.

Jünger thus described a world in which humanity was caught in a spiral of consumption and destruction, imprisoned in the hold of mechanical "dead time." Ostensibly rational and efficient, this world was in fact subject to irrational forces unleashed by technology. This situation was enabled by the interrelationships between the masses and technology. "Technical progress and the formation of masses are simultaneous," Jünger wrote. Likewise, "the mass is the most useful, the most pliant material of the technician." Technical organization of society operated on the "modern mass," and "to the extent in which the masses become subjected to rational organization, they become supercharged with blind elemental powers and bereft of all spiritual powers to oppose them. The masses are running berserk..."[21] With tunnel-visioned experts on the one side and dumb masses on the other, space was opened, in Jünger's account, for the primal "vengeance of the elemental spirits which the modern magician has conjured up." Taking Jünger at face value, there was literally a demonic force in technology: "The old superstitious idea...which held that some man-made apparatus might acquire demoniacal life, might unfold a will of its own, a rebellious and destructive will – this idea is by no means as erroneous as we now suppose."[22]

Jünger's book can be read as a critique of Nazism, and thus an accusation that technology and science were deeply implicated in the German political catastrophe. The description of masses running amok while led by organizers set only on their own narrow goals seems to be Jünger's description of the rise of Nazism. The chapter on war and technology can be read as an astute description, and condemnation, of the technical aspects of *Blitzkrieg*. (The English translation adds footnotes in a couple of spots making an explicit comparison to Nazism.)[23] But if the book is to be read as a critique of Nazism, then it is not so much of a critique of Nazism per se as of Nazism as a virulent aspect of an otherwise undifferentiated modern, technical civilization. This was what made it so troublesome in the context of re-construction, which depended inter alia on the application of modern technical know-how.

In short, Jünger maintained that technology, together with its cognitive basis in natural science, was a dangerous demonic force in recent history. He associated the dominance of scientific-technical thinking with the growth of mass society and the rise to importance of specialized experts and, in turn, the rise of Nazism. His dire vision of science and technology was the most widely discussed in West Germany in the late 1940s and early 1950s, but by no means the only one. During this same period Martin Heidegger, for example, become more and more interested in "The Question Concerning Technology." While there is a vast difference between Jünger's impressionistic essays and the linguistic subtlety and philosophical profundity in Heidegger's writings, the animus was similar: the sense that humans were trapped in a spiral of production and consumption driven by an autonomous technoscience.[24] Herein lies an irony, inasmuch as Heidegger, among other antimodernist intellectuals, had once seen Nazism as perhaps the *answer* to the problems of the industrial, technoscientific world.[25] The carry-over of these traditional, conservative tropes into post-war Germany was in some sense a sign of cultural and intellectual depletion and tiredness.[26] Indeed these tropes may have been tired, but in this new cultural, social, and political context, they took on new implications and connotations. In particular, whereas much traditional German conservatism was often quite explicit in the rejection of the values of the "West," now German conservatism was confronted with the necessity to re-assess that antipathy.

More moderate cultural critics writing in the new and revived journals of the literary renaissance, such as *Der Monat*, *Frankfurter Hefte*, and *Die Gegenwart*, denied the existence of a demonic force, but did see some dangers. Atomic physics served as a pervasive and potent symbol. An article by Hans Kudszus in the new literary journal *Athena*, for example, began with a depiction of Hiroshima after its destruction by the atomic bomb. The unleashing of atomic energy symbolized for Kudszus a current sense of "disorientation," paralleled by problems of social "massification," professional over-specialization, and mechanization of work. But he was resigned to the necessity of a scientific world for the sake of rebuilding "the deranged existence of Western humanity."[27] Here, as in many other cultural debates of the post-war era, the concept of Germany as part of the "the West" (often denominated *das Abendland*) or a "Western" cultural identification is quite prominent. As a lay observer, Kudszus thus perceived acutely the dilemma between power and purity faced by German scientists as part of this cultural tradition.

A pair of reviews in the journal *Merkur* came to differing assessments of *Die Perfektion der Technik*, but both agreed that responsibility for – and solutions for – the question of technology lay within human choice. Helmut Günther judged that although Jünger was wrong at almost every point of detail, his description as a whole nevertheless had pointed out an important problem: the "elementarization of civilization," the unleashing of primeval, seemingly demonic powers. But Günther located these "demonic and destructive forces in humans themselves." They had the freedom to shape their own fate.[28] Social theorist Max Bense argued that Jünger had neglected the benefits of science and recent theoretical developments. In Bense's view, the newest conclusions of science, notably modern physics, led to a renewed sense of human freedom and human responsibility. He conceded that this responsibility had not always been fulfilled. The solution he sought in a "renewed humanistic-spiritual education."[29] Bense affirmed the practical value of science but saw scientism as a shortcoming which required alleviation through philosophical immersion in the German humanist tradition.

Critiques of science thus sometimes functioned as apologias for the classical-humanistic educational tradition, in the context of the debate over education in the late 1940s. Such critiques of science reiterated traditional conceptions of intellectual purity, precisely because of the increasing practical power of science emblematized by its use in the Second World War. As noted above, defenses of a traditional structure and content for school and university curricula were relatively successful against American (and British) pressure for reform.[30]

Although apportioning blame for "the German catastrophe"[31] was a major subtext in critiques of science, the Cold War added another dimension to suspicions about the ideological susceptibility of science. One example of this kind of concern could be found at the Congresses for the Freedom of Culture in Berlin in 1950 – the already symbolic location in the divided city took on added significance with the coincident outbreak of the Korean War – and 1953 in Hamburg under the theme "Science and Freedom." These rallies of intellectuals to show solidarity against the Soviet bloc were, it turned out later, sponsored behind the scenes by the CIA. In Hamburg, the sociologist and philosopher Helmuth Plessner critiqued the "blinker effect of specialization," which resulted, he claimed, in a susceptibility towards particular ideologies. Historically oriented fields were susceptible to nationalism, he said, while physics and allied fields were easily prey to Marxism.[32] However, the Congress for Cultural Freedom meetings also were forums for promulgating the idea that the Soviet system was fundamentally inimical to free scientific inquiry, and vice versa, with the Lysenko affair serving as a prominent example.[33] As we will see below, defense of the democratic potentials of science was a key part of its post-1945 ideological reconstruction.

Promoting the power of science

In responding to these shifting, often contradictory visions of the role of science in the reconstructed Germany, members of the German scientific community deployed two main strategies. One was to claim science as ideologically pure, and therefore conducive to a nascent anti-totalitarian political culture. A second strategy was to call attention to the economic utility of scientific research and development and thus its projected contribution to long-term prosperity and stability. German scientists thus steered a course between the rhetorics of economic power and utility, and ideological or political purity. They sought thereby to maneuver through Allied restrictions on research, anti-modernist critiques of science in the literary and cultural press, the ambiguities of their professions' experience under Nazism, the anxieties of the Cold War, and the material necessities of reconstruction.

From the perspective of economic utility, the argument was that scientific research and technological development were the *sine qua non* of Germany's economic survival. This argument was compelling for American authorities as Allied policy swung towards integrating West Germany into Western Europe. It also appeared prominently in appeals for funding directed at both government and private (industrial) sponsors. One frequently repeated version of the argument was the assertion that West Germany's natural resources were too meager for its dense population. Long-term economic viability would only come from export trade, especially of highly engineered and technically advanced manufactures; exactly those in which research and development in optics, electronics, chemistry, metallurgy, and other branches of science would be crucial. As a major corollary, scientists such as Werner Heisenberg pressed for a heightened advisory role for scientists vis-à-vis policy makers.[34]

For physicists in particular, an important theme was energy production, a top priority for both Allied and German authorities. While coal production was the immediate focus, the dream of future supplies of atomic energy was profoundly appealing.[35] Even if nuclear research could not immediately be supported, cultivating the development of physics as a whole would at least hold the door open for that tantalizing possibility. Furthermore, by the early 1950s German participation in European Organization for Nuclear Research (CERN) was promoted as a safe way to allow Germans to undertake atomic research.[36]

Another commonly heard part of the argument from economic utility was the need to recruit and train, as quickly as possible, new scientific personnel. The subjective sense that a generation of young researchers had been lost was reinforced by statistical analyses in newspaper and journal accounts of the needs of German scientists. In addition to noting the Jewish and political "victims of the Hitler era," such articles also presented muted, but still noticeable, criticism of Allied recruitment of desired experts, removals of erstwhile Nazi scientists, and cases of "intellectual reparations."[37]

Claiming political virtue for science

Even while arguing strenuously for the utility of science in the new Germany, the leaders of the scientific professions and their defenders in other disciplines also asserted claims of science's ideological desirability. Countering the critiques of technical-scientific modernity such as Jünger's, they frequently portrayed science as a contributor to a political culture of freedom. In popular articles, university addresses, memoranda to officials asking for public support – and not a few Persilscheine (testimonials for denazification proceedings) – scientists argued that scientists, too, had been victims of Nazism and therefore now deserved better treatment. The "Hitler era" had been one of "sharp anti-scientific propaganda," argued a 1949 memorandum of the German Research Council to the federal government, and according to the Stifterverband für die deutsche Wissenschaft (Donors League for German Science), the scientific profession had suffered more losses than any other.[38]

The Nobel-prize winning chemist Otto Hahn argued to the American military government that the Kaiser Wilhelm Gesellschaft (now renamed the Max Planck Gesellschaft), of which he was president-designate, deserved their trust since various member institutes had during the Nazi era sheltered scholars threatened on grounds of race, world view, or politics.[39] From this position, it was a relatively short step to the implication or assumption that good science was, so to speak, the natural enemy of Nazism. The existence of the "Aryan physics" movement under the Nazis made this move easier. Since some of the leading and outspoken opponents of modern quantum and relativity physics were also outspoken Nazis, adherence to the standards of good professional discipline also counted as resistance to outside, viz. Nazi, interference in professional standards, and could further be construed (or, made to construe) as resistance to Nazism *per se*. Ways in which (some) German scientists may have collaborated with or benefited from Nazi policies were, by and large, left unnoticed. If science, too, had been victimized by Nazism, then science as a whole was free of responsibility for the political disaster which had overtaken Germany, and thus free to contribute to the democratization of the new Germany. Scientists in the post-war era thus tended to postulate a distinction between science in its ideal form which was at worst politically neutral and quite possibly positively aligned with democratic values, and the "misuse" of science by dangerous ideologies. Although the latter was a real possibility, scientists *per se* did not bear any more responsibility for it than any other segment of society.[40]

Conspicuously absent (at least in relative terms) in this line of discussion was the thorny issue of the potential usefulness of "useful" science for military purposes, raised so conspicuously by Jünger's critique of technological modernity. A disturbing possible corollary of the rhetoric of victimization was that if the Third Reich had better respected science, it would have been better equipped to fight a modern, technological war. However, as this argument directly undercut the emphasis on Demilitarization, it was more or less avoided in the immediate post-war years. This situation was to change somewhat later with German integration into the NATO military alliance; however, it remained touchy enough to set off the so-called "Göttingen Eighteen" controversy of 1957, in which eighteen leading atomic scientists, in response to what they saw as first steps by the Adenauer government towards nuclear armaments, issued a manifesto stating their refusal to work on nuclear weapons.

In the post-war era, many scientists also placed renewed emphasis on the humanistic, culturally uplifting aspects of their profession. These apologias shared the anxieties about specialist expertise and about massification which pervaded the critiques of science and technology from the humanistic portion of the cultural elite. Such apologias did not present a fundamental challenge to that viewpoint; rather, they attempted to placate it. One of the leading public spokesmen for science in post-war Germany was the theoretical physicist Carl Friedrich von Weizsäcker. In 1946 he presented a lecture cycle at the University of Göttingen on "The History of Nature" which admitted that scientists all too often lacked a sense of broader responsibility.

"One feels more and more," Weizsäcker stated, "the danger which lies in the specialization of sciences." The ideal scientist, he explained, shared a "responsibility for the whole," an obligation to contemplate and explain what "research means for the life of my fellow humans" in the broadest sense.[41] Interesting in this connection here is Weizsäcker's involvement in the 1947 *Gutachten zur Hochschulreform*, the so-called *Blaues Gutachten*, sponsored by the educational authorities of the British occupation government. Weizsäcker played a leading role in its composition. The report advocated, among other things, an increased attention to general studies – not only education within a disciplinary specialty – in university curricula.[42] Although the education reforms of the *Blaues Gutachten* failed, Weizsäcker's broader message was clear: science, understood properly and reformed so as to build upon the most valuable parts of its tradition, was quite compatible with the values of the new political culture.

A different kind of mediation between technologists, scientists, and the broader public was attempted by Joachim Leithäuser, a historian and popularizer of technology and science. Leithäuser's writings were straightforward cheerleading for the benefits of science and its technical applications. In a review essay in the cultural journal *Der Monat*, which was published under aegis of the Information Services Division of the U.S. High Commission for Germany, Leithäuser attacked Jünger and other "romantic pessimists" for their "sublimated animism." Leithäuser especially took umbrage at Jünger's linkage of technology with the power-hungry, dictatorial state. He acknowledged that "modern dictatorships" had indeed "exploited technology without restraint," but saw this as a degenerate case. Leithäuser asserted that contrary to the thesis of such a linkage, "the methodology of technical thought and its criteria of truth stand in crass contradiction to the methodology of systems of political compulsion." The Americans had provided clear models of this, e.g., as described in David Lilienthal's book *TVA: Democracy on the March*. Leithäuser thus associated the distrust of technology with a distrust of democratic political engagement.[43]

Leithäuser's article unambiguously dissociated scientific-technical thinking from dictatorship and associated it with democracy. The positive association was reinforced by a small note, taken from the *Observer*, placed by the editors at the bottom of the page at the end of his article. The note contrasted the fate of the Russian geneticist Sergei Vavilov, who resisted Lysenkoism and died in Siberian exile, with his brother N.I. Vavilov, who supported Lysenko and received a hero's burial in Moscow.[44] The implication was that totalitarian regimes could not be trusted to recognize scientific accomplishments fairly. Science and technology would foster democracy; contrapositively, dictatorship would thwart or skew science and technology.

Another clear but somewhat more nuanced argument that science went hand-in-hand with democratic political re-construction came from Clemens Münster, a former industrial physicist who was now one of the co-editors of the well known cultural-political journal *Frankfurter Hefte*, regarded in the late 1940s as the leading forum for left-of-center Catholic thought. Münster argued that scientists could have a constructive social role provided they cultivated connection with other cultural domains. Science at its best – i.e., at its most socially valuable – had an organization and ethos which counteracted the dangers posed by its enormous power. It was crucial for everyone to realize and act on the belief that science itself was a kind of socially-connected humanism, and not think of it solely as instrumental knowledge.[45]

Elsewhere, Münster pointed out the changing institutional structure of scientific research. It was no longer dominated by the ivory tower, he said. Rather, it now consisted of the coordination of work in theory and in technical application in many different subfields. In modern research and development, he wrote, "not much of the old specialization [*Spezialistentum*] is left. The bearers of research are groups, teams, work communities..." He believed, however, that German universities, in contrast to industrial laboratories, were behind the times. Indeed, in Germany "teamwork" – tellingly, he used the English word – was already the rule in industrial research and development, but in academia remained more the exception. Where Münster thought the future trend lay was clear. Furthermore, especially in the context of the Christian left *Frankfurter Hefte*, the social and political associations were hard to miss: here, the cultural meaning of modern research was cooperation between various professions and social groups.[46] Thus in Münster's conception, science at its best included cooperation among individuals and groups of various backgrounds and disciplines, and thus meshed ideally with a new culture of social democracy and active citizenship, in contrast to the elitism and subservience found in the old Germany.

Post-1945 Japan

Münster's eloquent pleas were an attempt to resolve the tensions between the economic necessities, the perceived need for self-criticism, and the vision of a reconstructed German political culture which all converged in post-war science policy in Germany. Turning to American-occupied Japan, we find that similar concerns – bracketing off a politically undesirable past, defending against a new ideological foe in the present, and ensuring a politically and economically stable nation – were behind shifts in science policy there. There were, however, significant contrasts to the German case, in large part due to differences in scientists' perception of their own social role and historical responsibilities. In particular, apparently more Japanese scientists than Germans expected to make a clear, radical break with the traditions of the past, as opposed to reformist visions such as those of Weizsäcker or Münster. Furthermore, the special sensitivities of the problem of atomic energy in Japan distinctly shaped the debates over science policy in post-war Japan. Somewhat ironically, given the Japanese experience, and in some contrast to the marked current of pessimism troubling many Germans, both the initial direction and final outcome of these debates were generally optimistic about the potentials of science and technology; the open question was how these potentials would be directed. A prime example of this orientation can been seen in the post-war writings and activities of the physicist Mitsuo Taketani.

By the end of the war, 3 to 4 percent of Japan's population of 74 million (c. 1941) had died, and sixty-six major cities had been heavily bombed. In the three largest cities there was much devastation, with 65 percent of homes destroyed in Tokyo, 57 percent in Osaka, and 89 percent in Nagoya.[47] It was against this backdrop that the Allied Occupation of Japan occurred. It involved not only U.S. forces but also what was known as the British Commonwealth Occupation Force (BCOF) which comprised of contingents from the United Kingdom, Australia, India, and New Zealand.[48]

The way the war ended had a major impact on the way the Japanese saw science. Underlying their belief in it was the understanding that scientific research, both fundamental and applied, would ultimately contribute to Japan's long-term economic prosperity. As in Germany, there was a belief that wartime regimes had ignored science at their peril. Immediately after the end of the war, many Japanese absolved themselves of any personal responsibility for what had occurred by accusing Japan's leaders of having misled them. The "irrational" militarists were "unscientific" in the way they waged war, and their nation's defeat could be traced to weaknesses in their science and technology, as symbolized by the atomic bomb, or rather the lack of one.[49]

Despite Japanese efforts to produce an atomic bomb, little real progress was made. Japanese physicists, including Mitsuo Taketani, have argued that there was passive resistance to the military leadership as exemplified by their less than active participation. Taketani himself carried out calculations for the Tokyo-based atomic bomb project while being detained by the Thought Police for subversive ideas. It is claimed that as there was little likelihood of producing a bomb, physicists were able to accommodate the demands of the military while maintaining their own values. The military request to build an atomic weapon was seen more as an assigned scholarly project,[50] a means by which related research could also be funded and conducted under the banner of building a bomb. The reality was a small-scale effort lacking in manpower, coordination and materials.[51]

The late mobilization of science in Japan during the war and backwardness in the military use of radar were seen as further examples of the nation's technological lag. Over-confidence in the *samurai* spirit, and the Shinto belief that the Emperor was a divine being, were all seen as ideologically-driven. Science was a quest for "truth" and in itself ideologically pure, but left-wing scientists sought to elevate Marxism to a science as well.

Instead of dwelling on the past (and interrogating the war responsibility of the Emperor), it was more convenient for both the Allied Occupation and many Japanese to look forward to the future, one in which science and technology would be mobilized for Japan's economic reconstruction. Both ends of the political spectrum were generally in agreement with this. Debate tended to focus on how science would be used and for what purposes. Left-wing activists argued that political reconstruction and a democratic administrative structure were vital. Socialism was portrayed as a type of scientific remedy to solve

Japan's problems, and society would need to change for science to prosper.[52] There was, thus, consensus in Japan about the need for economic reconstruction, and the importance of science and technology. But the Communist Party and left-wing scientists attached further ideological importance to the role of science. In Japan, as in Germany, reconstruction included both an economic and ideological dimension.

The Occupation was seen as an opportunity for great change and much of it occurred in the first year. In early 1946, a study mission of U.S. education experts visited Japan. The group suggested major changes based on the American system of education.[53] The Japanese educational system subsequently underwent complete restructuring with revision of curricula and decentralization of the system. The scientific community saw the Allied Occupation as a chance to create a space for themselves in the policy-making structure.

On January 12, 1946, the inaugural meeting of the Democratic Scientists' Association (abbreviated in Japanese as "Minka") was held. Minka has parallels with the Association of Scientific Workers in England and the Federation of Atomic Scientists in the U.S.[54] Minka sought to promote a "democratic" science which resisted feudalism and militarism. Its members called for the open publication of research and open access to scientific facilities. Science and technology needed to contribute to the enrichment of the lives of the Japanese people. With more cooperation and dialogue between natural scientists, social scientists and technologists in Japan, and liaison with like-minded groups overseas, Japanese scholars could forge a new identity for themselves. This involved cutting ties with their "feudalistic" past and reshaping existing institutions.

It was hoped to build a network of Minka branches throughout Japan. At the height of its popularity in late 1949, it had 110 branch offices, a specialist membership of 2000 and a student and non-specialist membership of 11,000. To gain a sense of the size and influence of Minka, one has only to peruse the list of periodicals the organization produced: *Shakai kagaku* (Social Science), *Shizen kagaku* (Natural Science), *Riron* (Theory), *Rekishi hyôron* (Historical Critiques), *Geijutsu kenkyû* (Studies in the Fine Arts), *Seibutsu kagaku* (Biological Science), *Ningen kaihô* (Liberation of Mankind), *Kagaku nenkan* (Science Yearbook), *Nôgyo nenpô* (Agricultural Annual Report), and *Nôgyô riron* (Agricultural Theory). The magazines *Kagakusha* (Scientists) and *Gakujutsu tsûshin* (Science News) appeared three times and twice a month, respectively.[55]

Minka's publications were part of the "literary renaissance" soon after Japan's surrender, not unlike that in Germany. There was a veritable explosion of "sôgô" (general) magazines which advocated progressive ideas, and Marxist intellectuals were prominent supporters. Notable magazines were *Kaizô* (Reconstruction), *Sekai hyôron* (World Review), *Shinsei* (New Life), and *Shisô no kagaku* (The Science of Ideas).

Postwar Japanese critiques of science and technology

The writings of physicist Mitsuo Taketani often appeared in such magazines, and had a major impact on the Japanese intellectual world. An essay on his Marxist-inspired theory of technology[56] appeared in the February 1946 edition of *Shinsei*. Taketani considered technology as the application of objective laws in productive, human practice. He distinguished between technology and skill: the former is objective and enriched by the bank of knowledge whereas the latter is subjective and acquired through experience. Both feed into each other in a dialectical process. New technology may require new skills. When skills are transformed into technology, production capacity and quality often improve. Taketani felt that the safety and welfare of workers had been neglected by industrialists who were more concerned with maintaining a supply of low-wage labor and using imported technology.

In terms of reconstructing Japan, Taketani wrote in his "An Analysis of Japanese Technology and the Rebuilding of Industry," published in *Gijutsu* (Technology) in March 1946, that there should be a proper place for unions in the process. Increasing the lot of the workers should be part and parcel of improvements to technology.[57] In the same issue, he likened Japan's plight to that of the U.S.S.R., a place where technologists were important in the construction of the socialist republic. They could fulfill their true potential by working for the welfare of the people and helping the nation overcome serious economic problems.[58]

In the opinion of Shôichi Sakata, a close colleague of Taketani's, the Japanese lack of creativity in basic science was not the result of national character but more of environment. It was time to work for an environment which was congenial to research of a fundamental nature.[59] Both Taketani and Sakata called for more research facilities, and greater funding of Japan's own research effort.[60] In this way, the physicists' own analyses of their history and the failings of Japanese science informed the debate on the role of science in Japan's reconstruction. On 10 April 1946, the first general election since the end of the war was held. The perceived importance of science and technology was such that each political party proposed policies for science and technology which were, in turn, commented upon and criticized by scientists such as Sakata and Taketani.[61] Overarching critiques of scientific-technical modernity or questions of the "demonic of technology," such as those offered in Germany by authors such as Jünger, were not prevalent in these discussions. Whereas in Germany, conservative cultural critics were eager to point to elements of a past still regarded as valid despite the rubble of Nazism, Japanese cultural critics – for the time being at least – focused on the future. Science could lead the way, though how it was to do so was open to debate.

Shifting American attitudes

The Occupation authorities sought to demobilize Japanese science, remobilize it to meet U.S. goals, and reorganize scientific bodies along more democratic lines.

As in Germany, the demilitarization of Japanese science and technology entailed the temporary control of all military research facilities and aeronautical laboratories, some of which were converted to establishments for peaceful purposes. Towards the end of 1945, after initial investigations had ascertained there was no Japanese atomic bomb, restrictions on Japanese nuclear physicists were relaxed, and orders to guard Japanese laboratories were lifted. But the prohibitions on research relating to atomic energy remained, and personnel with any knowledge of the topic were registered and kept under surveillance. Continued fear of Japan's atomic potential, especially given the high level of Japanese theoretical physics, led to the impounding of all uranium and thorium with zealous thoroughness.[62]

Whereas in Germany there were separate sections within the Allied command that had responsibility for labor, finance, and the economy, in Japan they were combined, along with science and technology in the Economic and Scientific Section of S.C.A.P. (Supreme Command for the Allied Powers).[63] Within this could be found the Scientific and Technical Division, headed by the Australian, Brigadier John O'Brien.[64] Harry C. Kelly, a young American physicist, was also a member of the Division with initial responsibility for basic research.[65]

The Division considered its most important achievement was the reorganization of Japan's national scientific organizations. Unlike progressive Japanese scientists, the Allied representatives viewed science and technology, not so much as a means of achieving the democratization of Japan, rather as a means by which to solve industrial problems and achieve economic growth. This included the improvement of quality control, and the enforcing of industrial standards, patents and trademarks. Japan was thus encouraged to carry out research and development in areas which might contribute to a "peaceful and productive nation." Meanwhile, maintaining prohibitions on weapons research, the Scientific and Technical Division carried out a surveillance function by reporting on individual scientists and their work, as well as carrying out inspections of laboratories.[66]

In an interesting parallel with Otto Hahn's role in defending the Kaiser Wilhelm Gesellschaft, the physicist Yoshio Nishina was given the task of fighting for the survival of the Institute of Physical and Chemical Research (abbreviated in Japanese as "Riken"), Japan's premier pre-Second World War scientific research organization, located in Tokyo. It was analogous to the Imperial Institute of Physics and Technology in Germany (established in 1887).[67] Riken was formed in 1917 with funds from government and the private sector. It combined a research institute with an industrial group of manufacturing companies which produced a wide range of products which included machinery parts and tools, synthetic liquor, chemicals, metals, pharmaceuticals, and optical equipment.[68] After the war, Nishina was called in to save Riken from financial disaster and possible dissolution by the Occupation authorities.

The attitude of Occupation authorities in both Germany and Japan are strikingly similar. Initial strict control of scientific research was followed by increasing relaxation once the Occupation authorities were able to ascertain via extensive reports and questionnaires what research had been carried out and what capacity remained. Even personnel such as the American scientist Roger Adams served as an advisor for both. In July–August 1947, a United States Scientific Advisory Group was sent to Japan by the U.S. National Academy of Sciences, at the urging of Kelly. Adams led the Group. Adams also took part in a second mission in November–December 1948.[69] The political climate had changed considerably by that time. I.I. Rabi, Nobel prize physicist for 1944 and professor at Columbia University, was also a member of that mission. He recommended the limited "importation" of Japanese scientific personnel to the U.S. for two or three years. This would help both the Japanese who were starved of funding and adequate equipment, and provide highly trained personnel for use in research in the United States.[70] On 10 December 1948, Rabi's proposal took the form of an official, "top secret" memorandum entitled "The Use of Japanese Research Facilities as an Advanced Base in the Event of Acute Emergency in the Far East." It related the potential usefulness of Japanese scientists, particularly in time of war.[71] Such recommendations reflected the rise of Cold War tensions around 1947–8. The Occupation authorities became less tolerant of the calls for social "revolution" and adopted what is sometimes called the "reverse course" in its policies towards workers' organizations and leftist groups. Japanese communism, not militarism, was now seen as the most urgent threat to East Asia.

Americans sought to exploit Japanese research and researchers, the most notorious case involving the biological warfare unit known as Unit 731. The Unit had been led by a Japanese Army surgeon, Lieutenant Shirô Ishii, and was active in Manchuria in the 1930s and 1940s.[72] It is clear that Americans were concerned that bacteriological warfare know-how not fall into the hands of the Soviets. Unit 731 research data was exchanged in return for war crimes immunity.

After initial suspicion of Japanese science, the attitude of Occupation authorities turned to helping to rebuild it. There were attempts by Japanese scientists and the scholarly community to establish a representative organization which would provide both a forum for issues relating to science policy and serve as a significant lobby group. Occupation personnel facilitated this process. A representative Renewal Committee for Science Organization (*Gakujutsu Taisei Sasshin Iinkai*) was elected, and this led to the establishment of the Science Council of Japan. A major issue was the relationship between the proposed Science Council and the government. Fearful of a left-wing takeover of the Science Council,[73] it was proposed that the Council should be independent. The government could seek the advice of the Science Council, but it was not bound to do so. The national parliament, the Diet, agreed in mid-1948 that the Science Council of Japan would consist of 210 elected members who, in turn, would belong to one of the following seven divisions: (1) literature, philosophy, and history; (2) law and politics; (3) economics and commerce; (4) fundamental science; (5) engineering; (6) agriculture; and (7) medicine, dentistry, and pharmacy. Each division would consist of thirty members. Furthermore, the Japan Academy would be set up within the Science Council. The election of members was held in December 1948. According to S.C.A.P., approximately forty of the 210 elected representatives were known to be communists or communist sympathizers.[74]

Promoting atoms for peace in Japan

During the Occupation, writing on the effects of the atomic bombs on Hiroshima and Nagasaki was heavily censored from 1945 to 1949. Discussion of the effects of the atomic bombs was discouraged.[75] The delayed coming-to-terms with the a-bomb experience has arguably contributed to a "nuclear allergy." The Japanese oppose the production of atomic weapons on the one hand, but have nevertheless enthusiastically pursued nuclear power on the other. The latter is viewed as being atomic energy for peaceful purposes.[76] As in Germany, a frequent argument in favor of pursuing nuclear power has been that the country is densely populated and lacking in natural resources. It is not surprising that in both countries, physicists would be involved in debates concerning energy production. While Americans were wary of allowing Germans and Japanese to become involved again in atomic research, President Eisenhower's Atoms for Peace plan in 1953 facilitated their reinvolvement.

At the same time as embracing nuclear power, there was a widespread perception that, courtesy of the atomic bomb, the Japanese were victims of the war rather than aggressors. The atomic bomb experience enabled the Japanese to fashion themselves as champions of a non-nuclear world.[77] This approach to the "atom" can be seen in Taketani's writing. During the Occupation, he wrote on nuclear war and suggested that the threat of mass annihilation was serving as a brake to further global conflict. He was also the first academic to push for the civilian use of atomic energy in Japan. In 1950 he published a book entitled Genshiryoku (Atomic Energy) (published by Mainichi Shinbun Sha)[78] which explored this idea. The end of the Occupation in 1952 saw many opportunities for Taketani to write on the atomic bomb and atomic energy. Particularly notable was his contribution to the special issue on the bomb in Kaizô (October 1952) in which he interviewed the director of the Atomic Bomb Casualty Commission in Hiroshima. A discussion of the "Smyth Report"[79] appeared in Asahi hyôron and he co-authored Genshiryoku (Atomic Energy) (Tokyo: Mainichi Shinbun Sha) with Seitarô Nakamura.[80] Soon after, in an article in the periodical Kaizô in November 1952, Taketani wrote that while he felt that Japan should pursue atomic energy, there were two problems: (1) insufficient domestic supply of uranium, and (2) the danger of ending up as a "subcontractor" for the U.S. atomic bomb program. Taketani felt that as the world's first victim country of the atomic bomb, the Japanese had a right to conduct research on the peaceful uses of atomic energy, with the assistance of overseas countries. He felt that such countries should unconditionally supply Japan with the necessary quantities of uranium, and that atomic energy research should not be of a secretive nature.[81] By his writings, Taketani was instrumental in educating the public about the facts and implications of atomic energy.

In February 1952, Taketani called for the Japanese to build their own nuclear reactors, in order to break the stranglehold on such know-how held by nations possessing nuclear weapons.[82] Although research into atomic energy was prohibited during the Occupation, Sakata and Taketani nevertheless felt that it was important to discuss means of ensuring that such research, when allowed, would remain autonomous. There were fears that the prohibition would remain in place even after a Peace Treaty was signed, but this was not so. The Peace Treaty came into effect in 1952, and the prohibition was removed.

Taketani was the first to propose three principles for the peaceful use of atomic energy, calling for public disclosure of atomic research, democratic management of research, and research autonomy. It was hoped that they would prevent atomic research from being used for military purposes. Science, in Taketani's mind, needed to be conducted in an open manner and made accessible to most people. Atomic energy became the focus of discussions in the Science Council of Japan, a representative assembly of scholars from a diverse range of fields and disciplines. Sakata's advocacy of the importance of the "three principles" ensured that public debate over atomic energy continued.

Ironically, the atom served to empower the Japanese. The dropping of the atomic bomb on Hiroshima and Nagasaki, and the subsequent push for the prohibition of nuclear weapons, helped shape the Japanese (especially physicists) as peacemakers.[83] It also provided a way of making their expertise relevant for economic reconstruction. While the bombs were dropped by Americans, science was generic and the sad outcome was, for some Japanese, arguably more the result of the backwardness of Japanese science and technology than evil Americans. Such beliefs facilitated the Americanization of both Japanese society and science. The U.S.–Japan relationship continues to remain one of the most important bilateral relationships in the world. Underpinning that relationship is the strength of the Japanese economy, the belief that Japan's future is tied to technological innovation, and the maintenance of a healthy export trade in value-added goods.

Concluding comments

We have seen that the calls after the Second World War for an approach to policy making that was at once more "scientific" and more "democratic," as well as the debates over science policy itself, contributed to the (re-)formation of a public sphere in West Germany and Japan. Underlying the project of ideological reconstruction was a largely American strategy of democratization, stabilization, and economic reconstruction. The public sphere was not only a realm of rational discourse where citizens could discuss matters of mutual interest, but it was also a place where scientists could pursue their own political goals.[84] The postwar literary renaissance in both countries enabled scientists to fashion themselves as both intellectuals and public watchdogs, people whose opinion and expertise mattered, and also to respond to more pessimistic assessments of the role of science and technology in the modern world.

It was against the backdrop of Cold War tensions that in both countries the economic and political deployment of science became of even greater concern to the occupation authorities. Starting from a policy model of controlling science, they became appreciative of the potential uses of science in the stabilization and international integration of the occupied countries, provided research was properly channeled. Simultaneously, however, the occupation powers were fearful of the ideological uses that science might be put to by ideologically suspect intellectuals – whether on the left wing (a concern relatively common in Japan) or the unreconstructed right (initially at least, the focus of "re-education" in West Germany). The subsequent relationship between scientific expertise and public policy was a sometimes turbulent one, with scientists actively seeking to make their expertise relevant, and to further their own political aims in the re-formed public sphere.

Notes

1 We focus here primarily on the complexities within the Americans' perspectives vis-à-vis Germany and Japan, the respective German and Japanese responses. Especially in the case of Germany, many of our conclusions would also apply a fortiori to the complex relationships between the Americans and the other Occupations powers. Apart from some comparison with British views and policies in the German case, however, a comparison among the various Allies is not within the scope of this paper.

2 Statistics cited by Jeffry M. Diefendorf, In the Wake of War: The Reconstruction of German Cities after World War II (New York: Oxford University Press, 1993), 15. For orientation on post-war West Germany and the American occupation see John Gimbel, The American Occupation of Germany: Politics and the Military, 1945–1949 (Stanford: Stanford University Press, 1968); Hermann Glaser, Die Kulturgeschichte der Bundesrepublik Deutschland, vol. 1, Zwischen Kapitulation und Währungsreform 1945–1948 (Frankfurt a.M.: Fischer, 1990); Jeffry M. Diefendorf, et al. (eds), American Policy and the Reconstruction of West Germany, 1945–1955 (Washington, DC: German Historical Institute, 1993); Rebecca L. Boehling, A Question of Priorities: Democratic Reforms and Economic Recovery in Postwar Germany (Providence, RI: Berghahn, 1996).

3 Adolf M. Birke, Nation ohne Haus: Deutschland 1945–1961 (Berlin: Siedler, 1998), 25, 30.

4 Statistics cited by Ullrich Schneider, "The Reconstruction of the Universities in American Occupied Germany," in Manfred Heinemann (ed.), Hochschuloffiziere und Wiederaufbau des Hochschulwesens in Westdeutschland 1945–1952, 3 vols. (Hildesheim: Lax, 1990–91), vol. 2, 1–8, here 1.

5 Thomas Stamm, Zwischen Staat und Selbstverwaltung: Die deutsche Forschung in Wiederaufbau 1945–1965 (Cologne: Verlag Wissenschaft und Politik, 1981); Alan D.

Beyerchen, "German Scientists and Research Institutes in Allied Occupation Policy," *History of Education Quarterly*, 22 (1982): 289–99; Maria Osietzki, *Wissenschaftsorganization und Restauration: Der Aufbau ausseruniversitärer Forschungseinrichtungen und die Gründung des westdeutschen Staates 1945–1952* (Cologne: Böhlau, 1984); John Gimbel, "U.S. Policy and German Scientists: The Early Cold War," *Political Science Quarterly*, 101 (1986): 433–51; Heinemann, *Hochschuloffiziere und Wiederaufbau* (1990–91); David Cassidy, "Controlling German Science, I: U.S. and Allied Forces in Germany, 1945–1947," *Historical Studies in the Physical and Biological Sciences*, 24 (1994): 197–235; David Cassidy, "Controlling German Science, II: Bizonal Occupation and the Struggle over West German Science Policy, 1946–1949," *Historical Studies in the Physical and Biological Sciences*, 26 (1996): 197–239.

6 On JCS 1067 see Cassidy, "Controlling German Science, I," 208–12; Boehling, *Question* (1996), 26–30; on the re-opening of university medical faculties, see James F. Tent, *Mission on the Rhine: Reeducation and Denazification in American-Occupied Germany* (Chicago: University. of Chicago Press, 1982), pp. 57–61.

7 "Gesetz Nr. 25 des Kontrollrats: Überwachung der wissenschaftlichen Forschung," *Physikalische Blätter*, 2 (1946): 49–53; for discussion, see Cassidy, "Controlling German Science, I," 221–35.

8 A comprehensive overview of American efforts to exploit German science after 1945 is John Gimbel, *Science, Technology, and Reparations: Exploitation and Plunder in Postwar Germany* (Stanford: Stanford University Press, 1990). On personnel "transfers," see Clarence G. Lasby, *Project Paperclip: German Scientists and the Cold War* (New York: Athenaeum, 1971); on Wernher von Braun specifically, see Michael J. Neufeld, *The Rocket and the Reich: Peenemünde and the Coming of the Ballistic Missile Era* (New York: Free Press, 1995), 267–79.

9 We will return to scientific issues connected to the educational reform effort below. On the rise and fall of American ambitions for "re-education," see Manfred Heinemann (ed.), *Umerziehung und Wiederaufbau: Die Bildungspolitik der Besatzungsmächte in Deutschland und Österreich* (Stuttgart: Klett-Cotta, 1981); Tent, *Mission* (1982); Glaser, *Kulturgeschichte* (1990), 147–82; Jürgen C. Hess *et al.* (eds), *Heidelberg 1945* (Stuttgart: Franz Steiner, 1996); Manfred Heinemann and Klaus-Dieter Müller, "Einleitung," in Manfred Heinemann *et al.* (eds.), *Süddeutsche Hochschulkonferenzen 1945–1949* (Berlin: Akademie Verlag, 1997), 1–28; Rebecca Boehling, "The Role of Culture in American Relations with Europe: The Case of the United States' Occupation of Germany," *Diplomatic History*, 23 (1999): 57–68.

10 See the discussions in Stamm, *Zwischen Staat und Selbstverwaltung* (1981); Osietzki, *Wissenschaftsorganization* (1984); Manfred Heinemann, "Der Wiederaufbau der Kaiser-Wilhelm-Gesellschaft und die Neugründungen der Max-Planck-Gesellschaft (1945–1949)," in Rudolf Vierhaus and Bernhard vom Brocke (eds.), *Forschung im Spannungsfeld von Politik und Gesellschaft: Geschichte und Struktur der Kaiser-Wilhelm/Max-Planck-Gesellschaft* (Stuttgart: Deutsche Verlags-Anstalt, 1990), 407–70.

11 By 1947 at the latest, the British were lobbying the Americans for an integrative science policy; an internal British military government report from January, 1948 stated that the Americans were coming around to the British view that "effective … control results from the active encouragement of the right kind of research." "Review of Scientific and Technical Research in Germany during 1947," memorandum, Science and Technology Research Branch, Control Commission for Germany [British Element] (12 January 1948) in FO 1005/1478, Public Record Office, Kew, London (hereafter PRO).

12 For example, science policy officers received cabinet-level orders to find secure employment for the nine German physicists (Werner Heisenberg, Otto Hahn, Max

118

von Laue, Carl F. Weizsäcker, *et al.*) interned at Farm Hall, in England, at the close of the war, according to "Control of Scientific Research in the British Zone of Germany," memorandum, Trade and Industry Division, Control Commission for Germany (British Element) (23 April 1946), in FO 1062/149, PRO.

13 Charles Frank (ed.), *Operation Epsilon: The Farm Hall Transcripts* (Berkeley, CA: University of California Press, 1993), 27, 89, 92, 112, 143, 177, 183, 199–203. It should be noted that these statements may also reflect a more or less conscious attempt of the interned scientists to position themselves favorably in the eyes of the Anglo-American Allies.

14 Richard H. Beyler, "From Positivism to Organicism: Pascual Jordan's Interpretations of Modern Physics in Cultural Context" (Ph.D. dissertation, Harvard Univ., 1994), ch. 6.

15 On post-war technological pessimism, see Jeffrey Herf, "Belated Pessimism: Technology and Twentieth Century German Conservative Intellectuals," in Yaron Ezrahi *et al.* (eds),*Technology, Pessimism, and Postmodernism* (Amherst, MA: University of Massachusetts Press, 1994), 115–36. Post-war cultural-political journals are surveyed in Ingrid Laurien, *Politisch-kulturelle Zeitschriften in den Westzonen 1945–1949: Ein Beitrag zur politischen Kultur der Nachkriegszeit* (Frankfurt a.M.: Peter Lang, 1991); see also Glaser, *Kulturgeschichte*.

16 For example, Hans Kudszus, "Dämon der Technik? Gedanken zur Philosophie der Technik," *Athena*, 2, 6 (1947–8): 20–8.

17 On the "mandarin" tradition and related concepts, see Fritz K. Ringer, *The Decline of the German Mandarins: The German Academic Community, 1890–1933* (Cambridge, MA: Harvard University Press, 1969); Paul Forman, "Weimar Culture, Causality, and Quantum Theory, 1918–1927: Adaptation by German Physicists and Mathematicians to a Hostile Intellectual Environment," *Hist. Stud. Phys. Sci.*, 3 (1971): 1–115; Russell McCormmach, "On Academic Scientists in Wilhelmian Germany," in Gerald Holton and William A. Blanpied (eds), *Science and Its Public: The Changing Relationship* (Dordrecht: D. Reidel, 1976), 157–71; Jonathan Harwood, *Styles of Scientific Thought: The German Genetics Community 1900–1933* (Chicago: University of Chicago Press, 1993).

18 On Jünger and the theme of technology, see Anton H. Richter, *A Thematic Approach to the Worlds of F.G. Jünger* (Bern: Peter Lang, 1982), 58–87.

19 Friedrich Georg Jünger, *Die Perfektion der Technik* (Frankfurt a.M.: Vittorio Klostermann, 1946); in English as *The Failure of Technology: Perfection without Purpose*, trans. F.D. Wieck (Hinsdale, IL: Henry Regnery, 1949), 5–25, 34–41, 94–6, 150–4.

20 Jünger, *Failure*, 168.

21 Jünger, *Failure*, 66–8, 126, 144–5, 177. Apropos the concept of "mass society," cf. the popularity in Germany at this time of the work of José Ortega y Gassett; see Ernst Robert Cutrius, "Ortega," *Merkur*, 3 (1949): 417–30; Laurien, *Politisch-kulturelle Zeitschriften* (1991): 149–50.

22 Jünger, *Failure*, 115, 176–7.

23 Jünger, *Failure*, 123, 133. The fact that publication was delayed may indicate that the work was meant – and immediately understood – as an attack on Nazism at the time it was written. Or maybe not: Richter, *Thematic Approach*, 59, implies that the work did not appear before 1945 because of disruptions of the war, namely, the destruction of the type of the first edition in a 1942 bombing raid, and later of a published first edition in 1944.

24 The classic sources from Heidegger are "Die Zeit des Weltbildes," in *Holzwege* (Frankfurt: Vittorio Klostermann, 1952), based on a 1938 lecture; and "Die Frage nach der Technik," based on a 1949 lecture, in Bayerische Akademie der schönen Künste (ed.), *Die Künste im technischen Zeitalter* (Munich: R. Oldenbourg, 1956),

48–72; translated in *The Question Concerning Technology and Other Essays*, trans. William Lovitt (New York: Harper & Row, 1977), 3–35, 115–54; discussed in Michael E. Zimmerman, *Heidegger's Confrontation with Modernity: Technology, Politics, Art* (Bloomington, IN: Indiana University Press, 1990). An irony in Heidegger's post-war statements is that he had once seen Nazism as perhaps the answer to the problems of the industrial, technoscientific world. See Jeffrey Herf, *Reactionary Modernism: Technology, Culture, and Politics in Weimar and the Third Reich* (Cambridge: Cambridge University Press, 1984), 109–15; Zimmerman, *Heidegger's Confrontation* (1990). Other contemporary anti-science critiques include Otto Veit, *Die Flucht vor der Freiheit: Versuch zur geschichtsphilosophischen Erhellung der Kulturkrise* (Frankfurt a.M.: Vittorio Kostermann, 1947); Robert Dvorak, *Technik, Macht und Tod* (Hamburg: Claasen and Goverts, 1948); Henrik de Man, *Vermassung und Kulturverfall: Eine Diagnose unserer Zeit* 2nd edn (Munich: Leo Lehnen Verlag, 1952). De Man was Belgian, but worked in Germany.

25 See Herf, *Reactionary Modernism*, 109–15; Zimmerman, *Heidegger's Confrontation*.
26 Jerry Z. Muller, "How Vital Was the *Geist* in Heidelberg in 1945? Some Skeptical Reflections," in Hess *et al.*, *Heidelberg 1945*, 197–200.
27 Kudszus, "Dämon", 20–21, 24–5, 28.
28 Helmut Günther and Max Bense, "Die Perfektion der Technik: Bemerkungen über ein Buch von F.G. Jünger," *Merkur*, 2 (1948): 301–10, here 301–6.
29 Günther and Bense, "Perfektion", 310.
30 See Stamm, *Zwischen Staat und Selbstverwaltung*, 60–70; Heinemann, *Umerziehung*; Tent, *Mission*; Glaser, *Kulturgeschichte*, 147–82; Heinemann, *Süddeutsche Hochschulkonferenzen 1945–1949*; Hess *et al.*, *Heidelberg 1945*.
31 Friedrich Meinecke, *Die deutsche Katastrophe: Betrachtungen und Erinnerungen* (Wiesbaden: Eberhard Brockhaus, 1946).
32 Helmuth Plessner, "Die ideologische Anfälligkeit der Wissenschaftler: Thesen zu einem Vortrag," in *Wissenschaft und Freiheit: International Tagung Hamburg, 23.-26. Juli 1953* (Berlin: Grunewald Verlag, 1954), 159–61, here 159–60.
33 Helmuth Plessner, "Die ideologische Anfälligkeit der Wissenschaftler: Thesen zu einem Vortrag," in *Wissenschaft und Freiheit: International Tagung Hamburg, 23.-26. Juli 1953* (Berlin: Grunewald Verlag, 1954), 159–61, here 159–60; proceedings of the 1950 Berlin congress appear in *Der Monat* 2 (1950): 341–493. On the meetings' background, see Peter Coleman, *The Liberal Conspiracy: The Congress for Cultural Freedom and the Struggle for the Mind of Postwar Europe* (New York: Free Press, 1989); Frances Stonor Saunders, *The Cultural Cold War: The CIA and the World of Arts and Letters* (New York: New Press, 1999).
34 See, for example, Hildegard Brücher and Clemens Münster, "Deutsche Forschung in Gefahr?" *Frankfurter Hefte*, 4 (1949): 333–344; the articles from the journal *Christ und Welt* assembled by the Stifterverband für die deutsche Wissenschaft as *Forschung heisst Arbeit und Brot* (Stuttgart: Steingrüben, 1950); Hellmuth Eickemeyer, *Abschlussbericht des Deutschen Forschungsrates* (Munich: H. Oldenbourg, 1953).
35 Wilhelm Moock, "Vorratskammer Erde," *Frankfurter Hefte*, 1 (1946): 87–8; Herbert Goldscheider, "Technik des befreiten Menschen," *Frankfurter Hefte*, 2 (1947): 70–7; Gimbel, *American Occupation* (1968), 95–7, 151–8.
36 Armin Hermann, "Germany's Part in the Setting-Up of CERN," in Hermann *et al.* (ed.), *History of CERN* (Amsterdam: North Holland, 1987), vol. 1, 383–429; Thomas Stamm-Kuhlmann, "Deutsche Forschung und internationale Integration 1945–1955," in Vierhaus and Brocke, *Forschung* (1990), 886–909, here 905–07.
37 Stifterverband, *Forschung*, 18–22.
38 "Denkschrift des Deutschen Forschungsrates über die Betreuung der wissenschaftlichen Forschung im Rahmen der Deutschen Bundesregierung," (1 September 1949), in B227/531, Bundesarchiv, Koblenz (hereafter BaK); Stifterverband, *Forschung*, 15.

39 Minutes, Wirtschaftsrat des Länderrates, Sonderausschuss Wissenschaftliche For-schung (16 January 1947), in Z1/1034, BaK.
40 See, for example, Robert Proctor, *Racial Hygiene: Medicine under the Nazis* (Cambridge, MA: Harvard University Press, 1988), 298–312; Herbert Mehrtens, "'Missbrauch': Die rhetorische Konstruktion der Technik in Deutschland nach 1945," in Walter Kertz (ed.), *Technische Hochschulen und Studentenschaft in der Nachkriegszeit* (Braunschweig: Universitätsbibliothek der Technischen Universität, 1995), 33–50; Mark Walker, *Nazi Science: Myth, Truth, and the German Atomic Bomb* (New York: Plenum Press, 1995), 59–61, 243–71; Ute Deichmann, *Biologists under Hitler* (Cambridge, MA: Harvard University Press, 1996), 290–335.
41 C.F. von Weizsäcker, *Die Geschichte der Natur* (Stuttgart: S. Hirzel, 1948), 5–7.
42 Studienausschuss für Hochschulreform, *Gutachten zur Hochschulreform* (Hamburg: n.p., 1948); see also Stamm, *Zwischen Staat und Selbstverwaltung* (1981), 65–7; and for Weizsäcker's role David Phillips, "Britische Intitiative zur Hochschulreform in Deutschland: Zur Vorgeschichte und Entstehung des 'Gutachtens zur Hochschulreform' von 1948," in Heinemann, *Umerziehung* (1981): 172–89, here 187.
43 Joachim G. Leithäuser, "Im Gruselkabinett der Technik: Kritische Bemerkungen zur Mode des romantischen Pessimismus," *Der Monat*, 3 (1951): 474–86, quotes on 478, 480. Leithäuser rather conveniently skips over the controverted status of the TVA, and Lilienthal's leadership of it, in the American political arena.
44 "Die Brüder Wawilow," *Der Monat*, 3 (1951): 486.
45 Clemens Münster, "Naturwissenschaft ist Humanismus," *Frankfurter Hefte*, 4 (1949): 484–92.
46 Hildegard Brücher and Clemens Münster, "Deutsche Forschung in Gefahr?" *Frankfurter Hefte*, 4 (1949): 333–44, here 334–5.
47 John W. Dower, *Embracing Defeat: Japan in the Wake of World War II* (New York: W.W. Norton, 1999), 45–6.
48 James Wood, *The Forgotten Force: The Australian Military Contribution to the Occupation of Japan, 1945–1952* (St Leonards, NSW: Allen and Unwin, 1998).
49 Dower, *Embracing Defeat*, 490–2.
50 For example see Yôichi Yamamoto, "Nihon genbaku no shinsô" (The Truth about Japan's Atomic Bomb), *Daihôrin*, 20, 8 (August 1953): 6–40.
51 For example see Nobuo Yamashita, "Ma ni awanakatta Nihon no genbaku" (Japan's Uncompleted Atomic Bomb), *Kaizô*, special number, 15 November (1952): 162–5; Mitsuo Taketani, *Taketani Mitsuo chosakushû: 2, Genshiryoku to kagakusha* (Collected Works of Mitsuo Taketani: Volume 2, Scientists and Atomic Energy) (Tokyo: Keisô Shobô, 1968), 355; Morris Low, "Japan's Secret War?: 'Instant' Scientific Manpower and Japan's World War II Atomic Bomb Project,"*Annals of Science*, 47, 4 (1990): 347–60.
52 Translation by Allied Translator and Interpreter Service, Military Intelligence Section, G.H.Q. of Department of Science and Technology of the Japan Communist Party, "Shortcomings of Science and Technology in Japan and the Duty of the Communists," *Zen-ei* (Vanguard), (November 1946), G.H.Q./S.C.A.P. records, ESS (E) 06393, National Diet Library, Tokyo.
53 Thomas P. Rohlen, *Japan's High Schools* (Berkeley, CA: University of California Press, 1983), 65–6.
54 See Alice Kimball Smith, *A Peril and a Hope: The Scientists' Movement in America, 1945–47* (Chicago: University of Chicago Press, 1965); William McGucken, *Scientists, Society and State: The Social Relations of Science Movement in Great Britain, 1931–1947* (Columbus, OH: Ohio State University Press, 1984).
55 Confidential intercept of Kôichirô Ichikawa (Minka) to Kansei Inoue (Imperial Petroleum), 15 June 1949, Civil Censorship Detachment, CIS 03977, G.H.Q. /S.C.A.P. records, National Diet Library, Tokyo.

56 For an explanation of Taketani's concept of technology see Yoshirô Hoshino, "On Concepts of Technology," in Shigeru Nakayama, David L. Swain and Eri Yagi (eds), *Science and Society in Modern Japan* (Tokyo: University of Tokyo Press, 1974), 39–50.

57 Mitsuo Taketani, *Shisô o oru* (The Interweaving of Ideas) (Tokyo: Asahi Shinbunsha, 1985), 122, 126–7.

58 Mitsuo Taketani, "Democratic Revolution of Japan and Technologists," *Gijutsu* (Technology), 5, 2 (1946), 3–5, abridged translation, G.H.Q./S.C.A.P. records, ESS (E) 06393, National Diet Library, Tokyo.

59 "Gakumon to shisô no jiyû no tame ni," 1950, in Shoichi Sakata, *Kagakusha to shakai, Ronshû 2* (Scientists and Society, Collected Papers, Vol. 2) (Tokyo: Iwanami Shoten, 1972), 73.

60 "Shortcomings of Science and Technology," G.H.Q./S.C.A.P. records, ESS (E) 06393, National Diet Library, Tokyo.

61 Mitutomo Yuasa, *Kagakushi* (The History of Science) (Tokyo: Tôyô Keizai Shinpôsha, 1961), 292.

62 Supreme Commander for the Allied Powers (S.C.A.P.), General Headquarters, *History of the Nonmilitary Activities of the Occupation of Japan 1945–1951, Volume 54: Reorganization of Science and Technology in Japan* (hereafter referred to as S.C.A.P. *History*), National Diet Library, Tokyo, 2.

63 Dower, *Embracing Defeat*, 529.

64 Gordon Rimmer, *In Time for War: Pages from the Life of the Boy Brigadier, The Biography of John O'Brien* (West Ryde, NSW: Mulavon, 1991); Hiroshi Ichikawa, "Technological Transformation of Occupied Japan: The Implications of the Policies and Activities of the Scientific and Technical Division of the Economic and Scientific Section of GHQ/SCAP," *Historia Scientiarum*, 5, 2 (1995): 183–97.

65 Hideo Yoshikawa and Joanne Kauffman, *Science Has No National Borders: Harry C. Kelly and the Reconstruction of Science and Technology in Postwar Japan* (Cambridge, MA: MIT Press, 1994).

66 "Report on Missions and Accomplishments of Scientific and Technical Division, Economic and Scientific Section," (16 September 1949), G.H.Q./S.C.A.P. records, ESS (A)-09796, National Diet Library, Tokyo.

67 For further details of the Imperial Institute see David Cahan, *An Institute for an Empire: The Physikalisch-Technische Reichsanstalt, 1871–1918* (Cambridge: Cambridge University Press, 1989).

68 Eikoh Shimao, "Some Aspects of Japanese Science, 1868–1945," *Annals of Science*, 46 (1989): 69–91, here 85.

69 Yoshikawa and Kauffman, *Science Has No National Borders*, 50, 56.

70 Rabi to Kelly, 8 December 1948, G.H.Q./S.C.A.P. records, TS 211, National Diet Library, Tokyo.

71 Rabi to Marquat, "The Use of Japanese Research Facilities as an Advanced Base in the Event of Acute Emergency in the Far East," (10 December 1948), stamped top secret, G.H.Q./S.C.A.P. records, TS 210, National Diet Library, Tokyo.

72 Keiichi Tsuneishi, *Kieta saiken-sen butai: Kantô gun dai-731 butai* (The Germ Warfare Unit that Disappeared: Unit 731 of the Kwantung Army) (Tokyo: Kaimeisha, 1981); Keiichi Tsuneishi and T. Asano, *Saikin-sen butai to jiketsu shita futari no igaku-sha* (The Germ Warfare Unit and the Two Doctors Who Killed Themselves) (Tokyo: Shinchosha, 1982); Keiichi Tsuneishi, *Ishii: 731 butai to Bei-gun chôhô katsudô* (Target, Ishii: Unit 731 and U.S. Army Intelligence Activities) (Tokyo: Otsuki Shoten, 1984). In English, see Peter Williams and David Wallace, with an introduction by John Pritchard, *Unit 731: The Japanese Army's Secret of Secrets* (London: Hodder and Stoughton, 1989).

73 Shigeru Nakayama, "The American Occupation and the Science Council of Japan," in Everett Mendelsohn (ed.), *Transformation and Tradition in the Sciences: Essays in Honor of I. Bernard Cohen* (Cambridge: Cambridge University Press, 1984), 353–69.

74 *S.C.A.P. History*, 72.

75 Monica Braw, *The Atomic Bomb Suppressed: American Censorship in Occupied Japan* (Armonk, NY: M.E. Sharpe, 1991).

76 See Glenn D. Hook, "The Nuclearization of Language: Nuclear Allergy as Political Metaphor," *Journal of Peace Research*, 21 (1984): 259–75. Hook examines how the metaphor of nuclear allergy was used in the late 1960s as a means of branding those opposed to Japan's nuclearization as abnormal.

77 Dower, *Embracing Defeat*, 493.

78 Mitsuo Taketani, *Genshiryoku* (Atomic Energy) (Tokyo: Mainichi Shinbun Sha, 1950).

79 The "Smyth Report" is the official account of the history of the development of the American atomic bomb, written by the experimental physicist Henry DeWolf Smyth, distributed to the press on 11 August, 1945 and later sold in book form to the public. It has been reissued as: Henry DeWolf Smyth, *Atomic Energy for Military Purposes: The Official Report on the Development of the Atomic Bomb Under the Auspices of the United States Government, 1940–1945* (Stanford, CA: Stanford University Press, 1989).

80 Taketani, *Interweaving of Ideas*, 132–4.

81 Japan Atomic Industrial Forum, *Nihon no genshiryoku: 15 nen no ayumi, jyô* (Atomic Energy in Japan: A 15 Year History, Part 1) (Tokyo: Japan Atomic Industrial Forum, 1971), 12–14.

82 Taketani, *Interweaving of Ideas*, 144.

83 Dower, *Embracing Defeat*, 494–5.

84 This idea is discussed in Arthur Edwards, "Scientific Expertise and Policy-making: The Intermediary Role of the Public Sphere," *Science and Public Policy*, 26, 3 (1999): 163–70.

6

THE TRANSFORMATION OF NATURE UNDER HITLER AND STALIN

Paul Josephson[1] and Thomas Zeller[2]

Introduction

Adolf Hitler and Joseph Stalin dreamed of the creation of eternal regimes more powerful than any the world had ever seen. Employing the advantages of closed political systems which tolerated no dissent, they used coercion, arrest, and execution to push magnificent plans for the transformation of their countries' economic, social, and cultural systems into unbeatable world powers. They took advantage of central planning mechanisms to bring the economy into sync with transformationist visions. Those visions included the alteration of nature itself to operate according to the ideological precepts of those two regimes. Nature transformed served symbolic purposes of demonstrating the personal visions of regime leaders, as well as economic and social goals. To achieve the ends of transformation, in both systems the political authorities involved scientists and engineers, in the Nazi case in the elaboration of a notion of *völkisch* nature, and in the Soviet case a kind of "proletarian" nature.

Nazi and Soviet scientists and engineers, like those in other industrial powers of the twentieth century, studied natural resources to develop means of ensuring their availability for present and future generations. Many of them welcomed government attention to their research in terms of increased funding and kind words. Many of them also believed that they could remain apolitical in the conduct of their research, merely providing expertise necessary to ensure access to resources and provide the power needed to harvest, excavate, transport, and process those resources, whether ore, fossil fuel, or forest. But because of the fact that their scientific efforts were bound tightly to state programs for economic development, cultural transformation, and military preparedness, they discovered it was an impossibility to be apolitical. In fact, those scientists who welcomed Nazi power in Germany and Soviet power in the USSR celebrated the fact that their research meshed nicely with the overarching ideological and political aims of their regimes.

A comparison of Nazi and Soviet forestry practices, and Nazi and Soviet engineering, in this chapter of highway design and construction (the *Autobahn*) in the former and hydroelectric stations in the latter, reveals that there were significant differences in the experiences of scientists and engineers in the two countries, too. The most obvious devolved from differences in the economic systems. In Nazi Germany private ownership of industry, and a public-private mix of resource ownership limited the size of nature transformation projects to those where the regime could build consensus. And once the Second World War commenced, many projects fell on hard times because of the need to concentrate on military expenditures. In the USSR, state ownership of the means of production enabled the state to embark on projects for canals, hydropower stations, and the like that rivaled Egypt's pyramids in scale. Germany had a longer tradition of forestry, hydrology, and other fields than the USSR. The political and ideological differences were pronounced as well. While both regimes were authoritarian, one was fascist and the other socialist. The most obvious aspect of this difference was the Nazi proto-organic vision of nature and the role of scientists in its mastery as opposed to the Soviet materialist vision of nature based on mechanistic understandings of biology. But whether the *Autobahn* or the Dnepr hydropower station, the goal remained the same: to bring nature within state plans for economic and military might, while at the same time appropriating its symbolic qualities.

The Third Reich and the forest

"German forestry is crestfallen; the German forests are on the down grade, their stands are badly reduced, their growth is diminished, research is at a standstill. German forestry is in despair, is in a desperate condition. German forestry has lost its leadership."[3] An aging expert, Carl Alwin Schenck, issued this simultaneously sad and condemning assessment in 1948. Three years after the liberation of Germany from the Nazis, it was apparent that the legacy of the dictatorship for forestry had been devastating. Schenck, a German sylvan academic who had founded the first American forest school more than fifty years before, had witnessed the growing demands on the German forests between 1933 and 1945. The very concept of sustainable forestry, once a token of pride for the academic community of foresters, had to be renegotiated under the auspices of the Nazi regime. Woods were hacked down at increasing speed; reforestation could not keep up with the new pace of cutting. Yet, the advent of the regime had signaled a turn towards a more "natural" forest, an ecologically more stable and aesthetically more attractive arrangement of trees and shrubs, with less stress on economic exploitation. These two contradicting tendencies formed the backdrop of Nazi forestry attitudes and politics. Scientists, administrators, foresters, and party officials fought out inevitable contests over political, financial, and ideological resources. More often than not, the former aligned their rhetorical strategies with what they perceived to be the core ideology of the latter.

In this respect, it would be too facile an undertaking to present the Nazi era as the culmination of a long and uniquely German relationship with the forests as expressed in literary romanticism, public attitudes, and forestry. Rather, Nazi Germany seems to stand for the antagonistic confluence of utilitarian and romantic attitudes and practices; the tension between these strands rather than their proclaimed harmony characterized the collective history of tree groupings in this era.[4]

One of the contested issues was sustainability. Since the eighteenth century, the meanings of *Nachhaltigkeit*, as it was called in German, oscillated from the exertion of centralized control over the woods to a mathematical care for future generations and their anticipated lumber needs.[5] With no apparent problem, it could be integrated into the prevalent Nazi ideology of the *Volksgemeinschaft*, a racially pure community transcending class, denominational, and political divisions. In this vein, the Nazi slogan *Gemeinnutz vor Eigennutz* (common good comes before individual gain) could justify experts' central control resulting in steady growth of trees.[6] (The Soviets, of course, had in mind common good and common gain, but the state was the beneficiary in both cases.)

Forestry was even presented as the forerunner of this ideology. Introducing statistical evidence on the fairly evenly balanced age-class composition of German high forests, a professor at one of the leading forestry schools commented in 1938 to an American audience: "It shows clearly that the socio-economic principles which only now are coming to be recognized as necessary in other economic activities have been consistently followed in the forest for more than a hundred years. If the principle of unrestricted economic egoism, i.e., *laissez-faire*, had dominated forestry no such even distribution of the age-classes would be present."[7] In the field of forestry, anti-market rhetoric struck a tone with overarching Nazi ideology; what is more, this data could be interpreted as evidence that past generations considered themselves as "only the temporary members of a perpetual folk organism," thus making German forestry "a shining example for other countries."[8] A skewed version of history played a crucial part in this argument. The past performance of foresters in Germany and the planting of trees, some of which had developed roots years before there even was a *second* Reich, only helped to underscore the teleological nature of German history leading inevitably to the Third Reich. On another level, this forest academic displayed a tendency to portray his discipline as a paragon of Nazi values, as a particularly assiduous branch.

There was a reason to do so. As it was the case in many other fields, new institutions and practices of governance challenged the established administrative and academic structures. Claiming rhetorical allegiance with the Nazi's core ideology through past practices would facilitate access to resources. In the case of forestry, the new regime enlarged its own power base by establishing a new political powerhouse for forestry and hunting on the Reich level. A pertinent law of July 3, 1934 gave Hermann Göring the illustrious titles of *Reichsforstmeister* and *Reichsjägermeister* and, for the first time in recent

German history, created a centralized office supervising all German forests, the *Reichsforstamt*. Power was appropriated both from the federal states and the Reich's Department of Agriculture.[9] All public woods were now under Göring's direct or indirect control; private forests continued to be administered by the Department of Agriculture.[10] The latter, predominantly owned by small farmers, were subsequently subject to even higher levels of exploitation than public woods.[11]

The total control over public lumber in the hands of the *Reichsforstamt* created a potential for faster, further-reaching political management of the forests. This attracted the attention of different actors, among them proponents of sustained forestry as well as government administrators in rivaling agencies. The foresters staked their claims early. Since the 1920s, a proto-ecological movement among German foresters had called for a lowering of annual cuttings and for replacing monocultures with polycultures. Academic foresters such as the Munich professor Heinrich Mayr pointed at the heavy tree cutting during the First World War. This overcutting exemplified the dangers of rash economic exploitation. What is more, academics diagnosed Norwegian spruce monocultures in Saxony as harmful for the forest soil and blamed them for diminishing yields. The combination of ecological and economic problems was contrasted with the vision of a "virgin forest" as an ecologically stable and thus economically productive community. The re-establishment of a primal forest became the final goal of sylviculture, and *Dauerwald* (permanent forest) was the buzzword of the debate. Especially in Prussia, where the poor sandy soils showed signs of exhaustion after generations of pine trees, this movement first took hold in the 1920s. Yet, since reaching this goal required selective cutting in order for differently aged trees to grow, foresters relinquished a degree of control to the "forces of nature." This proved to be an obstacle for the implementation of these principles before the advent of the Nazi regime.[12]

Since an ardent proponent of the *Dauerwald* was Göring's appointee for the position of head of the Prussian forest administration, this idea was officially propagated in the first years of the Nazi regime.[13] Clear-cutting in the Prussian forests was now discouraged; the division of every forester's domain into "improvement units" (*Pflegeblöcke*) would provide for ecologically diverse communities. Yet, this did not change the outlook of the forest administration at large; the proto-ecological advocate failed to round up broad support and was discharged in 1937.[14] As a trend, these efforts tended to jeopardize Göring's more imminent objective as plenipotentiary for streamlining the economy into preparedness for war. Ecological sensibilities and experiments were costly not only in their reallocation of administrative resources – a price that no one in power was willing to pay. Obviously, they threatened to lower the timber production in Germany, as Göring realized very soon. As early as in the fall of 1934, Göring ordered an increase in the pace of cutting for Reich forests by 50 percent on the average, a directive bluntly contradictory to these of his subordinate.[15] Even if the economic crisis of the early 1930s had decreased the demand for timber, the economic recovery after 1933 translated into higher cutting rates even without added interference by the state. Yet, the planners of the Nazi economy made it clear that timber production was an integral part of the politics of economic autarky. Timber production, at a low in 1933, exceeded the levels before the economic crisis as early as 1934 and grew by 85 percent until 1938.[16]

This was the desired result of a coordinated effort to create a corporatist marketing structure, epitomized in an October 1935 law establishing an oxymoronic *Marktordnung* (market order). Timber producers, sawmill owners, and lumber merchants were now part of a timber board supervised by the Reich government, which exercised its power to set prices.[17] Centralized control of wood supply and prices also became a part of the Four-Year-Plan of 1936, whose secret part aimed at preparing the German economy for war. Significantly for the German woods, Hitler appointed Göring as the chief executive of the Four-Year-Plan administration. The corporative spiderweb of control intensified in the fall of 1936, now covering tree nurseries as well as plywood and fiberboard producers. Starting in December of that year, forests owned by towns, cities, and large private estates were required to increase their cutting rates by 50 percent as well. This internal expansion, which soon tested the liminal stages of sustainability, was supplanted by increased cutting in the annexed and occupied territories as the Nazi conquests of the Second World War progressed.[18] In this way, timber typifies the mixed economy of the Third Reich, where private ownership of production was mostly upheld, yet with varying and progressively intensifying levels of centralized economic planning. It would be easy to understand this process as the industrialization of wood, although characterizations such as a heightened flow of production seem more apt.

The administrators of the Four-Year-Plan sought to redefine the applicability of wood as well. Artificial silk had been a staple since the turn of the century. Under the shibboleth of autarky, further synthetic usages were introduced. Cellulose produced from timber provided the basis for the production of spun rayon in a viscose process; the state provided major economic incentives for setting up new plants.[19] Indeed, scientists and engineers propagated cellulose derived from lumber as a source of myriad applications, from motor fuel to food, from uniforms to planes. Friedrich Bergius, who had shared the 1931 Nobel Prize in chemistry with Carl Bosch for the hydrogenation of fuel, rallied successfully for financial resources with a vision of solving Germany's raw material problems. The chemical giant IG Farben as well as Reich institutions supported his industrial plant, which succeeded in obtaining sugar from cellulose in wood.[20]

It is doubtful whether a master plan elevating wood to the status of a backbone of the German economy was drawn up and followed. More likely, the polycratic governing structure of the Reich with competing agencies, different funding sources and political rivalries rendered the implementation of master plans as impossible for wood as for other economic sectors.[21] Still, increasing self-sufficiency in the textile sector relied on the exploitation of lumber in Germany and the occupied territories, decreasing the share of firewood and introducing hardwoods such as birch and beech into production processes. Sugar, alcohol, and *Holzbenzin* (wood gas) were manufactured on an industrial scale; and the chief of the forest products division of the Four-Year-Plan has been reported prancing around in a light green suit made entirely of artificial wool called *Wistra* or *Wollstra*.[22] State regulation demanded that garments produced for consumer markets contained 20 percent synthetic fibers, which made their production economically more feasible.[23]

With the preparation for war, ecological concerns for the woods took a back seat in Nazi Germany. Research and development of synthetic uses for lumber garnered state support, leaving the forests with little more than a role as provider of raw material. The foresters seemed to have complied with these exploitative techniques. One indication explaining this behavior was their almost insuperable membership level in the Nazi party, reaching between 88 percent and 93 percent in the public sector and 78 percent in the privately owned sector.[24] Forest academics, eager to partake in the new research opportunities, increasingly studied the applicability of the resource and founded new research institutes at their universities.[25]

For this goal, they integrated their work into a framework of expansion and rigorous resource utilization, with the forests serving as a seemingly augmentable homegrown resource. The transformation of nature in the realm of forestry in Nazi Germany amounted to thinning out the woods instead of replenishing them as the supposedly "green" rhetoric of the peacetime years had indicated. The self-imposed pressures on the German economy in preparing for war found one outlet among others in utilizing lumber resources on a wider scale and scope and with an increasing pace of cutting. The personal factor of Göring as both head of the Reich's forest service and head of the Four-Year-Plan captured this dilemma in a nutshell. The proto-ecological efforts of the first years left few lasting effects on the forest landscape as preparation and execution of war became paramount goals of the Nazi dictatorship. As much as conservationists had welcomed the advent of the Nazi regime, they quickly realized that political and economic factors overruled their concerns as a rule.[26]

Rather than giving the conservationists a real voice in forestry matters, the Nazi regime created an opportunity for furthering forestry as a science-based enterprise in the hands of academically trained experts who had swiftly reoriented their work towards economic self-reliance at home and economic and military expansion abroad. As the demands of the war machine spiraled upwards and included growing amounts of timber, Eastern European forests were conquered, appropriated and exploited. While a few forests were set aside as hunting resorts, forestry scientists worked on a globalization of the lumber industry and trade under German tutelage.[27] Thus, the rhetorical transformation of forests into a mere resource was accompanied by constant over-cutting in the countryside. According to one estimate, by 1945 overcutting amounted to some 65 out of the initial 350 billion board feet of timber; the growing stock was reduced by roughly 25 percent.[28] In the later part of the Nazi years, the cutting rates set by the regime were often not reached. Instead of introducing labor-saving technology, administrators increasingly relied upon forced laborers in their ranks.[29] The pace of overcutting increased in the ensuing years, reaching a high in 1947 and requiring the complete recovery of the German forests.[30]

In the life of a forest, twelve years is a short time. However, Nazi forestry policies aimed at changing the parameters for the woods thoroughly and simultaneously from two opposing ends of the spectrum. On the one hand, a movement within academic forestry aiming at reconciling aesthetics, proto-ecology, and production perceived of the new dictatorship as a vehicle for attaining their goals. These proponents of the *Dauerwald* recognized the economic pressures on the forest, yet sought to answer them while at the same time attaining a more diverse, stable, and more beautiful forest. This stance received initial sympathies from the forest administration under the Nazis' aegis. On a political level, the centralization of the forest service offered a chance to implement these ideas more forcefully. Ideologically, they could be tied in with a particular streak of the Nazi movement stressing the link between nature and culture in a racially determined way.[31]

As the agenda of forestry changed from including these new approaches towards concentrating on forestry as ancillary for autarky and the preparation for war, these proto-ecological ventures declined. Instead, another group of academic foresters answered the call for economic self-reliance by pointing to the manifold usages of wood. They increasingly portrayed lumber as a replenishable and local resource whose range of applicabilities only waited to be tested. Ironically, an industrially developed nation such as Germany now spent considerable money and effort to replace coal and oil-derived gasoline as the fuel of the economy with wood as an energy source, a feature usually assigned to pre-industrial economies. This circumstance, together with thinner forests and a scientific redefinition of the woods, was one of the major results in this realm.

The people's automobile

If the National Socialist transformation of the forest failed, in transportation the legacy of Nazi rule is – even in today's Germany – widely apparent. The road network known as the *Autobahn* and the *Volkswagen*, the people's car, were supposed to represent a technologically advanced Germany with a broad allure, advancing economic modernization and gaining symbolic hegemony. Massive propaganda efforts hammered these technological icons into the collective memory, presenting them as symbols and means of an economically and technologically advanced nation. Hitler seized the opportunity to leave his dictatorial imprint on these artifacts. The propagandists never tired of labeling the

Autobahn "Adolf Hitler's roads" and celebrating his supposed contribution to the construction of a Central European version of Ford's Model T.[32]

Hitler announced the road-building program in connection with introducing an affordable car for the masses, the *Volkswagen*, during the first year of his rule. Germany was a poorly motorized country, at least in comparison with Britain or France, not to mention the United States. A series of policies such as tax breaks, direct subsidies to potential car buyers and increased investment into road construction enabled both the car-producing and the road-building sectors to grow much faster than the rest of the economy during the 1930s.[33] Cars and roads possessed a distinct appeal: they could be presented as successful public work programs, as they were highly visible and simultaneously carriers of a modern transportation technology. Consequently, the regime exaggerated the number of workers who found jobs building the *Autobahn* and the *Volkswagen* and stressed how increased auto production would make them affordable.[34]

When Hitler introduced a "massive road-building" program in February 1933, he described the advantages of an affordable car for the masses. The first known feature was its cost, which Hitler established at 990 Reichsmarks against the advice of the car manufactures. In a speech in March 1934, the dictator asked the companies to produce such a car within four years; the "common man" should thus partake in the grand scheme of motorization. Car manufacturers were taken aback by Hitler's sudden maneuver and labeled his plans as fantastic, preferring to hold onto established profit margins by selling expensive cars to upper middle-class customers.[35] The maverick engineer Ferdinand Porsche, however, captured the moment and presented Hitler with a prototype of his *Volkswagen* in 1936. The successor of the trade unions, the Deutsche Arbeitsfront was subsequently put in charge of building a production plant utilizing Fordist principles of mass production since 1937.[36]

After long deliberations, a site near Fallersleben in Lower Saxony was chosen. Due to its proximity to the Reichswerke Hermann Göring, another state-run company, planners envisioned the "greatest industrial plant in the world" when both companies would finally merge into one big spatial and industrial conglomerate.[37] This was a direct challenge to Stalin's planners and to industrialists in the United States, for it shared elements of the size of both, and promised to take better care of the workers than either. The production site changed the geography of the region from rural farmland to an industrial center built from scratch on 2,000 hectares. Ford's River Rouge plant was the example to be both emulated and surpassed; on this side of the Atlantic, it was state power combined with a vision for technology that would enable such a gargantuan project, not a market economy. Without relying on the established car manufacturers, the plant was built at an amazing speed, starting in February 1938. The new town of Wolfsburg was to house the workers; its architecture, as the one of the factory itself, included elements of the Bauhaus *moderne* as well as inflated regionalist styles.

The "German River Rouge" was never finished; the mainstay of production proved to be war vehicles instead of civilian cars. Forced laborers and racial enemies of the regime worked in the half-finished halls until their liberation. The new spatial order set in motion by the Nazis continues to exist, however. The industrial geography of heavy industry, centered around the Ruhr valley, was complemented by this new, remote center of car production that necessitated more transportation for the production of vehicles. On another level, the idea of the *Volkswagen* proved to be culturally pervasive; consumers' demands for small, affordable cars were ignited in the 1930s and only fulfilled in the 1950s and 1960s in West Germany.[38]

In what seems to be a peculiar way to motorization, the roads built under Nazi rule had to await sizable traffic for some time. The *Autobahn* came to express this discrepancy most vividly. In retrospect, the size of the network – 3,600 kilometers by 1945 with thousands more planned – raises questions about the planning capabilities of Nazi Germany.[39] Hitler overruled concerns within the military which objected to creating enemy targets and relied on the railway for its logistics. The *Autobahn*'s potential for propaganda clearly was its most immediate appeal for Hitler. Unknown to anyone not associated with the *Autobahn* were the brutish working conditions in the camps for the formerly unemployed, even leading to unrest. Later, Jews were forced to work on these roads.[40]

The official *Autobahn* propaganda took place on different levels. Bombastic radio shows, movies, board games, and cigarette collectibles brought home the message to German workers that this piece of German engineering was only possible through the advent of Hitler.[41] Theater plays, books, commissioned *Autobahn* paintings, and newspaper articles conveyed to educated middle-class audiences that the transformation of nature entailed by the roads was beneficial. Instead of destroying the surrounding countryside, the roads were touted as enhancing their beauty. Presenting these roads as an organic extension of the landscape aimed at nothing less than overcoming the long-established chasm between nature and technology. Upholding certain building styles was offered as the solution.

The engineers in charge of the *Autobahn* did not have to go to great lengths to integrate their beliefs into Nazi ideology; rather, they accommodated core elements of the newly established doctrines as much as the Nazis accommodated their profession. The first engineer ever to head a Reich agency was Fritz Todt, who supervised the *Autobahn* project. When he presented his proposals to Hitler in January 1934, the building style was already of importance. Todt contrasted the new roads with railways and their physical appearance. "The railway has mostly been an alien element in the landscape. A motorway, however, is and will be a street, and streets are an integral part of the landscape. German landscape is full of character. Therefore, the motorways must assume a German character."[42]

By summoning landscape rather than nature, Todt alluded to a powerful, normatively charged concept. German geographers in the nineteenth century had established cultural landscapes (*Kulturlandschaften*) as an heuristic tool as well as a collective embodiment of the work of previous generations.[43] (Agri-) Culture had formed nature, and the result was a culturally loaded landscape. It was exactly this essentialist link between nature and culture that made landscapes so valuable and so contested by conservationists. While earlier environmentalists in Germany had been quick to condemn industrialization *per se* as the destroying force of landscapes, after the turn of the century a growing number of them sought ways to incorporate technology into their agenda.[44] Regionalist designs of power plants, the dressing of factories with trees and shrubs, and finding the appropriate location for other technological artifacts figured prominently for this group.[45] While assuring the proper function of these plants, their design itself would make sure that the landscape's features would be upheld and not be overruled. Mere decoration was frowned upon; inner harmony and outward beauty had to go hand in hand. An avid student of this approach, the landscape architect Alwin Seifert, approached Todt as the chief engineer for the *Autobahn* and persuaded him to install a system of landscape advocates (*Landschaftsanwälte*) to ensure that a landscape-friendly style was realized.

Todt supported this step and deemed it appropriate for the national technological style which he was advocating. *Deutsche Technik* (German Technology) was the ideological tool with which a number of leading German engineers aligned themselves with the Nazi regime.[46] According to these ideas, the capitalist profit motive had led to a detrimental influence on technology that had served individualistic needs and furthered social unrest. An economic and social system governed by a more egalitarian mode of production, possibly under the guise of technocracy, would lead to the development of improved technologies and benefit the social status of engineers. What is more, this trope was increasingly racialized: capitalism as the buzzword of criticism was supplanted with "Jewish influence" upon technology, which, as this argument would have it, had turned it into an egotistical and short-sighted endeavor.

With the advent of the Nazi regime, technology's role increased considerably. According to one contemporary writer, *Deutsche Technik* was linked to the inner core of National Socialist beliefs, namely "the biological, the nature-given thinking and acting, which now attains central focus in all political, *völkisch*, and cultural fields."[47] Technology was no longer peripheral and engineers' work no longer marginal or even detrimental to society. They could now feel themselves to be part of the vanguard of culture in what they perceived as a new era.

Thus the stakes were high for the engineers involved with the Nazi *Autobahn*. As much as Todt welcomed the assistance of the landscape advocates, he stipulated that "no petty landscape dreamers" should be appointed. Only if they could appreciate the "technical problems" involved, would they be capable of ensuring that the *Autobahnen* would attain a scenic character in accordance to the "German spirit."[48] Landscape, at this early stage, was portrayed as an object worthy of preservation or embellishment. On this almost trivial point, all the actors involved in *Autobahn* building could agree, but the specifics of altering the landscape, that is the specifics of road design, were very much under debate.

Todt's cadre of landscape architects consisted of fifteen professionals selected by Alwin Seifert, who was the liaison to Todt. Their job was to consult civil engineers in questions of design. Many of them, however, were employees of the national railway, the Reichsbahn, thus habitually committed to building streets with mostly the same parameters as used for railways. That is, engineers traditionally avoided curves wherever they could, situated roads on dams, and paid little or no attention to roadside plants. The ensuing conflicts between civil engineers and the landscape architects were therefore inevitable. These altercations have been portrayed as ultimately a success story for the landscape advocates, who slowly overcame resistance with Todt's help.[49] However, central issues of road design and thus the transformation of nature were not resolved during the building phase of the *Autobahn*. Ideological as well as methodological differences prevailed and even intensified.[50]

For one thing, the actual alignment of roads remained contested. Seifert abhorred straight roads, and recommended curvilinear designs as appropriate to the values he saw incorporated in German landscapes. Nature was the guiding authority for him, since in his eyes a straight line was unnatural: "The straight line is of cosmic origin, it is not from this earth and is not found in nature. No living thing can move itself forward in a straight line."[51] Todt, in his reply, did not shy away from invoking nature as well. Branching out into the animal kingdom, he declared a comparison between the car and the rabbit's or deer's sweeping jumps as invalid. "Rather, the car resembles a dragon fly or any other jumping animal that moves shorter distances in straight lines and then changes its direction at different points."[52] In line with the reorientation of technology towards nature's ultimate authority, these men invoked whatever rhetorical resources they could use. In 1935, this conflict was decided to Todt's liking, as he simply declared his design to be appropriate. No controlling authority or public discussion could deter him from doing so. It was only after large stretches of the *Autobahn* had been built that Todt reconsidered his stance. Adhering to the contours of the landscape, he surmised in 1941, would make the ride a livelier experience due to the higher number of curves. He also believed that this road-building style was less expensive.

On a related matter, the landscape advocates sought to use the *Autobahn* as a showcase of nature restoration. Some of the areas traversed by their roads had become overused or changed their face through capitalist exploitation. Building the highways, in their eyes, gave them a chance to reintroduce native plants that had been obliterated. In the confusing jumble of agencies that made up the highway bureaucracy, the landscape architects garnered support for a team of ecologists called "plant sociologists." They studied the local climax communities and produced lists of species that had once been indigenous to the area. The architects increasingly relied on these lists of autochthonous trees and shrubs when placing orders. Yet, the Todt administration questioned the immediacy of the undertaking and – more pressingly – was averse to the high cost of too many plantings. In the end, a compromised version of this green revisionism prevailed.

While the landscape advocates reached some of their objectives in individual cases, general differences between them and the civil engineers over the transformation of nature remained. The latter aimed at creating a speedy, impressive, and varying driving experience in which nature would be transformed into a visual commodity for the drivers' consumption. The landscape advocates, however, presented a mixture of aesthetic agendas and ecological restoration that they thought to connect to *Deutsche Technik* and to Nazi ideology at large. Instead of relying on the merely aesthetic qualities of the undulated line recommended since the days of William Hogarth, the landscape architects did not tire to point out that it was in fact nature's laws that they were about to implement and not a passing artistic fashion. In invoking this final authority, they aimed at tapping rhetorical resources of seemingly insurmountable force.

In this case of nature transformation, nature itself was the object invoked by all the parties involved. It was through the prism of *Deutsche Technik* and its close alignment with core tenets of Nazi ideology that building roads could be presented as a boon to the countryside. These highways were subordinated to the goal of establishing a vision of German landscapes that was compatible with the racial and expansionist goals of the regime. Nature, in this respect, was appropriated and groomed as an object for visual consumption. On another level, the new highways contributed to the creation of a more closely knit and more spatially homogeneous country, one in which geographical distance could be replaced by driving time.

In Nazi Germany, the large-scale transformation of nature, increasingly a prerogative of the modern nation-state, thus took on a new quality, both in the name of science and of the dictatorship.[53] Increasingly aware of their role as social systems, forestry and road building were able to attain different levels of attention, funding, and symbolic capital during the relatively short time-span of twelve years. The millennial promises and chiliastic pretensions of January 1933 offered distinct attractions to the scientific discipline of forestry, whose rise had been closely connected to that of modern state institutions and whose temporal perspective transcended daily operations. The nascent movement within German forestry calling for more ecologically diverse and thus aesthetically more pleasant woods expected the dictatorship to support and fund its efforts. The centralization of sylvan administrations furthered this hope, since it offered the chance to multiply those efforts.

After some initial shows of sympathy, Nazi planners in the Four-Year-Plan administration quickly realized that these experiments would threaten, if not thwart, their goals of refashioning the nation's economy into a war machine. Indeed, lumber supply became one of the pillars of this martial conversion, both in its traditional usages from construction to heating and in the heavily funded efforts to synthesize fabrics from cellulose. Lumber's qualities as a renewable resource strengthened the positions of both chemists and some forestry scholars eager to redefine their objects of study as a widely applicable resource. Thus, even the hallowed principle of sustainability was overruled in practice and increasingly ignored in theory. With the expansion of the German war efforts, trees fell faster than they could be replenished. This depletion stands in sharp contrast to the declarations of respect for nature and the creation of forest preserves which some scholars tend to take as the essence of Nazi forest politics. The gap between oratory salutations and practical exploitation of the forests could hardly have been larger than in the field of forestry.

This discrepancy, although a different level, also holds true for the example of motorization. It appears that the Nazi regime chose the internal combustion engine and its corollaries as emblems of its politics for reasons of propaganda and their potential for public displays rather than for economic benefit. While Nazi propaganda wrongly credited Hitler with "inventing" the *Autobahn*, it is hard to imagine the sudden start of the massive road construction program without his personal interest in this technology or, to be more precise, their potential for groundbreaking and opening ceremonies. These roads were to literally solidify the economic and political resurgence of Germany and thus became instrumental in the constantly ceremonial public life of this dictatorship. Civil engineers and technical universities were ill prepared for this jumpstart, both in numbers of engineers and in scientific expertise. The regime's demand to build "landscape-friendly" roads only augmented the confusion over the design of the *Autobahn*, which proved to be more contradictory than organic as the accompanying propaganda claimed.

During the hurried first years of road building, railway design, appropriations of the early American parkway and landscape architects, and demands for sweeping, "organic" roads imbued the construction style simultaneously. The resulting conflicts were not resolved, only overshadowed by a debate on grounding these roads in a specifically national style of technology, the *Deutsche Technik*. While the parameters of this discourse allowed for conflicting viewpoints and technological designs under one umbrella, the racial core of this ideology united these conflicting views. Under the same auspices, the *Volkswagen* was foisted upon a grudging car industry by the regime. The production plant for the car was eventually run as a public enterprise and modeled after Henry Ford's River Rouge factory, thus deliberately appropriating the American model of mass production and consumer orientation. The consumer, to be sure, was racially defined in 1930's Germany, and instead of experiencing yearly model changes, Germans adopting the savings plan for the car were denied consummation of their dreams. Still, the motorization of Germany's economy and society was more than a chimera. The *Autobahn* became both a symbol of Nazi rule and an almost empty shell of future transportation needs, while the *Volkswagen* embodied a state-centered approach to traffic, a system of tight regulation and symbolic exploitation.

As implausible as it may sound, the German *Autobahn* actually contributed to the imprisonment of a Red Army soldier in Russia. In 1947, the student Mikhail Isayevich Tanich was arrested in Rostov-on-Don. Tanich was interrogated and admitted that he had praised the German roads in a conversation with his co-students. He had seen these roads as a soldier in the Red Army. Tanich's crime qualified as "praising life abroad," thus implicitly disdaining life in Stalinist Russia. He was jailed for six years.[54] Many of Tanich's countrymen had fared as poorly in their efforts to bring technology to bear on the transformation of nature to meet the needs of the proletariat.

Nature transformation under Stalin and beyond

The Soviet effort to tame nature in the name of the proletariat gathered strength after the death of Lenin under Stalin, Khrushchev, and Brezhnev. Both the urban and rural landscape were included. It became more bold as Moscow's reach, in the guise of engineers and their forest, fish, and water resource research institutes, extended from the European USSR to Siberian rivers and forests and the pelagic waters well beyond the country's borders. No tree, no river, seemingly no molecule of water escaped the planner's designs for a regular, orderly nature.

Because of national ideology based on state-defined notions of the common good – in this case, the good of the proletariat as opposed to the *Volk* – Soviet resource management programs resembled those in Nazi Germany. The importance of central planning organizations in bringing natural resources within plans defined as rational, the role of scientific and engineering institutes in fulfilling those plans, and the symbolic importance of such artifacts as dams, canals, and highways in demonstrating control over nature were also important similarities. The obvious differences between the two nations – in politics, economics, physical geography, and historical experience – make the similarities all the more striking, and are one indication of some of the universal attributes of science, technology, and engineering, when the overused analytical categories of "pseudo-science" and "totalitarianism" often postpone in-depth analysis of science under Hitler and Stalin.

Perhaps the single greatest difference between Soviet and Nazi efforts to alter the natural landscape was the size of the countries and scale of the effort. The Soviet Union covered eleven time zones, with climate and vegetation ranging from permafrost and tundra in the far north to the semi-arid regions of Central Asia; from the mountains of the Caucasus to the deep pine forests of Karelia and Arkhangel; and to the productive agricultural regions of Ukraine. Germany lacked this variation and size, and also lacked the USSR's extensive natural resources. Germany also had historically strong universities and a technologically advanced industry, whereas Stalin's universities and industry had had less than ten years of peaceful development when he came to power in 1929.

Nature transformation found ample support among Bolshevik leaders and engineers from the first days of the Russian Revolution. The engineers, as part of the enlightenment tradition, fully believed in their ability to identify scientifically-based ways of managing resources to ensure their availability for present and future generations. They welcomed the new government for its stated willingness to support science and technology for the masses, for the Tsarist regime had left them searching high and low for funding to pay for new equipment and for oceanographic, cartographic, and geological expeditions. Such scientists as N.M. Knipovich, who managed to chart the fish resources in the North and Barents Seas at the turn of the century, spent years organizing their research and often ended up creating from scratch or importing from abroad even basic measuring instruments.[55] With Lenin announcing that scientists were "naturally" materialists in philosophical viewpoint and crucial to the success of the state, there was little doubt among scientists or engineers in the 1920s that it was easier to do research than before the revolution.

Indeed, Lenin was a technological utopian. He believed that technology, in particular electrification, would change agrarian Russia into an industrial superpower overnight. He argued that electrification would transform agriculture and industry. Since the time that Lenin advanced a program for rapid electrification, large-scale systems have been at the center of Soviet economic development programs.[56] The Bolsheviks popularized the program through the famous slogan "Communism equals Soviet power plus electrification of the entire country!"

Lenin asked a colleague, Gleb Krzhizhanovskii, to compose a State Program for the Electrification of Russia, known usually by its acronym GOELRO, which served as the basis of Soviet electrification efforts for decades. Hydroelectric power had a major place in initial GOELRO plans. This is not surprising, given the poor state of Tsarist coal industry in the Don basin of Ukraine and oil industry in the Caspian Sea, near Baku, Azerbaidzhan, and also the cost of the extraction of oil and coal. Further, Tsarist hydroelectric engineering was in an infantile state. By 1915, Russian engineers had built a few hydropower stations at 150 kW, 300 kW, and up to 2,800 kW. If underutilized, hydroelectric power seemed to be less costly and, engineers believed, could be deployed on any moving body of water. No longer ignored by the state, the engineers gained political support to build a series of modest stations and conduct surveys of hydro potential as far east as the Enesei, Angara and Amur Rivers in Siberia. Yet this support was inadequate to finish projects on even a modest scale until Stalin came to power, in part because of the backwardness of Russian industry.

Soviet engineers, egged on by party officials who were eager to turn the proletarian USSR into a mighty power, adopted brute-force approaches to nature transformation. At first, they turned to America, and Germany in particular, to rebuild industry on a modern technological basis in the 1920s. Siemens, General Electric, and other corporations provided know-how, turn-key plants, engineers, and various technologies to rebuild Russian industry (for example, the Elektrosila factory for turbogenerators and Izhorsk Works for magnets, pressure vessels, and the like) which in 1926 only had reached pre-war – 1913 – production levels). But in the 1930s, as Stalin pushed the country toward so-called socialism in one country, as show trials against scientists and engineers – foreign and suspect foreign sympathizers – as "wreckers" were held to squash any nascent, and largely imagined, technocratic tendencies, and as industry was forced to innovate in conditions of autarky, engineers and scientists developed new technology along Soviet rails and with Soviet style in form and content. These were generally large-scale technological systems which, whether power-generating, irrigating, harvesting, or processing technologies, were known more for their ability to alter the natural landscape rapidly and for the purpose of gathering the masses in one site where they were subject to political control rather than for their efficiency, as this discussion of the evolution of water and forest management technologies will show.

The effort to alter nature's very essence, to turn something perceived to be irrational and capricious, and in the lexicon of the time an "enemy of the people," into something planned, orderly, and proletarian through and through, had its first major manifestation in the 560,000 kW Dnepr hydroelectric power station (known as DneproGES and built by a construction trust of thousands of poorly skilled workers known as Dneprostroi). At DneproGES enlightenment visions, political backing, economic support, and Marxian ideological certainty came together. Between 1920 and 1928 the Soviets had installed only 100,000 kW of hydroelectric power stations, but they were confident they would finish DneproGES in short order, even if preliminary geological surveys had showed the Dnepr's flow might not justify the station. (The political authorities ordered that these surveys be disregarded.) Taking five years to complete, DneproGES became one of the largest hydroelectric power stations in the world. With the construction of three locks, it made the Dnepr River navigable to the Black Sea. The decision to build DneproGES delayed other projects such as the Volga–Don canal whose supporters claimed would be less expensive and of more immediate benefit to agriculture.[57] But form was more important than style in the effort to showcase socialist technological verve.

Stalin focused more attention on the so-called productive relations than Lenin had. He wished to see modern technology in agriculture, industry and the military enable the USSR to survive hostile capitalist encirclement. But in addition to technology, he sought socialist economic relationships, abandoning Lenin's New Economic Policy of small-scale capitalism and a moneyed economy for state control of the means of production. He initiated the "Great Break" with previous economic development and political policies. He pushed a break-neck pace for industrialization and forced peasants into collective farms, in part to serve as a source of investment capital for industrialization, and in part to bring the tractor and electricity to the countryside. The human costs of the Great Break were immoral by any measure: millions died of famine, millions perished in the purges, and engineers and managers whose projects failed to reach targets were arrested, and some executed for "wrecking."

139

Many engineers conflicted with the Communist Party over industrialization policies even before Stalin's rise to power. They criticized the growing tendency of the state to "gusher psychology" – the emphasis on short-term spectacular achievements as opposed to long-term more economical use of resources. For them, Soviet planning ignored local geological conditions by favoring functional, national or sectoral planning; it emphasized large scale projects at the expense of handicrafts, office materials, tableware, and clothing industries might have been more efficient owing to their small scale; and it stressed the value of capital, but abused humans and the environment. Yet political pressures on the engineer made it impossible for individual engineers to practice the profession with integrity. Inevitably, political judgements predominated. The engineer was responsible for raising production at any cost. The "human" factors in production that might be encouraged by higher pay, good housing, good food and health, and that might lead to sensible environmental considerations within the limits of scientific uncertainty were ignored.[58] For engineers, the result was a lesson learned well. To put it in the words of one of the engineers involved in Dneprostroi, Alexandr Vinter, "The gigantic construction on the Dnepr was the first serious school for our Soviet builders."[59]

Increasingly coopted by the regime, engineers themselves advanced large scale projects to change the face of the USSR from agrarian to urban and industrial. Logically, large-scale urban construction projects were the centerpiece of Stalinist programs. The Moscow metro, part of the "socialist reconstruction" of Moscow undertaken in the 1930s, which served as a flagship of the Stalinist system,[60] and Magnitogorsk at the south of the Ural mountains, a factory based on the most modern steel mills in the world in Gary, Indiana, which was intended to be an exemplar of Stalinist technological might,[61] are two examples of the effort to change the urban landscape.

From the urban landscape, Stalinist planners turned their attention to the transformation of nature itself. One author likened the reconstruction of rivers, whose flow changed seasonally, into machines operating according to the dictates of man as similar to the rational rebuilding of streets and cities.[62] "Bulldozer" technology grew to epic proportions during the Stalin era as major scientific, engineering, and ministerial organizations were put to the task of improving upon nature.[63] For the Soviet Union this meant harnessing rivers and lakes through a series of ever more grandiose canals and hydroelectric power stations. By the 1930s, Soviet engineers were projecting an entire series of hydroelectric power stations on Siberian rivers – ten stations at 2 million kW each – to develop rich bauxite deposits.[64]

What the Soviet Union lacked in technological sophistication, its engineers and policy makers made up for with their unbridled enthusiasm. Without the impediments of public opposition or the legal requirements of environmental impact statements, they quickly moved to change forever the face of European rivers – the Volga, Don and Dnepr – with attendant human, economic and ecological costs that continue to be felt. During the first five-year plan of forced industrialization and collectivization, larger hydroelectric power stations were being built including the Nizhne-Svirskaia, Rionskaia, Gizel'donskaia and others; in addition to DneproGES, by 1933 there were ten new ones totaling 345,000 kilowatts capacity. During the second plan (through 1938) another 745,000 kilowatts of capacity was added, and on the eve of the Second World War, including the installation of the Uglichskaia station on the Volga, over ten percent of the country's electrical energy was now generated by hydropower. The thirty-seven stations built from 1928 to 1941 had a total capacity of 1.5 million kW.[65]

Canals were another important cog in the machine of Soviet environmental management. Most were for irrigation, with the others for city water supply (civil and industrial purposes). Seven were built before Second World War, seven from 1946–60, and twenty-two in the next twenty years.[66] At its seventeenth Party Congress, the Communist Party approved the construction of the infamous 127 km Belomor canal, built between 1931 and 1934, and the Moscow–Volga canal (1932–7), intended to bring Volga water to the Kremlin and facilitate waterway shipping between Leningrad, Moscow, and Gorky.[67] Slave laborers built Belomor with pick axes and shovels. Tens of thousands of them froze or starved to death. The canal turned out to be too shallow for barge traffic. But its purpose was political education of alleged state enemies, not shipping. In 1933, at a special session of the Academy of Sciences, engineers discussed partial diversion of waters from northern European rivers into the Caspian. The Leningrad-based institute Gidroenergoproekt offered a Siberian rivers diversion canal project as early as the 1940s to irrigate Central Asia and Kazakhstan.[68]

If DneproGES was the Soviet Hoover Dam, then the Volga–Don was its Panama Canal. Its goal was to unite all of the great rivers of the European USSR, primarily through the tributaries of the Volga: the Kama, Oka, Viatka, and Belaia rivers, so that all major ports – Moscow, Leningrad, Belomorsk, Iaraslavl, Kuibyshev, Saratov, Stalingrad, and so on – were part of one waterway. Construction began in earnest in 1948 only after the damaging interruptions of the war. In building the Volga-Don canal, 152 million cubic meters of earth was excavated, 57 million cubic meters of concrete and reinforced concrete was poured, and 45,000 tons of metal devices and mechanisms were employed. Construction was faster, the sources claim, than any effort the west could muster, and increasingly mechanized too, using 900 graders, 300 bulldozers, and 350 excavators – including several Soviet monsters with bucket capacity of fourteen cubic meters – three dozen suction dredges, and thousands of trucks, tractors, cranes, and winches.[69] The size of the hero projects and the machines used to build them clearly had much to do with Cold War competition for ideological superiority: this was technology envy, plain and simple.

Stalinist canals were important for their display value. An early publication intended for western audiences touted the virtues of socialist canals, while foreshadowing Stalinist transformation of nature. Its author wrote, "On the bank of what once was a small stream called Khimki, just a few miles outside of Moscow, towers a magnificent structure built of granite and marble." A five-pointed, gold star sat on top of stainless steel spire 262 feet in the air. The Moscow-Volga canal brought clean water for Moscow's burgeoning population, to the Moscow Sea, created at Volga terminus and holding nearly 40 billion cubic feet of water. The canal itself was modest at 80 miles, but the authorities patriotically announced that only Soviet machines took part in the construction. This was quite a change from DneproGES, where American and German methods and expertise were crucial to the project.[70]

Interrupted by the Second World War and the Nazi invasion, nature trans-formation regained immense importance in the postwar years. The German armies destroyed much of what had been achieved in Ukraine and western Russia, taking special glee in dynamiting the Dnepr hydropower station. The postwar, fourth Five-Year Plan therefore would make up for lost time, rebuilding what had been destroyed, building what had been merely a vision, and turning the European USSR into a unified transportation, hydroelectric, agricultural system based on the Volga and Don Rivers, hydropower stations, irrigation ditches, canals, locks, and forest protection belts. Within a few years, thirty stations were built including the Khramskaia, Ozernaia, Farkhadskaia on the Syr-Darya River, and Shcherbakskaia, plus several smaller ones in the Ural, Caucasus, and Central Asian regions. By 1947 DneproGES had been rebuilt and upgraded. But this was merely the appetizer. The crowning event in the all-out Soviet commitment to these projects was the Party's promulgation of what came to be known as the "Stalinist plan for the transformation of nature."

In October 1948, in the Great Hall of the Soviets in Moscow the Communist Party faithful gathered to pass this plan to transform nature to its full proletarian potential unanimously and with great fanfare. Within days, a schematic map of the European USSR appeared at kiosks throughout the city. According to the fantastic map, all major rivers had been dammed; vast irriga-tion systems spread into fertile but arid land of the southern steppe; hydroelectric power stations were distributed liberally; huge reservoirs backed up behind them; canals and locks guaranteed ease of inland water transport; scores of "forest belts," dozens of kilometers long and several hundred meters wide, ringed land that had been plagued by constant dry winds, but now would be fertile and lush. Nature, like the Russian peasantry, the bourgeoisie, and society itself, would succumb to the Party's will and the planner's magic wand. Not thwarted by small landholders (the state owned all property), planners could grab a map and draw with impunity. Once tamed by the scientist under the watchful eye of Stalin and the party elite, nature would operate according to plan, provide water and electricity to the masses, produce agricultural goods in clock-like fashion, and facilitate their rational distribution by barge and river.

Scientists and engineers contributed to the promulgation of the Stalinist Plan for the Transformation of Nature, having studied natural resources of water, timber, fish, and minerals under Soviet power for two decades. Building on the work of government and Academy of Sciences commissions with pre-revolutionary roots, they did more than catalogue the extensive resources. Many of them welcomed a government which proclaimed itself to be scientific in essence, and promised to lavish their research enterprise with funding they had not known in the Tsarist era. Many of them had never lived under any non-Soviet regime and were entirely Soviet-educated, and virtually none had traveled abroad. They could now embark on the dream of scientific management of natural resources, ensuring their rational and efficient development and use for generations to come.

Using a series of locks, canals, irrigation systems, and hundred-mile long belts of trees, planted by Soviet workers to protect newly reclaimed farmland from drought and wind, engineers intended to use their expertise to ensure that the USSR never again fell prey to the forces of nature. A series of government resolutions in 1950 called for the construction of the massive Kuibyshevskaia, Stalingradskaia, and Kakhovskaia hydropower stations, with hundreds of miles of canals (the Main Turkmen, the South Ukraine, the North Crimean, and several smaller ones associated with them), and hundreds of thousands of miles of irrigation channels. Construction began in 1951, and was intended to be completed by 1957.[71] For the engineer Aleksandr Vinter, who had served well his two masters, electrification and Soviet power, the hero projects of late Stalinism reflected "a scale and tempo of economic and cultural construction hitherto unseen in human history" which had turned the USSR from abackwards, poverty-stricken nation into a "an unbeatable industrial socialist power."[72]

Having freed the country of the dangers of internal enemies lurking among engineers and scientists in the 1930s, and having beaten the hated fascists during the Second World War, only one obstacle stood between Soviet power and the achievement of socialism: the capitalist West, in particular the United States. So postwar reconstruction involved every effort to create economic might at all ends of the empire, far from potential invaders. Under Khrushchev, the development of Siberia commenced; under Brezhnev it picked up steam. The engineers in Gidroproekt, Ukrgidroproekt and other organizations who studied river flow, turbidity, seasonality, speed, and temperature, and the planners who had only dreamed of the world's largest hydropower stations to develop Siberian resources, now saw their dreams realized in the assembly of huge construction trusts of fifty, sixty, even one hundred thousand men who fulfilled plans to turn nature into a machine. In the 1970s and early 1980s alone they poured 27 million cubic metres of concrete, moved 164 million cubic metres of earth, and installed more than 463,000 tons of metal works in water works. Huge economic complexes formed as a result: Irkutsk-Cheremkhovskii, Bratsk-Ust-Ilimski, Siansk and others with hundreds of thousands of workers and their families and tens of thousands of km of high tension power lines.[73] The gigantic hydroelectric power stations of the Soviet period resulted in tremendous human dislocation and submerging of tens of thousands of square miles of towns, homes, cemeteries, farmlands, and forests. But this was apparently a small cost to pay for the achievement of socialism.

As the centerpiece of the Stalinist plan for the transformation of nature, the "Big Volga" project saw thirteen major hydropower stations on that river completed which produced approximately 60 billion kWh annually. For each of these major stations, a dam roughly five kilometers was constructed across the river, including 1,000 meters or so of reinforced concrete, 70 meters high with spillways and locks,[74] using serially-produced Electrosila turbogenerators, often at 105,000 kW, and fitted with three-phase 400,000 to 500,000 volt transmissions lines, to send up to 60 per cent of electrical energy hundreds of kilometers to the Moscow region grid. Indeed, Ministry of Power and Electrification personnel siting in Moscow directed electrical power from their control rooms. There was great human cost, from the workers who died during construction who included tens of thousands of prisoners, to the persons who had lived for generations near the Volga and were forcibly moved into unfamiliar, poorly-constructed prefab homes, and saw their churches and town centers submerged. For example, the Rybinskoe reservoir, once the largest artificial lake in the world, is 4,550 square kilometres, and covered 663 inhabited areas including six cities. In all, Soviet dams flooded 2,600 villages, 165 cities, almost 30,000 square miles (the area of Maryland, Delaware, Massachusetts, and New Jersey combined), including nearly 12,000 agricultural and 12,000 square miles of forest land.[75] Soviet communism tolerated no dissent from nature or squeamish engineers.

Seeing the forest for the socialist trees

Another correctable misfortune associated with the Volga basin concerned rainfall and its impact on agriculture. In the upper Volga, annual average rainfall is 300–400mm, but in the lower Volga it is half that much at 200–300mm. Hot dry winds that raise air temperature to 95° or 100° and push the humidity as low as 15 per cent plague the entire lower Volga basin. The most audacious aspect of the Stalinist "Big Volga" plan for the transformation of nature involved the planting of forest belts and grasslands, the creation of a network of ponds and lakes, and the construction of irrigation channels. After all, Lenin himself had referred to the glories of irrigation, albeit once, in his collected works.[76]

Forests covered 1.2 billion hectares or more than half the USSR. Perhaps 30 percent of the forest resources in the world were in the Soviet Union, primarily in Russia, in the Urals, Siberia, and to the northeast of Leningrad and Arkhangel regions. Wood, pulp, and cellulose from trees was used in over 20,000 different products: building materials, in the shipping, transport, and fishing industries, of course in books, furniture, sports, and so on. Beginning in the 1920s scientists set out to map, age, type, and categorize every forest and each tree for rational use. Forestry specialists studied typology, cartography, selection, and hybridization, phytopathology, chemistry (with applications for paper and cellulose production, pollution control in the wood products industry,

144

and fertilizers, herbicides, and pesticides for reforesting after clear-cutting), phytopathology, and the special field of forest drainage management science (draining swamps and bogs to prepare them for harvest since roughly 15 percent of Soviet forests are swampy land). Scientifically determined norms of selection, thinning, and clear-cutting stood at the center of Soviet forestry to improve the quality of the tree and speed up growth and harvest.

Russian forestry was an undermechanized and underpowered cottage industry through the early 1930s. It was slow to recover from revolution. In 1920 it operated at one-tenth of its 1913 level. But through the organization of logging trusts that filled the forest with men and noise such as Severoles, Zapadoles, and Verhnevolgoles, production increased ninefold by 1927. In 1928 the Supreme Economic Council decreed thinner boards to lessen waste and overcome a growing shortage in keeping with Russia's economic recovery. The first Five-Year Plan saw great increases in production; lumber and wood products were needed for the burgeoning industrialization effort in building materials, wagon and farm machinery, aircraft, light and food industry (packing and shipping crates, etc.). Fifty-two new mills were opened. But most of the new equipment was imported. An exception was the RLB-75 log frame. Many of the mills had to take on machine building them-selves because heavy industry received the lion's share of Stalin's investment, so that on the eve of the Second World War there was little standardized machinery. Experiments with Taylorism in the 1930s had little impact on production.

Few mills had more than one saw, and fewer still were powered by internal combustion engines or electricity until the late 1930s. Economic planners and political commissars, if not scientists, rejected efforts to conserve forest through scientific management as "bourgeois," for such conservation practices would slow the construction of the major technological artifacts of early Soviet power: the famous Dneprostroi, Belomor, and Magnitogorsk. Wood was needed for paper for propaganda, dormitories and barracks for workers, and for factories. Yet the planners did not come close to meeting targets of first Five-Year Plan through 1932 because of poor mechanization There were only 1,000 tractors working in forests, they broke down frequently, and horses were unavailable to replace them, for peasants had slaughtered one-half of Soviet livestock rather than collectivize it. Investment flowed into iron, steel, and concrete, but not into the forest where the Soviets needed to build ice and dirt roads, railroads, and trunk lines.[77]

The Second World War set back the development of forestry again. As for the Soviet people, so for the trees, the war was a great loss, with one-quarter of the forest industry lost to occupation or destruction, so that in 1945 production had recovered to only 40 percent of the 1940 level. Only by 1949 had the logging industry recovered, but it suffered from a labor shortage; twenty million Soviet citizens perished during the war. Throughout the 1950s and 1960s it embarked on extensive, albeit wasteful growth, largely through the taming of

Siberian forests. The war had exposed the USSR's strategic weaknesses. The political leaders determined to develop ore, fossil fuels, water, fish, and forest resources east of a natural geographic barrier, the Ural mountains. Further, European forests had been decimated by clear-cutting practices and extensive soil erosion. At first under the auspices of the Committee for the Study of the Productive Forces, largely a cartographic and surveying organization, Siberian development proceeded under the direction of the Siberian division of the Academy of Sciences, and its scientists and institutes in Akademgorodok, Krasnoiarsk, Vladivostok, Irkutsk, and in Buriatia and the Komi regions. The Main Administration for the Manufacture of Woodworking Equipment (Giprodrev) and the Research Institute of Woodworking Machine Building (NIIDrevmash) were to provide the modern, indigenous equipment needed for the effort.[78] But recovery from the Russian Revolution and from the Second World War did not mean modern efficient technology, but rather increasing expanses of forest under assault by armies of men equipped with axes and chain saws.

For all the fascination with big technology, the Soviet forestry and paper industries labored under the constraints of a low level of mechanization, both in terms of numbers of machines and their horsepower. This, and the USSR's continental climate, may be the most important factors in limiting the extent of deforestation. But the engineers had great dreams for mechanizing forestry. The goal was easiest where agricultural machines might be slightly modified, especially when the question was working the soil, for example cultivators, planters, and stubbing machines based on tractors. It was more difficult when questions of harvesting and removing cut wood were the issue. But machines such as skidders provided more challenges, and these were produced only in small numbers until after the Second World War; the general level of mechanization of forest work remained very low through the 1950s.[79] If modern technology was intended to help him be more circumspect, the psychology of the lumberjack remained the same: he wasted one-third of what he cut, in part with encouragement of the Ministry of the Forest Industry (Minlesprom) which ordered him to takes what was closest and cheapest to cut. It compensated through plans to harvest generally one-third more than needed.[80]

As for other regions of natural resource management, so in forestry the industrial paradigm was an central aspect of Soviet forestry, nowhere more clearly than in the Stalinist plan to transform nature. Stalinist climate change would increase the harvest in the steppe and forest-steppe regions. Planners had their eyes on 14 million hectares (35 million acres) of the Sarpinskaia and Norgaiskaia Steppe, land with rich soil but low rainfall in the lower Volga and northern Caspian regions, as well as other land in Ukraine, Kazakhstan, the North Caucasus, and some land west of the Ural mountains. (The Norgaiskaia steppe, at 6.5 million hectares, had more land than the Netherlands and Belgium together.) On basis of extensive scientific research on forests, Soviet agricultural and forestry ministries embarked on an effort to plant twelve million hectares of forest defense belts. The advantages were as follows: forests help manage water flow into rivers, leaves retain moisture on the ground, in spring and summer they release moisture into the atmosphere, they slow erosion, and they increase the quantity of ground water, retaining the water that "uselessly flows into the Caspian sea." The forest belts would moderate the temperature, humidity, and frequency of winds that hurt agriculture. The Volga, Dnepr, Amu-Darya and Don rivers would be harnessed for irrigation to served primarily wheat grains, and to a lesser degree vegetable and fruit farming and animal husbandry.[81] The Kuibyshev and Stalingrad GES provided the electricity to pump the water.

By 1965 eight forest belts had been built on shores of Volga, Ural, Don, northern Donets and other rivers, each from 170 to 1100 km in length, with a total length of 5320 km stretching over 118,000 hectares. Each belt consisted of two to six strips of 30 to 100 meters diameter, standing at a distance from one another of 200–300 meters. Planners acknowledged that irrigation and forest belts would affect climate for a significant part of the country. This did not trouble them; they had studied local impacts, and were comfortable making substantial changes in plant cover, soil composition and moisture level through melioration, irrigation, and planting. They also recognized that "Only under Soviet power under the leadership of the great Stalin" was the implementation of a great plan to transform nature possible. One popularizer of Stalinist gigantomania noted "Here there will be gardens, vineyards, melonfields, and plantations of technical culture."[82]

After Stalin, nature transformation and scientific management fell prey to the difficulties encountered in managing huge forests that extended far beyond the reach of modern technology, accessible only to workers on foot or in open trucks, armed with rudimentary tools – axes, low-power chain saws, pries, and awls – whose every step was overwhelmed by rapid changes in the weather that should have been anticipated. They cut far more than needed, for they could count on losing much of it on the forest floor, in rivers during the spring float, or at factories which threw away one-third of the tree during processing. The float produced over 100 million cubic meter/year on more than 700 rivers and lakes. But the industry seemed unable to manufacture safe and efficient machinery. From Arkhangel to the Volga, they lost huge quantities of timber to miscalculation about the onset of ice in rivers that could be removed only in the spring by cranes and tractors, and much of it still sank to the bottom, or was lost in log jams. Many logs spilled over onto flood plains, and disappeared into the mud. The raft crews handled the task poorly, for the rafts themselves were unstable. So much wood was lost that foremen wrote that it "sank" when it had not. To overcome anticipated problems in many regions (for example, Kalinin province) they cut five to ten times more lumber than indicated by plan, even when the district attorney threatened to prosecute.[83]

How could a country with one-third of the world's forests fail to meet targets? Paradoxically, the destruction of forest and stream could have been much more extensive. In all areas of industrialized nature, workers failed to meet scientifically determined yields and targets. One reason is that in the forest, on the farm, and at sea the worker remained underpaid, mistreated, and unmotivated to improve his unhappy lot. He realized that in comfy offices in Moscow sat cartographers, compilers, and codifiers who had no clue what life in a dump truck, on a tractor, or on a boat was like. The lumberjacks would tell the Moscow bureaucrat that their slovenly performance was linked not only to the low level of mechanization, but to their miserable conditions: the dorms in which they lived were spartan and filthy, with broken windows and no shades. The workers dropped their clothes on the floor at the end of the day, drank vodka, and fell asleep exhausted. Mice and cockroaches loved these new homes, especially because the clothes were rarely washed. Of course, there were no laundry facilities. Dining halls were breeding grounds for intestinal disorders, if the workers could stomach the long lines that stretched far from the door into the mud. In the 1920s and 1930s the turnover of personnel ranged from 40 to 70 percent annually. The lumberjack who stayed in one place for three years was the exception. This led to miserable conditions at camp sites, for no one cared, and no one was responsible for living and working conditions. Not surprisingly, given these conditions, turnover increased in the 1960s and to the end of the Soviet empire.

The men spent far too much time on bad roads that were icy and unsafe in winter, muddy, mosquito infested, and impassible in the spring. They did not have bulldozers and graders to fix roads. The trains broke down. In Kerki in the Komi Republic, workers of the Ukhtales enterprises harvested 5,240 million cubic meters of wood products during the tenth five year plan. They claim the achievement is the result of mechanization and socialist competitions. But they built less than two thousand 150 square foot apartments for them and spent little else on *sotskultbyt* (the Soviet term for apartments, stores, schools and entertainment and cultural infrastructure). Housing lagged far behind targets for all regions of the country. Because of low wages and hard work, it was nearly impossible to retain workers. Thousands of seasonal workers hardly made a difference for the forestry enterprises.[84]

Each leader of the USSR stamped the country's economic development program with his own preferences, seeking to distinguish himself from his predecessor and establish the legitimacy of his rule before the people of the nation and the world. From Lenin and electrification to Stalin, Magnitogorsk, and the "Big Volga," and from Khrushchev's hydropower and agricultural programs to Brezhnev's many officially proclaimed "projects of the century," the Soviet people learned by direct participation in nature transformation projects that there would no obstacles to the effort to transform nature from a capricious enemy into a socialist machine that operated according to plan.

In many ways, there was, however, little to distinguish features of hydropower under Stalin and Khrushchev or forestry under Brezhnev and Khrushchev. As the Soviet system evolved into a centrally managed economy beholden to planners' preferences and to one-party rule, large-scale projects, which were seen as best able to organize workers and ensure access to machinery, equipment, materials, and natural resources, became the standard. In spite of significant technological improvements and vastly increased scale in the projects over time, more crucial were the similarities: the utopian, Leninist belief in the efficacy of scientific socialism to change nature for state purposes, improving the lot of the worker along the way; the search for economies of scale; the pursuit of and standardized approaches; and a legacy of an altered landscape of rivers and forests forced to succumb to Soviet political will and ideological precepts.

Authoritarian regimes and nature transformation reconsidered

A comparison of Nazi Germany's and Soviet Russia's record on land transformation reveals surprising similarities. Both dictatorships harbored a specific and nationally charged unifying ideology, the inner contradictions of which at times seemed to outweigh the similarities. But the existence of one single ideology with identifiable core elements – proletarian rule in the Soviet case, racial goals in the Nazi case – made it both possible and increasingly likely for scientists and engineers to connect their work to the prevalent beliefs. Our case studies indicate different scales of behavior, ranging from grudging concessions to the new regime to willful self-mobilization and jumping at the possibilities to better one's social, ideological, or political status within the given society. But Nazi and Soviet engineers and scientists realized that economic and political considerations might overrule their concerns about designs or environmental degradation. They found it simpler to comply with state programs rather than risk an interruption in their career paths. And perhaps more to the point, the state provided them with research funding.

Both regimes also realized the potential of symbolical gain that large-scale technologies offered to them. In both countries they likened dam building and road building to the pyramids of ancient Egypt, thus invoking the stature of world empires at the height of their power. The awe, sublimity, and resourcefulness of the technologies were powerful resources to exploit. The dictators themselves utilized them consciously, for example at commemorative ceremonies as political reassertions in a technologically charged arena.

Central planning attained new roles during the Soviet and Nazi efforts to achieve economic autarky. While Soviet authorities could continue a tradition of strong rule from the center with complying bureaucracies in the peripheries, Germany's young and federal nationhood had left a legacy of diversity on many levels, including scientific funding. The Nazi efforts at coordination on the Reich level included the spheres of science and technology with a vengeance. Both in the case of road building and in the case of forestry, new nationwide agencies replaced older systems of regional control. For the scientists and engineers associated with these fields of research and practice, this development opened up a chance to reach a wider audience, receive increased funding, and to bolster egos. In the Soviet case, however, the massive transformations set into motion by the acronym-laden new agencies amassed power from the center and were accountable strictly to the central powerholders. It is almost a tautology to claim that the scientific expertise that might call for a rethinking of some of the grandiose plans for building dams and highways or diverting rivers would be silenced by the ideology of steadily increasing production, of big and bigger numbers, and of state power that prevails more often than not.

The role of the state diverged on a number of counts. While the absence of private ownership of land and of production facilities made the Soviet Union an authoritarian planner's dream land, Nazi Germany, although inconsistent in its course, left private ownership of production usually in place. In the case of forestry, the ownership structures remained more or less the same, whereas in the transportation sector an impatient Hitler gave the jump start to a state-run company that would make him his country's own Henry Ford. More typically, however, the mixed system of profit-oriented privately owned companies and the increasing efforts at controlling the economy by central state agencies was characterized by inefficient altercations, double work and petty conflicts.

Nature itself remained an adversarial object in both countries. While some of the Germans propelled into power during the first years of Nazi rule expressed proto-ecological beliefs, the tendency to appropriate nature economically was stronger in many respects. In Nazi Germany, this latter development was inextricably bound up in the preparation for the Second World War. Ultimately, science, ideology and nature itself were merely parts of general mobilizations, in spite of any claim of their organic relationship with the proletariat or Volk.

Notes

1 Department of History, Colby College.
2 Department of History, University of Maryland, College Park.
3 Carl Alwin Schenck, *Forestry in Germany: Present and Prospective* (New York: Newsprint Service Bureau, 1948), 1. Between 1895 and 1909, Schenck (1868–1955) had been manager of the Biltmore estate forests in Tennessee and founded their forestry school in 1898. During the winter of 1945/6, the American military government hired Schenck as administrative head of the forest service for Greater Hesse. Carl Alwin Schenck, *Cradle of Forestry in America: The Biltmore Forest School, 1898–1913*, ed. O. Butler (Durham, NC: Forest History Society, 1998); Edward Stuart Jr., "German Forestry During the American Occupation: Dr. Schenck's Pivotal Role," *Journal of Forest History*, 29 (1985): 169–74.
4 For a critical assessment of the allegedly essential German romantic love for the woods, see Joachim Radkau, "Wood and Forestry in German History: In Quest of an Environmental Approach," *Environment and History*, 2 (1996): 63–76, esp. 72–4. On the other hand, Simon Schama, in his *Landscape and Memory* (New York: Alfred A. Knopf, 1995), 118–19, is not always free from continuing this long-lasting and all-too-rarely empirically challenged trope. His assertion that "arguably, no German government had ever taken the protection of German forests more seriously than the Third Reich and its *Reichsforstmeister* Göring" (119) is questionable and needs to be balanced with a look at actual Nazi forestry politics.

5 Henry E. Lowood, "The Calculating Forester: Quantification, Cameral Science, and the Emergence of Scientific Forestry Management in Germany," in Tore Frängsmyr, John L. Heilbron, and Robin E. Rider (eds), *The Quantifying Spirit in the 18th Century* (Berkeley, CA: University of California Press, 1990), 315–42; Heinrich Rubner, *Forstgeschichte im Zeitalter der industriellen Revolution* (Berlin: Duncker und Humblot, 1967); Radkau, "Wood and Forestry".

6 Michael Burleigh and Wolfgang Wippermann, *The Racial State: Germany 1933–1945* (Cambridge: Cambridge University Press, 1991).

7 Franz Heske, *German Forestry* (New Haven, CT: Yale University Press, 1938), 113–14. Publication of this volume was supported by the Oberlaender Trust and the Carl Schurz Memorial Foundation, two major organizations of German-Americans. Heske's chair was located at the Forstliche Hochschule Tharandt, close to Dresden. W. Keuffel and M. Krott, "Forestry Tradition in the Long Term. A Review of Franz Heske (1927/28): 'A General Review of Post-War Forestry in Central Europe,'" *Forestry*, 70 (1997): 351–8.

8 Heske, 114.

9 Heinrich Rubner, *Deutsche Forstgeschichte 1933–1945. Forstwirtschaft, Jagd und Umwelt im NS-Staat* 2nd edn (St. Katharinen: Scripta Mercaturae, 1997), 95–109.

10 Although private forests accounted for 46 percent of all woods in 1934, it seems reasonable to accept Rubner's argument that noble estate owners followed official lines more closely out of self-interest. Rubner, *Deutsche Forstgeschichte*, 102.

11 Rubner, *Deutsche Forstgeschichte*, 103. The author of this richly detailed study states that the funds available for educating these farmers to avoid exploitation were not sufficient.

12 Heinrich Rubner, "Sustained-Yield Forestry and its Crisis during the Era of Nazi Dictatorship," in Harold K. Steen (ed.), *History of Sustained-Yield Forestry: A Symposium* (Santa Cruz, CA: Forest History Society, 1983), 170–75.

13 Walter von Keudell, a conservative politician who allied himself with Göring in 1932, had practiced a more ecologically oriented approach in his own estates since 1903. Rubner, *Deutsche Forstgeschichte*, 66–8.

14 Rubner, *Deutsche Forstgeschichte*, 136–7, 139–140. There is also evidence of resistance on the side of foresters who opposed the "confusion" in the forests.

15 Rubner, *Deutsche Forstgeschichte*, 120, 137.

16 Rubner, *Deutsche Forstgeschichte*, 115. The figures are for million *Festmeter* of untreated lumber. 1927: 42; 1933: 35; 1934: 44; 1938: 60.

17 Rubner, *Deutsche Forstgeschichte*, 121.

18 Rubner, *Deutsche Forstgeschichte*, 124.

19 Rubner, *Deutsche Forstgeschichte*, 125.

20 Peter Hayes, *Industry and Ideology. IG Farben in the Nazi Era* (Cambridge: Cambridge University Press, 1987), 36–7, 145–7; Harald Beck, "Friedrich Bergius, ein Erfinderschicksal," *Abhandlungen und Berichte des Deutschen Museums*, 50, 1 (1982): 1–37; Thomas P. Hughes, "Technological Momentum in History: Hydrogenation in Germany 1898–1933," *Past and Present*, 44 (1969): 106–132; Anthony N. Stranges, "Friedrich Bergius and the Rise of the German Synthetic Fuel Industry," *Isis*, 75 (1984): 643–67.

21 Egon Glesinger, *Nazis in the Woodpile: Hitler's Plot for Essential Raw Material* (Indianapolis: Bobbs-Merrill Company, 1942), 23, mentions seeing such a blueprint upon visiting Berlin after 1935. Rubner, based on archival studies, makes no mention of such a plan, but points to the many competing institutions researching and producing synthetic applications. On the other hand, Glesinger's position as secretary of the clearing-house "Comité International du Bois" enabled him to gather insights only a few outsiders could gain. Yet, his hyperbolic tone, "Wood has become the backbone of the German war effort," (25) points to the main goal of the

book, which was to raise awareness among the U.S. public for the "story of Nazi wood-scheming" (10) by way of a conspiratorial tale and to incorporate lumber topics into post-war Allied planning. As of today, a systematic, comparative analysis of the relationships between science, industry, and the state in the field of lumber in Nazi Germany is still lacking.

22 Rubner, *Deutsche Forstgeschichte*, 129–31, Glesinger, 24–39 (for the suit anecdote 32).

23 Hayes, *Industry and Ideology*, 146ff.

24 Rubner, *Deutsche Forstgeschichte*, 151. Rubner surmises that the Nazis' millenial ideology was more attractive to foresters who were professionally conditioned to think in long cycles. Also see Joseph C. Kirchner, "The Forests of the U.S. Zone of Germany," *Journal of Forestry*, 45 (1947): 249–52.

25 Rubner, *Deutsche Forstgeschichte*, 118.

26 Karl Ditt, "Nature Conservation in England and Germany 1900–70: Forerunner of Environmental Protection?" *Contemporary European History*, 5 (1996): 1–28.

27 Rubner, *Deutsche Forstgeschichte*, 159–160, on occupied Poland, 196–204; Glesinger, *Nazis*; Schama, *Landscape*.

28 Kirchner, "Forests of the U.S. Zone," 249.

29 Ulrich Herbert, *Hitler's Foreign Workers: Enforced Foreign Labor in Germany under the Third Reich* (Cambridge: Cambridge University Press, 1997).

30 Keuffel and Krott, "Forestry Tradition," 353.

31 Adalbert Ebner, *German Forests: Treasures of a Nation* (New York: German Library of Information, 1940).

32 Karl Lärmer, *Autobahnbau in Deutschland 1933 bis 1945. Zu den Hintergründen* (Berlin: Akademie Verlag, 1975); Karl-Heinz Ludwig, *Technik und Ingenieure im Dritten Reich* (Königstein: Athenäum, 1979); Erhard Schütz and Eckhard Gruber, *Mythos Reichsautobahn. Bau und Inszenierung der "Straßen des Führers" 1933–1941* (Berlin: Links, 1996); Rainer Stommer (ed.), *Reichsautobahn. Pyramiden des Dritten Reiches. Analysen zur Ästhetik eines unbewältigten Mythos* (Marburg: Jonas, 1982).

33 Hansjoachim Henning, "Kraftfahrzeugindustrie und Autobahnbau in der Wirtschaftspolitik des Nationalsozialismus 1933–1936," *Vierteljahresschrift für Sozial- und Wirtschaftsgeschichte*, 65 (1978): 217–42; Richard J. Overy, "Cars, Roads, and Economic Recovery in Germany, 1932–1938," in Richard J. Overy, *War and Economy in the Third Reich* (Oxford: Oxford University Press, 1994), 68–89; Richard J. Overy, *The Nazi Economic Recovery 1932–1938* 2nd edn (Oxford, Oxford University Press, 1996).

34 Dan P. Silverman, *Hitler's Economy: Nazi Work Creation Programs, 1933–1936* (Cambridge, MA: Harvard University Press, 1998), 147–74.

35 Neil Gregor, *Daimler-Benz in the Third Reich* (New Haven, CT: Yale University Press, 1998); Mark Spoerer, "Die Automobilindustrie im Dritten Reich: Wachstum um jeden Preis?" in Lothar Gall and Manfred Pohl (eds), *Unternehmen im Nationalsozialismus* (Munich: Beck, 1998), 61–8.

36 Hans Mommsen and Manfred Grieger, *Das Volkswagenwerk und seine Arbeiter im Dritten Reich* (Düsseldorf: Econ, 1996).

37 Mommsen and Grieger, *Volkswagenwerk*, 162.

38 Heidrun Edelmann, *Vom Luxusgut zum Gebrauchsgegenstand. Die Geschichte der Verbreitung von Personenkraftwagen in Deutschland* (Frankfurt am Main: Henrich, 1989).

39 Christopher Kopper, "Modernität oder Scheinmodernität nationalsozialistischer Herrschaft. Das Beispiel der Verkehrspolitik," in Christian Jansen, Lutz Niethammer, and Bernd Weisbrod (eds), *Von der Aufgabe der Freiheit. Politische Verantwortung und bürgerliche Gesellschaft im 19. und 20. Jahrhundert. Festschrift für Hans Mommsen* (Berlin: Akademie, 1995), 399–411.

40 Wolf Gruner, "Juden bauen die 'Straßen des Führers.' Zwangsarbeit und Zwangs-arbeiterlager für nichtdeutsche Juden im Altreich 1940 bis 1943/44," *Zeitschrift für Geschichtswissenschaft*, 44 (1996): 789–808.

41 Schütz and Gruber, *Mythos Reichsautobahn*.

42 Statement of Todt (January 18th, 1934, R 43II/503, Federal Archives Koblenz (hereafter: FAK).

43 Gerhard Hard, *Die "Landschaft" der Sprache und die "Landschaft" der Geographen. Semantische und forschungslogische Studien zu einigen zentralen Denkfiguren in der deutschen geographischen Literatur*, (Bonn: Dümmler, 1970).

44 For a sometimes overstated revision of the older historiography of these groups, see William Rollins, *A Greener Vision of Home: Cultural Politics and Environmental Reform in the German Heimatschutz Movement, 1904–1918* (Ann Arbor, MI: University of Michigan Press, 1997).

45 Paul Schultze-Naumburg, *Kulturarbeiten*, Parts I–IX (Munich: Callwey, 1902–1917). Vol. VII to IX deal with "Die Gestaltung der Landschaft durch den Menschen," the designing of landscapes by man.

46 Helmut Maier, "Nationalsozialistische Technikideologie und die Politisierung des 'Technikerstandes': Fritz Todt und die Zeitschrift 'Deutsche Technik,'" in Burkhard Dietz, Michael Fessner and Helmut Maier (eds), *Technische Intelligenz und "Kulturfaktor Technik". Kulturvorstellungen von Technikern und Ingenieuren zwischen Kaiserreich und früher Bundesrepublik Deutschland* (Münster: Waxmann, 1996), 253–68.

47 Alf Gießler, "Natur und Technik," *Deutsche Technik*, 4 (1936): 7–11.

48 Todt to Reichsbund Volkstum und Heimat, Reichsführung, 12–11–33, 46.01/1187, Federal Archives Potsdam (hereafter FAP).

49 Alwin Seifert, *Ein Leben für die Landschaft* 2nd edn (Düsseldorf: Diederichs, 1962).

50 Thomas Zeller, "Landschaften des Verkehrs. Autobahnen im Nationalsozialismus und Hochgeschwindigkeitsstrecken für die Bahn in der Bundesrepublik," *Technikgeschichte*, 64 (1997): 323–40; Thomas Zeller, *Straße, Bahn, Panorama. Verkehrswege und Landschaftsveränderung in Deutschland von 1930 bis 1990* (Frankfurt am Main: Campus, 2002).

51 Alwin Seifert, "Schlängelung?" in Alwin Seifert, *Im Zeitalter des Lebendigen. Natur – Heimat – Technik* (Planegg: Langen Müller, 1942), 114–17, quote from 114.

52 Todt to Seifert, 6–26–1935, NS 26/1188, FAK.

53 James C. Scott, *Seeing like a State: How Certain Schemes to Improve the Human Condition Have Failed* (New Haven, CT: Yale University Press, 1998).

54 "Highway to Siberia," *Newsweek* (European edition), 83, 11 (March 15th, 1999), 62.

55 N M. Knipovich, *Ekspeditsiia nauchnopromyslovykh issledovanii u beregov murmana*, 3 vols (St. Petersburg: Khudozhestvennaia pechat', 1902).

56 Johnathan Coopersmith, *The Electrification of Russia, 1880–1926* (Ithaca, NY: Cornell University Press, 1992). See also P.S. Neporozhnyi (ed.), *40 let plana GOELRO. Sbornik materialov* (Moscow-Leningrad: Gosenergeticheskoe izdatel'stvo, 1960).

57 Anne D. Rassweiler, *The Generation of Power* (New York: Oxford University Press, 1988).

58 Loren Graham, *The Ghost of the Executed Engineer* (Cambridge, MA: Harvard University Press, 1996).

59 A.V. Vinter and A.B. Markin, *Elektrifikatsiia nashie strany* (Moscow-Leningrad: Gosenergoizdat, 1956), 79.

60 On the history of the Metro, see Lazar Kaganovich, *Za sotsialisticheskuiu rekonstrukt-siiu Moskvy i gorodov SSSR* (Moscow-Leningrad: OGIZ 'Moskovskii rabochii,' 1931), 6–20 and *Pobeda metropolitena – pobeda sotsializma* (Moscow: Transzheldorizdat, 1935); *Rasskazy stroitelei metro* (Moscow: Izdatel'stvo 'istoriia fabrik i zavodov', 1935);

I. Kattsen, *Metro Moskvy* (Moscow: Moskovskii rabochii, 1947) and N. Bliskovskaia, editor, *Dni i gody metrostroia* (Moscow: Moskovskii rabochii, 1981).

61 For an eyewitness account of the construction of Magnitogorsk, see John Scott, *Behind the Urals: An American Worker in Russia's City of Steel* (Bloomington, IN: Indiana University Press, 1989). For a discussion of the fate of Magnitogorsk in the Gorbachev years, see Stephen Kotkin, *Steeltown, USSR* (Berkeley, CA: University of California Press, 1991).

62 M.M. Davydov, M.Z. Tsunts, *Ot volkhova do amura* (Moscow: Sovetskaia rossiia, 1958), 61.

63 For example, on the impact of the enlightenment tradition on the U.S. Army Corps of Engineers, see Todd Shallat, *Structures in the Stream: Water, Science, and the Rise of the U.S. Army Corps of Engineers* (Austin, TX: University of Texas Press, 1994).

64 N.P. Bakhtin, *Reka Enisei* (Leningrad: Gidrometeorologicheskoe izdatelstvo, 1961), 105.

65 Vinter and Markin, 80–1.

66 G.V. Voropaev, D.Ia. Ratkovich, *Problema territorial'nogo pereraspredeleniia vodnykh resursov* (Moscow: IVP AN SSSR, 1985), 252–3.

67 Vinter and Markin, 81–9.

68 See *Trudy noiabr'skoi sessii AN SSSR 1933 goda: problemy volgo-kaspiia* (Leningrad: Izdatel'stvo Akademii nauk SSSR, 1934).

69 E. Riabchikov, *Volga-don* (Moscow: Geografgiz, 1954), 44–50, and V. Kudriavchikov (ed.), *Trudovye podvigi stroitelei volgo-dona* (Moscow: Gosizdatpolit, 1954), 8–9.

70 "American Methods Win Fight to Control Russian River," *Engineering News-Record* (June 23, 1932), 877–83.

71 F.M. Loginov, "Velichaishee sooruzhenie stalinskoi epokhi," *Pravda* (September 5, 1950), 2, and I. Grishin, "Na Stalingradskoi zemle," *Literaturnaia gazeta* (September 19, 1950), 2. "Glavnyi turkmenskii kanal," *Pravda* (September 12, 1950), 1; S. Babaev, "Velichaishee sooruzhenie stalinskoi epokhi," *Izvestiia* (September 15, 1950), 2; G. Mukhtarov, "Kanal schast'ia," *Sotsialisticheskoe zemledelie* (September 17, 1950), 2; and I. Basalaev, "Pustynia budet sluzhit' nam," *Zvezda*, No. 1, (1951), 141–5. L. Mel'nikov, "Po stalinskomu planu," *Pravda* (September 22, 1950), 2; A. Palladin, "Rodnoe delo nashego naroda," *Pravda* (September 26, 1950), 2; and A. Kozlov, "Velichaishaia stroika sovremennosti," *Pravda* (September 23, 1950), 2.

72 A.V. Vinter, *Velikie stroiki kommunizma* (Moscow: Izdatel'stvo Akademii Nauk SSSR, 1951), 10.

73 V.V. Alekseev, *Gidrostantsii Sibiri – unikal'nye pamiatniki energeticheskoi revoliutsii XX veka* (Ekaterinburg: n. p., 1992).

74 On the development of Soviet practices in hydropower construction, see for example M.M. Zherbin (ed.), *Razvitie stroitel'noi nauki i tekhniki v Ukrainskoi SSR, II* (Kiev: Naukova dumka, 1990), 48–50.

75 Murray Morgan, *The Columbia, Powerhouse of the West* (Seattle, WA: Superior Publishing Co., 1949), 156–8, 162–3.

76 Lenin, *Sochineniia*, vol. 32 (Moscow: Gosizdatpolit, 1950), 295–7.

77 S. Gorelik, *Lesnaia promyshlennost' SSSR ko 2-oi piatiletke* (Moscow-Leningrad: Goslestekhizdat, 1932), 22–44, and S. Lobov, *O lesnoi promyshlennosti* (Moscow-Leningrad: Gosizdat, 1930), 16–18, 27–34.

78 V.A. Popov (ed.), *Lesnaia promyshlennost' SSSR, 1917–57, II* (Moscow: Goslesbumizdat, 1957).

79 After all, Lenin himself had referred to the the glories of irrigation, albeit once, in his collected works. On forests, forest belts, and the environmental movement in the postwar USSR, see Douglas Weiner, *A Little Corner of Freedom* (Berkeley, CA: University of California Press, 1999). On forestry and forestry science in the USSR,

see the journals *Lesnoe khoziaistvo* and *Lesnaia industriia*; A.I. Bovin (ed.), *Lesnoe khoziaistvo SSSR, 1917–1957* (Moscow-Leningrad: Goslesbumizdat, 1957); V.N. Sukachev (ed.), *Dostizheniia nauki v lesnom khoziaistve SSSR za 40 let* (Moscow-Leningrad: Goslesbumizdat, 1957); V.I. Rubtsov (ed.), *Lesnoe khoziaistvo SSSR za 50 let* (Moscow: Lesnaia promyshlennost', 1967); and I.S. Melekhov (ed.), *Lesokhoziaistvennaia I lesomeliorativnaia nauka v SSSR (1917–1967 gg.)* (Moscow: Lesnaia promyshlennost', 1967).

80 V. Gribanov, "Les rubit', no I shchepki berech'," *Izvestiia* (August 9, 1980), 2.

81 According to Douglas Weiner, the goal of increasing cereal crops in the postwar famine had not only practical but symbolic importance, for it signaled the attack on anarchic nature and its replacement by planned, socialist nature. If bourgeois specialists worried about natural disasters resulting from human interference and disasters, Soviet biologists would defend grain from predators with trees. See Weiner, *Freedom*.

82 Kasimovskii, *Velikie stroiki*, 50–58, and L. Ognev and P. Serebriannikov (eds), *Velikie sooruzheniia stalinskoi epokhi* (Moscow: Molodaia gvardia, 1951), 119–20, 206–7.

83 A. Semenov, "Bagor v kolese avtomatiki," *Sovetskaia rossiia* (March 25, 1972), 2; .*Lesnaia promyshlennost'* (July 13, 1961); *Sovetskaia rossiia* (June 27, 1960); "Lazeiki dlia beskhoziaistvennosti," *Ekonomicheskaia gazeta* (December 7, 1963); and *Sel'skaia zhizn'* (July 1, 1961).

84 A. Terent'ev, "Est rezervy na deliankakh," *Pravda* (January 23, 1981), 3; *Lesnaia promyshlennost'* (January 25, 1962, March 22, 1960, January 21, 1960, and August 14, 1962); *Izvestiia* (January 4, 1962); *Komsomolskaia Pravda* (May 13, 1960); *Trud* (July 14, 1961); and "Lespromkhoz-ne prokhodnoi dvor," *Sovetskaia rossiia* (March 29, 1969), 2.

7

LEGITIMATION THROUGH USE[1]

Rocket and aeronautic research in the Third Reich and the USA

Burghard Ciesla and Helmuth Trischler

Introduction

At the beginning of the 1990s, with grave concern, the Western press registered the emigration of Russian military researchers, as well as the dire future prospects for the collapsed R&D system of the dissolved Soviet Union. In Russia, depression was widespread among scientists and engineers. After the end of the Cold War and the internal decay of the "military-industrial-academic-complex," day-to-day research was determined by catastrophic work and living conditions. The money and material needed for the simplest maintenance and providing the necessities of life were not there. On top of this came galloping inflation, which, considering wages were often not paid and living conditions were poor, practically forced researchers to take second jobs. Many of the best minds of the younger generation of scientists moved over to the newly created professions in banking, trade, and finance. This internal brain drain was accompanied by a wave of emigration, particularly to the U.S.A., Israel, and Western Europe.[2]

However, the so-called "threshold countries" (on the threshold of nuclear weapons) also sought after the Russian scientific and technical elites. The trade in know-how from former Soviet high-tech armorers flourished. Western security services, military establishments, and politicians were especially concerned about the danger of emigration to Iraq, Iran, and Libya, as well as the sale of modern armaments to these countries. In January 1992 the *Washington Post* referred in this connection to the acute danger of nuclear weapons proliferation. At this time there was "a 10 percent chance that former Soviet military officials – or others in a position to secure access to nuclear weapons in the commonwealth – may become so demoralized by deteriorating conditions there that the sale of nuclear weapons to foreign bidders will emerge as a live danger. The purchaser might be a foreign state – Libya, Iraq, Iran – or it could even be a terrorist organization."[3]

The U.S.A., as victor in the Cold War, faced a fundamental question: how should it deal with the military-relevant technological innovations of their former Cold War opponent? One thing was generally clear to the State Department, Pentagon, and the defense industry: "We can get real advantages. And if we don't, someone else will." Support for the reorganization of Russia's worn-out R&D system appeared very worthwhile. An American-Russian cooperation "would help American industry, save taxpayer dollars, enhance national security and, at the same time, keep Russian scientists employed."[4]

From the perspective of the historian, this discussion is hardly new. The historical parallels to the beginning phase of the Cold War are clear. Less than a half century before, at the end of the Second World War, people had already made similar arguments in the United States. At that time it was a matter of utilizing scientists and engineers from defeated Germany and the resulting benefit for the American defense economy. It was in the national interest to deny the German "Specialists" to the other victorious powers. Over the course of the Cold War, the policy of denying relevant knowledge was concentrated more and more on China, the Soviet Union, and other east European countries.

However, in December of 1945 the American Secretary of Commerce Henry A. Wallace justified this political concept to President Truman by first of all emphasizing the domestic benefit:

> The transfer of outstanding German scientists to this country for the advancement of our science and industry seems wise and logical. It is well known that there are presently under U.S. control eminent scientists whose contributions, if added to our own, would advance the frontier of scientific knowledge for national benefit....A positive program along the lines described is essentially "intellectual reparations" and may well be the most practical and enduring national asset we can obtain from the prostrate German nation.[5]

This advantage was made even clearer a few months later by Colonel Donald L. Putt, who as chief of the Air Force Technical Center at Wright Field was responsible for the German aeronautic specialists working there. Before a hand-picked audience at the Country Club in Dayton, he bluntly described the operations of the Air Technical Intelligence at Wright Field, as well as the significance of the Germans and their experience for the United States:

> I think it will be clear...that the many tons of material, documents and the scientists which have been brought to Wright Field represent the end product or results of the expenditure of millions of dollars for research facilities, research work, and millions of man hours expended in the construction of these facilities and their operation. The material, documents and scientists were screened and selected from those fields in which the Germans were ahead of our own state of research and development. The developments in these fields are now of first order importance to us because they provide the simplest means of

obtaining previously unheard – of speeds in air transportation with the possibility of flight high in the stratosphere and some day, perhaps, interplanetary transportation. So, I might ask, with this information in our possession, should we or the American tax-payer spend additional time and money to supplant facilities and work which has already been accomplished by German science and is now available for our use with the expenditure of a relatively small amount of money and effort. If we are not too proud to make use of this German-born information, much benefit can be derived from it and we can advance from where Germany left off.[6]

However, the advantage for the United States of using the German scientists, engineers, and technicians emphasized by Putt remains a controversial episode in American immigration. Initially Germans were brought to the U.S.A. by going around the official immigration regulations with the silent agreement of the American government. Moreover, a series of scientists with dark Nazi pasts also came to the U.S.A. As early as the end of 1946, the American press discovered the secret transfer of scientists.[7] The American government had placed the usefulness of a scientist before faithfulness to the principle of democracy, and the resulting public outcry was at first loud. Among others, the Society for the Prevention of World War III and the Federation of American Scientists protested against the fact that government officials had deliberately broken American laws and accepted principles.

But the Cold War quickly quieted down the protests, and there were many ways for the government to justify its actions to the public.[8] The new enemy, the Soviet Union, played into its hands. On October 22, 1946 the Soviets forcibly brought more than 2,000 scientists and technicians, along with their families, to the Soviet Union.[9] Strengthened by the Soviet actions, the American government, military, and industry could push a pragmatic line: because of the leading position the U.S.A. played in the world, it had to take advantage of the technical know-how of the Germans because of the potential financial advantages and military superiority. In the discussion held within the political system, the justification was, and still is, "legitimation through use."

This chapter will use the example of rocket and aeronautic research to demonstrate how the institutional structures of Big Science created during the Third Reich and the experts operating within it were transferred from a dictatorial to a democratic system. At the center of our interest stand the Army Experimental Institute in Peenemünde-Ost, the center of German long-range rocket development, and the Aeronautics Research Center "Hermann Göring" in Völkenrode near Braunschweig, both founded in the 1930s. The construction and operation of these research installations were thoroughly examined by the Allies in 1945, and they served as models for similar research centers in the victorious Allied countries. Among others, the Air Force's Arnold Engineering

Development Center (AEDC) in Tullahoma and the Army's Redstone Arsenal in Huntsville were built in the 1950s as "replicas" of German Big Science institutions.

These Germans experienced two political systems along with their different scientific institutions. In order better to understand the behavior of the German rocket and aeronautic research elite in the Third Reich and subsequently in the U.S.A., this chapter will examine their socialization before 1933. Building upon this, the transfer of German scientists and engineers to the U.S.A. and its effect on the German experts will be examined by means of the case study of the "Paperclip" engineer, Tasso Proppe.

Structures I: rocket and aeronautics research in the Third Reich

In the middle of the 1930s German rocket and aeronautics research experienced a historically unique boom. The construction of the Air Force and Army Experimental Station in Peenemünde, the expansion of the well-established aeronautic research installation in Berlin-Adlershof (Deutsche Versuchsanstalt für Luftfahrt, DVL) and in Göttingen (Aerodynamische Versuchsanstalt, AVA), as well as the founding of new research centers like, in particular, the Aeronautic Research Center "Hermann Göring" in Braunschweig (LFA) signaled the will of the regime to become world leaders in the militarily important field of aeronautics as well as a field of military technology with potential for future importance, rockets as weapons. On one hand, this expansion was an expression of an explicit research and technology policy directed towards armaments and war.

On the other hand, this was the result of the polycratic structure of the National Socialist State (NS State), which encouraged the construction of competing institutions under the control of different powerful individuals in the state, working in parallel toward similar goals. The development of aeronautic research in the Third Reich clearly illustrates two interconnected processes, which also played a similar role in other areas of the NS system: (1) the secular development of Big Science; and (2) research fields achieving institutional independence while simultaneously being connected to other parts of the social system, especially the military, politics, and the economy.

If the first process is examined closely, then Big Science is clearly based on developments which reach far back in history. These include secular trends like the following: the increasing role played by science in parts of society; the increasing interdependence of science and technology; the exponentially increasing demands for financial, manpower, and apparatus for research; the increasing number of research results and their relevance for the state, and its ever-increasing costs for administration, standardization and norms, provisions, and future insurance. In the incubation period of Big Science, which dates back to the period between the world wars, new factors arose which first

159

established this type of institutional research. The following characteristics were ideal types for Big Science:[10]

1   Binding together different scientific-technical disciplines (multi-disciplinarity) in one project, with a large apparatus often standing at the center.
2   Binding together extensive resources in manpower and finances (intensity of resources).
3   Predominant financing by the state (involvement of the state).
4   Orientation towards concrete, middle to long-term projects (project orientation).
5   Connecting basic and applied research in an industrial context.
6   Orientation towards goals, which are considered politically and socially especially relevant (goal orientation).
7   Dualism of a clear political goal and far-reaching autonomy of scientists in the establishment of concrete work goals.

The secular process inherent in research developing towards Big Science manifested itself in Germany especially in the rocket and aeronautic research, and was politically supported through a fundamental transformation of the State research structures in the 1930s. In the course of rearmament, the NS regime made extensive resources available, which provided a dynamic push towards the development of disciplinary structures and tasks for the development of rockets and aircraft, which were already tending towards Big Science. The following developments can be identified:[11]

1. The degree of multidisciplinarity of rocket and aeronautic research increased. Research fields like gas dynamics, aeroelastics, engine development, and piloting electronics extended the disciplinary research matrix. The work of teams composed of scientists, engineers, and technicians from very different disciplines became the norm. Specialization and functional internal differentiation increased.

2. Large apparatus more and more determined the research process. In the rocket and aeronautic research, the secular trend towards increased size can be demonstrated by the example of a wind tunnel as one of the most important experimental instruments. Up until 1945 the power of the tunnels grew at a factor of ten in a period of around twelve years, before a stagnation arose for technical reasons.[12] The example of the Aeronautics Research Center in Munich (LFM), founded in 1938, shows in particular how decisively the location of the research can be influenced by high-performance wind tunnels. In order to be able to provide the energy required, the center found itself forced to situate its high-performance wind tunnel in the scientific diaspora of the nearby Ötztal. The increasing determination of the research process by large apparatus alienated conservative (in the scientific sense) scientists. From the retrospective of the postwar period, the aerodynamicist Albert Betz, head of the AVA,[13] even

160

identified the danger of a "mechanization of research" in experimental installations of "intolerable orders of magnitude," through which a "true intellectual culture" was being pushed to the wall.

3. The expanding research also required a high level of resources. The factor limiting growth was, in contrast to the Weimar Republic, not the lack of financial means but rather the availability of qualified personnel. On one hand, academic aeronautics was considerably expanded in 1935 by the construction of Aeronautic Education Centers at the Technical Universities of Berlin, Braunschweig, and Stuttgart, as well as the partial expansion of Technical Universities in Aachen, Darmstadt, Dresden, and Munich. However, the universities could not come close to finding sufficient research personnel to meet their needs. In 1937, only sixteen aeronautics engineers earned their *Diplom*, and in the two following years respectively only fifty-six and fifty-seven. These numbers barely made a dent in the yearly figure given out by the Reich Aviation Ministry (RLM) of 3,600 new engineers needed each year.[14] During the war, the personnel shortage grew further. The shortage of construction capacity was an additional limiting factor. Although the totalitarian regime usually is considered friendly towards infrastructure, here the constraints of the Third Reich show themselves. In the end, the number of construction projects running in parallel within the polycratic chaos of the NS regime forced scientists, engineers, and administrators again and again into competitions for resources.

4. Over the course of the 1930s, rocket and aeronautics research were aimed at goals which were considered politically very relevant. The militarily and technically important areas of high-altitude flight research, jet engine development, high-velocity aerodynamics, high-frequency research, ballistics, and rocket research were systematically expanded and the project orientation generally strengthened: high-altitude motors, jet engine, sweptback wing and tailless-aircraft, radar and radio navigation, ramjet and rocket engines all became development projects on which the research oriented itself first and foremost.

5. Over the course of the Third Reich, research grew more dependent on the state and politics. In this connection the removal of the AVA from the Union of Kaiser Wilhelm Institutes in 1937 is an example of how those in power sought to go beyond the classical instrument of financial control in order to fix research goals directly by influencing its content. During the opening years of the war, the regime also attempted to control not only the global setting of goals, but also the detailed goals of research.

The process by which aeronautic research became independent was greatly simplified by the fact that it was placed under the power bloc dominated by Reich Aviation Minister Hermann Göring. Until well into the war, he was the most powerful individual after Hitler within the NS hierarchy. Even influential institutions like the Reich Education Ministry (REM), Army Ordnance (HWA), or the Reich Research Council (RFR, created in 1937) were unable to influence the research organization and activities of the RLM. Conversely, the development of army research shows how the Armed Forces used the HWA and REM to create their own powerful research structures.[15]

The Big Science complex at Peenemünde also illustrates the dialectical process of double-structure and institutional independence running along the internal fragmentation of the dominant power blocks under the NS system. At Peenemünde, lying on the Baltic Sea peninsula of Usedom, the Air Force and Army founded in April 1936 what was at first a common experimental station. However, after only a short period of time the direction of research was divided up into two independently acting areas: Peenemünde-West (Air Force) and Peenemünde-Ost (Army). A common administration of the installation remained, but with regard to research and development tasks, the two experimental stations went their own way. While Peenemünde-Ost used the principle "everything under one roof" when implementing research and development tasks, Peenemünde-West employed the principle "everything farmed out." In other words, despite the immediate proximity and far-reaching compatible goals (development of repulsion technologies), the two Peenemünde experimental stations developed independent research organizations.[16]

This process of developing an independent niche so typical of the NS regime went along with an intensification of the interconnections between the State, the military, science, and the economy. In 1935 this manifested itself for aeronautics with the founding of the Lilienthal Society for Aeronautic Research, in which the state, science, and industry all came together, and in 1936 with the creation of the German Academy of Aeronautic Research. The latter illustrated the spectacular increase in the significance of these areas of research. Despite the bitter resistance of the traditional "pure" academics of science, the ministerial bureaucracy, with the support of Göring, was able to establish an academy of technical sciences, a novelty in the historically developed German scientific landscape.[17] However, it has rightly been pointed out that the interconnection of partial systems could only partially overcome historically developed barriers. The relationship between the military and the scientists was still plagued by considerable communication problems, which during the Second World War proved to be a great obstacle to the implementation of research results for innovative armaments.[18]

Rocket development at Peenemünde-Ost

The complex Peenemünde-Ost, with its research, development, and production of long-distance rockets, can be seen as the NS project which illustrates most clearly the formation process of Big Science. A comparison with the American Manhattan Project would be useful, but up until now neither this, nor a comparison with the German uranium project, have been done. Along with other important knowledge, these comparisons would help establish the significance of the respective actors and power constellations in the polycracy of the

NS regime. When the development of rockets in Germany is examined in this context, the following developments are revealed.

The 1919 Treaty of Versailles prohibited Germany from developing and possessing heavy artillery. Moreover, during the First World War the artillery had reached its technical boundaries. The German military searched for alternatives, and they found them at the end of the 1920s in the rocket. When the space flight film *Frau im Mond* (Women in the Moon), produced under the direction of Fritz Lang at Ufa's Babelberger Studio, had its premiere in October, 1929, the military had also been swept up by the "rocket craze" that had touched large parts of the German public.[19] HWA recognized that rockets could not only replace tactical weapons of limited range, but also heavy artillery with greater range. Moreover, the possibility of using rockets to send chemical weapons[20] over great distances appeared especially promising. Still in the Great Depression, the HWA provided a young group of enthusiastic rocket experts gathered around Wernher von Braun with enough resources to allow systematic research for the first time after years of tinkering.

After Hitler came to power, military research and, along with it, rocket technology experienced a further boost. This occurred not least because the Air Force, which on March 1, 1935 was elevated to an independent branch of the armed forces, cast an eye on the rocket development, which up until that time had been exclusively supported by the Army. The Air Force offered to support the construction of a new center for rocket technology with five million Reichmarks (RM), which would replace the rather unsuitable research and development installation at the Kummersdorf shooting range near Berlin for the test of experimental rockets. The HWA reacted to the "shamelessness of the young Air Force" with annoyance and made clear that the Air Force could play only the role of a junior partner in the rocket business.

The Army outdid the offer of the Air Force with a further six million RM and thereby fulfilled the financial precondition for the construction of a one-of-a-kind research establishment. After a little searching, in the summer of 1936 the regime began building the Army Experimental Institute Peenemünde-Ost and the Air Force Experimental Institute Peenemünde-West near the small fishing village of Peenemünde on the Usedom peninsula in Vorpommern. When on October 3, 1942 the world's first large ballistic rocket was successfully launched from Peenemünde-Ost and the institute signaled its readiness for serial production, this result not only represented a new dimension of human destructive technology and military strategy, but also a new degree of fusion between science and the military.[21]

Even before the Manhattan Project, Peenemünde signaled the spectacular breakthrough of Big Science as a new form of institutionalized production of knowledge. At Peenemünde, the tightly woven "triple helix" of science, the economy, and state politics expanded over the course of rearmament and war to a new quadruple helix.

Aeronautic research at Völkenrode

On March 21, 1935 the head of the Research Section of the Reich Aviation Ministry, Adolf Baeumker, gave a presentation in Göring's presence on the organization, goals and tasks, and the international level of development of aeronautic research. Baeumker used a type of argument that scientists and science policy makers worldwide used over and over during the twentieth century to push through their goals successfully: he referred to the scientific-technological advantage of other countries and the great efforts, which above all the U.S.A. had undertaken, in order to supplant Germany as the leading scientific country. As expected, Göring then demanded that the Reich should build all the types of research installations which successfully serve the leading aeronautics nations. He used the following maxim: compensate for the superiority of other countries within a few years, in order subsequently to take the lead in all areas and to have the best facilities in the world.

The full discretionary power of the head of the powerful Aviation Ministry and second-in-command to Hitler opened new opportunities to aeronautic research to push through its interests in the complex machinery of the NS rearmament effort, even against the resistance of the military. After Baeumker's presentation and Göring's directives, in the following weeks a series of meetings took place with the leading figures in German aeronautic research, including Ludwig Prandtl, *eminence grise* and representative of the scientific community, Albert Betz, head of the AVA, and Friedrich Seewald, director of the DVL.[22]

In the internal administrative negotiations over the implementation of Göring's directives, the men who did not consider the further expansion of the DVL in Berlin and the AVA in Göttingen sufficient were successful. They argued for the construction of a completely new type of research institution. Since the DVL and AVA lay within the precincts of Berlin and Göttingen, they thereby did not offer the possibility of large-scale expansions, nor could they be protected against enemy air attacks. The new research center was to include, along with the "usual" aerodynamic research, the fields of weapons research and gas dynamics, which should not be located near densely settled areas. The location would be determined by the following criteria:

1 sufficient distance from all of Germany's boundaries;
2 contact with a mid-sized city with a university or technical university;
3 good camouflage against observation on the ground or from the air;
4 space for the construction of its own air field.

Finally the village Völkenrode, northwest of Braunschweig, was chosen. In May of 1935 Erhard Milch, State Secretary for Aeronautics, gave his approval for the construction of the center, which was officially opened in February, 1936 as the German Research Institute (LFA). In the autumn of 1938 the first buildings and research institutes were finished. The construction site comprised approximately 485 hectares, facilitated the generous laying-out of an airfield of

1.2 kilometers diameter and, with its dense forest, promised an excellent camouflage of the widely dispersed individual buildings. By the end of the war there were fifty-six buildings, which included laboratories for ballistics, aerodynamics, gasdynamics, and power plant technology. After the end of the war a Canadian investigative report established the following about the LFA:

> was the most spectacular and impressive establishment seen in Germany...If the discovery of the establishment was a surprise, the realization of the scope and lavishness of its equipment was dramatic. Comparisons may be odious, but the L.F.A. has been described as the most magnificent aeronautical research establishment the world has ever seen. Certainly the planning and conception and the extravagant expenditure of money on the buildings, their equipment and supplies was most impressive.[23]

The historical assessment of this judgment must take into account the effect of surprise and note that the overflowing words of praise from the Allied experts also served as arguments in the debates with their own governments over the realization of similar high-quality research and experimental installations. The founding, construction, and operation of the LFA represented a decisive fundamental model for the research policy of the NS regime, and after 1945 served as model and orientation for the allied victorious powers and the realization of their own plans for expansion. The approach of the RLM was hereby carefully analyzed. Basic research was forced in the areas especially important for aeronautic rearmament: high-velocity flight, power plant research, and weapons research. Like the Third Reich, when the Allies built similar research centers they were very concerned about secrecy and camouflage against air attack, while the precise determination of the work goals was left up to the scientists. Moreover, the Allies recognized that control of the boundary conditions, while simultaneously giving science autonomy in the definition of concrete research goals, was an effective instrument for implementing policy.

First Actor: towards the socialization of German rocket and aeronautic researchers

In order to understand the behavior of the German rocket and aeronautic experts in the Third Reich and in the United States, it is important to recognize the collective mentality of the socialization of scientists and engineers within the authoritarian society of the German Empire, beginning in the late nineteenth and early twentieth centuries. Only with this background in mind does an interpretation or judgment of the individual conduct and behavior make sense. Of course, it is difficult to determine the effect of individual persons or groups of persons within complex historical connections. The thought and perception of individual personalities preclude, as Margit Szöllösi-

Janze recently noted with justification, "beyond a given point an adequate textural portrayal."[24] However, such an approach facilitates connections between the general and the particular, and can illuminate individual and collective freedom of action under the respective social boundary conditions. In contrast, the efforts, often called for by biographical research, to achieve an intuitive opening up of the mind to manifestations of historical actors, appears to be a historiographic dead end. The actions of individual persons should be understood much more as part of a social, cultural, economic, and political process and investigated in a multi-perspective analysis.[25]

With regard to the generation that laid the foundation for aeronautic research, as well as the subsequent rocket development in Germany at the beginning of the twentieth century, Thomas Mann's statement about himself and his generation in his "Observations of an Unpolitical" also holds: "Spiritually, I am essentially a proper child of the century, in which the first twenty-five years of my life fell: the nineteenth." Mann did acknowledge in the same book that elements, necessities, and instincts of the new century drove him. For him it was decisive, however, that his "spiritual focal point" lay "on the other side of the turn of the century."[26]

Thomas Mann confronts us with the double question: what effect the "spiritual focal point" of the nineteenth century had on the "turn-of-the-century generation," and how the academics, engineers, technicians, and civil servants socialized in the "golden age of security"[27] reacted to the social upheaval after the First World War? These questions can in no way be satisfactorily answered by this essay, but they nevertheless play a central role in judging this generation of scientists. Their collective character was formed by the duality of Wilhelmian civil society, which coupled the rules of behavior dictated by a strict Christian ethic with service to the state or to the nation. The combination of religious and political responsibilities yielded the "upright Wilhelmian professor, the decent and responsible state servant."[28] A high regard for the law, trust in the established institutions, and fulfilling responsibilities with regard to the state and bureaucracy were characteristic of the majority of the academics, engineers, and state civil servants belonging to this generation.

For them and the next generation, the end of the First World War destroyed not only the state system of salaries and benefits, but also the collective model of norms and values. An entire world, their trusted world, began to fall apart and what they had taken for granted about life from the period before 1914 became invalid. "The fear of losing one's position, the feeling that one no longer presented a good figure, no longer to count for anything and no place to go, the annoyance and resentment over being superfluous or one of too many, increasingly pushed aside and cast out – all that foments hate and jealousy."[29] After the Treaty of Versailles, the collapse and thereby the end of the familiar world were generally received with helplessness and bitterness.

The transfer from the German Empire to the Weimar Republic shocked the old elites of the German Empire and especially on the state-oriented academics.

Majority rule and the seemingly chaotic political debates of the early and late Weimar period led many Germans to look back at the world of the German Empire, apparently ruled purely by experienced bureaucrats on the basis of reason, as a state of paradise. It was typical "to dream of the glorious past of the German Empire, the Emperor, and bureaucracy and to hope, also in the area of science, for a restoration and reconnection to these traditions. In emotional patriotism this was at once demanded and depicted, and with the young Weimar Republic contrasted."[30] Within the context of this fundamental general feeling, most scientists, in agreement with the majority of Germans, perceived the 1919 Treaty of Versailles as a collective humiliation, a kind of national trauma.[31]

In particular, the generation born between the turn of the century and the First World War was very disoriented by the helplessness of their teachers and role models, and was susceptible to political radicalization. Collective disillusionment and individual resignation influenced everyday life and formed fertile ground for anti-democratic ideologies and movements. During the Great Depression, political discussion was led with increasing intensity, as the aircraft builder Ludwig Bölkow remembered: "What did we not discuss at that time! We were dominated by a social and national longing for justice internally, but also from the outside."[32] Many saw a possibility for experiencing the desired justice and recognition in the enthusiasm for the assumed objectivity of science and technology. On the search for orientation, technology, and technocracy offered an effective help because of its impressive clarity. Success in science and technology could strengthen the battered national self-confidence of the Germans. They became compensation for the loss of national power or, as Michael Neufeld has formulated it: "Germans tended to seize on almost any sign of their technological superiority or their rapid recovery from the humiliations of the war and Versailles. Despite bitter political and ideological divisions in the country, technological progress was desired by almost everyone."[33]

If Thomas Mann, and with him a good portion of German intellectuals, understood himself before the background of the collective experience of the First World War as "unpolitical," so did the model, drawn from the educated middle class, of the unpolitical hold even more for the scientific and technical elites of Germany. In particular, they understood conduct within their profession as a space free from politics. As Mark Walker and others have shown, this understanding was certainly compatible with active support for the goals of the NS regime, which in the meantime can be described as self-mobilization. Research in the service of the Fatherland was understood as unpolitical within this model, while to withdraw from this responsibility was understood as political. After the "revaluation of all values" through the collapse of 1945, a good part of the scientific elites clung to this model, which well into the 1970s could be sustained as the collective lie of the educated German middle class.[34]

After the end of the Second World War, the victorious Allies often turned to German specialists because under the Third Reich the latter had experienced a unique education and their specialist knowledge was indispensable. Thus, for example, after 1933 the German aeronautic research and aircraft industry sought out suitable candidates for subsequent careers as test pilots. These pilots were given scientific and technical training in order to improve the practical trials of new types of aircraft. After the end of their studies, those selected began a three-year education (1936–9) for becoming aircraft construction leaders. The graduates took exams equivalent to those for a Ph.D. in engineering and for an aircraft construction technician. They thereby belonged to a group of engineer-pilots who were regarded as pilots with special training in engineering and science. Moreover, they had to have practical knowledge in order to be in a position to make themselves understood among scientists, builders, and technicians within a broad spectrum of disciplines. Their main task was, first of all, to gather data during the flight, to register the flight performance, and to test extreme flying conditions. Second of all, on the ground they should help assess the results of the experimental flights. Their education included several months in different aircraft enterprises, as well as practical training in research centers.[35]

Precisely the goal-oriented bringing-together of practice and theory was one of the strengths of the German engineering education before 1939. In the 1930s an engineer usually had to have worked for two years as a machine builder or mechanic in order to be admitted to a technical university or engineering school. Many even finished a third year and took the journeyman's examination before their studies. This study path was called the "path of the dirty fingernails." This led to Germany having a large number of "practical" engineers who also had solid theoretical knowledge. They were encouraged during their studies to design technical products "which could be operated, serviced, and repaired as easily as possible; they learned to make realistic designs as a function of the expected use of the product."[36] This ability was also valued very highly worldwide after 1945.

Returning to the collective mentality of German scientists, the overwhelming majority of aeronautic researchers had received the impression in the 1920s that the democratic state was unable to satisfy their professional needs and desires. Parliament and the state bureaucracy did not appear to appreciate the necessity of expanding aeronautic research. In general, scientists tend to orient themselves to the working conditions of their colleagues inside and outside of their country. German aeronautics researchers looked first of all to the U.S.A., which in the course of the 1920s had developed excellently equipped and well-organized research facilities. While in Great Britain, France, and especially in the U.S.A. the infrastructure requirements for high-quality research had been fulfilled, the Germans operated with out-of-date equipment which did not even allow model experiments on new types of aircraft. Like large sections of Weimar society, scientists refused to recognize the republic as a suitable form of government and began to favor alternatives.[37]

The National Socialist "seizure of power" was therefore generally welcomed by the scientific community of aeronautic researchers. With unconcealed pleasure, aeronautic scientists registered the fact that their discipline was now receiving much financial support from the new government. The appointments of the previous director of the Lufthansa airline company, Erhard Milch, as State Secretary and of Adolf Baeumker to head of the Research Section of the newly-formed Reich Aviation Ministry raised hopes that the significance of aeronautic research for the well-being of the nation had finally received the recognition it deserved. Baeumker, in particular, had in the 1920s gained the confidence of scientists as the official in the Reich Transportation Ministry responsible for aeronautic research. The son of a well-known Munich philosophy professor, he was seen as the guarantor of the autonomy of science who, in an unbureaucratic way, sought new ways of supporting research.

The German aeronautic research elite were not disappointed. Just a few weeks after Hitler came to power, the AVA received permission to build the large wind tunnel it had been trying to construct for many years. Stocking the aeronautic budget with 40 million RM taken from the job creation fund allowed the DVL to take the expansion plans it had developed at the end of the 1920s out of the drawer and realize them on an even larger scale. The scientists quickly learned that their requests could not be big enough. Research goals which were unthinkable before 1933 were now realized. The financial question, which up until then had been the limiting factor for expansion and scientific advance, was suddenly only of secondary significance.[38]

Structures II: the replication of a principle of organization

When the Allied commission for investigating German research and technology visited the aeronautic research facilities in Germany and the territories formerly occupied by Germany after the war, they were very impressed by the level of development and research conception. Among other things, their analysis makes clear that the German rocket and aeronautic research was far more advanced than that of the Allies at the end of the war, precisely in the areas especially important for the future development of military technology: supersonic aerodynamics, ballistics, jet engine technology, and long-range and maneuverable rockets. The Allies quickly realized that, if a better coordination and technological realization of the research results had been achieved in Germany during the war, then the Third Reich would have presented a dangerous challenge.

In the end, this threat was not effective mainly because of the polycratic structure of the NS regime, discussed above, which doomed every attempt to build bridges of communication and interconnectedness between the systems of science, technology, and the military. Under the conditions intensified by the war, polycracy meant the communicative isolation of the systems and their permanent involvement in Darwinist competitions for political influence and resources.[39] For example, the high level of aeronautic research was contrasted with the far-reaching failure to convert these research results into armaments for the Air Force. At the end of 1945, the Allied side noted in this connection that: "The thought remains that a certain brilliance and daring was associated with their basic conceptions of the application of science to aeronautics and aeronautics to warfare. That they came so close to success but failed in the ultimate struggle may be due to many separate factors, but lack of close co-coordination and poor timing played a large part."[40]

Similarly, beginning in the middle of the 1930s American scientists registered the new quality of the rapidly expanding aeronautic research in Germany. National Advisory Committee for Aeronautics (NACA) director George W. Lewis had already received the impression during his European trip in the autumn of 1936 that the German aeronautic research facilities were ahead of American centers with regard to size, depth of research and education, as well as the technological competence of their personnel.[41] What Lewis saw at its beginning, presented itself at the end of the war to his colleague Theodore von Kármán, former student of Ludwig Prandtl and now director of the Scientific Advisory Group of the Army Air Force (AAF-SAG), as an exemplary organizational model for the desired reorganization and expansion of American aeronautic research. In particular, the center at Völkenrode fascinated him with regard to size and structure. What he saw there "was incomprehensible" and needed to be transferred to the U.S.A. as a guiding principle of organization.[42]

If Völkenrode stood out through its connection of broadly conceived basic research and innovative experimental and testing facilities (high-velocity wind tunnels, power plant testing installations, and ballistics laboratories), then Peenemünde was impressive because of its focusing of research and development on one technology: long-distance rockets. In August, 1945, von Kármán formulated the conclusion of the Scientific Advisory group he led as a "lesson to be learned" from the activities of the Peenemünde team and the Big Science centers of the German Air Force:

> Leadership in the development of these new weapons of the future can be assured only by uniting experts in aerodynamics, structural design, electronics, servomechanisms, gyros, control devices, propulsion, and warhead under one leadership, and providing them with facilities for laboratory and model shop production in their specialties and with facilities for field tests. Such a center must be adequately supported by the highest ranking military and civilian leaders and must be adequately financed, including the support of related work on special aspects of various problems at other laboratories and the support of special industrial developments.[43]

The polycracy of the NS research was replicated with the transfer of the organizational model of Big Science to the American rocket and aeronautic research. The latent tendency of military research, to split itself off into systems leaning towards autonomy and bound to a particular arm of the Armed Forces, was intensified by the transfer of scientists and technology from Germany. Underneath the race of the Allies for the know-how carried by individuals, the different arms of the Armed Forces within the U.S.A. also competed for the service of German scientists. Instead of the central development laboratory which von Kármán and other scientists and military officers called for, in the postwar years the Army rocket research center at Redstone Arsenal, near Huntsville, and the Air Force Arnold Engineering Development Center were founded. The Navy also tried to go its own way, if not on this order of magnitude.

Replicas: Huntsville and Tullahoma[44]

Rocket development in Huntsville

On December 28, 1948 the U.S. Army founded its new research center for rocket development, Redstone Arsenal, near Huntsville, Alabama. A few years later, in the middle of April 1950, Wernher von Braun and a number of his co-workers moved from Fort Bliss in Texas to Huntsville. With the arrival of the Germans, a modern rocket development complex arose, just as in 1937 with the transfer from the Kummersdorf shooting ground near Berlin to Peenemünde-Ost. The similarity of Huntsville to Peenemünde was, as a co-worker of von Braun remembered, "in many respects almost unbelievable." In fact, Huntsville was the resurrection of Peenemünde-Ost. Americans and Germans quickly began calling the place Peenemünde-South.

A significant reason for Huntsville's success was that it followed the organizational principle established in Peenemünde-Ost of "everything under one roof." This did not mean the physical concentration of the various research, development, and production facilities at one place, but rather that the coordination of the different areas and branches of science, technology, and production lay in the "firm" hand of those in Peenemünde, and now in Huntsville. As previously in Germany, it was again Wernher von Braun who, as its motive force and talented planning manager, drove the development forwards. The "replication" of Peenemünde in Huntsville was also the most visible expression of the exemplary international effect of the former German rocket center. The Peenemünde "rocket trail" also led to the Soviet Union, France, England, the Middle East, China, and – as has recently been made public – Australia.[45]

Aeronautic and rocket research in Tullahoma

Even before the end of the war, Army Air Force General Henry H. Arnold was pushing to bring German technological advances to the U.S.A. and apply them there. In September, 1944 Arnold got in touch with von Kármán and convinced him to take over the leadership of the AAF-SAG. The group was composed of thirty-one hand-picked experts from diverse areas of rocket and aeronautic research. A specialist, among others, on wind tunnels was Frank L. Wattendorf, who during his visit to Europe in 1945 was stunned, like von Kármán, by the technical aeronautic innovations of the Germans. He immediately sent a "transatlantic memo" back home which proposed "a new engineering development center to consolidate the test facilities and the talents of the nation's best civilian and military scientists – a center to properly test and evaluate weapon systems needed to guarantee superiority of American airpower and thereby the national security."[46]

After years of preparation, American president Harry S. Truman formally opened the Arnold Engineering Development Center near Tullahoma in Tennessee. Truman's speech fit well into the hot phase of the Cold War, and the president emphasized that the center had to make a significant contribution to the American "effort to make our air power the best in the world."[47] The Center was in part a result of the "Unitary Wind-Tunnel-Plan" of the NACA, which drew upon the precedent-setting recommendations of the Air Policy Commission in 1947–48. The plan included the construction of high-velocity wind tunnels in universities and industrial firms, as well as civilian and military research facilities, costing the gigantic sum of a billion dollars. The expansion program, finally approved by Congress at the end of July, 1950, benefited the Tullahoma Center in particular, which quickly developed into the "largest complex of flight simulation test facilities" in the Western world.[48]

It was no coincidence that the choice fell to a location in the state that had the most water and energy in the U.S.A. The Third Reich had already experienced how dependent Big Science installations were on the availability of large amounts of energy when it planned a new generation of wind tunnels on the eve of the Second World War. Ötztal, a valley south of Munich, used its natural resources of water power to provide favorable conditions for the planned high-velocity wind tunnel, with a engine unit capable of 75,000kW of power, in which experiments on high-performance engines were to be carried out up to full flying velocity. The NS leadership hoped to use tunnels of this magnitude to compensate for the threatening scientific and engineering superiority of the U.S.A. This experience helped the U.S.A. with the construction of the center near the Tennessee Valley, not least because they brought Walter Dornberger, the former head of Peenemünde-Ost, into the planning. "The result was that the AEDC combined Peenemünde, Völkenrode, Kochal, and Ötztal windtunnels."[49] It is significant that the Center was led by Bernhard Goethert, who during the Second World War was one of the leading high-velocity aerodynamics experts in Germany.

Second actors: transfer of scientists and technology to the U.S.A.

Goethert belonged to those German specialists who came to the U.S.A. after the end of the Second World War through the American military as part of

"Project Paperclip."[50] Originally a program code-named "Overcast" was created in July, 1945, which was supposed to help shorten the war in the Pacific. However, the American atom bomb led to an unanticipated quick Japanese capitulation. The bottom thus fell out of the market value of the German weapons experts, scientists, and experts coordinated by "Overcast", and now only hand-picked experts were needed. The group selected first included the German rocket researchers gathered around Wernher von Braun, and especially aeronautics specialists. They were highly qualified experts who were ranked by the Americans as "leading authorities" in their specialty and whose ability was rated "outstanding." However, the German and Japanese capitulations caused serious difficulties for "Overcast." The *normal* American state bureaucracy took back control, so that immigration regulations and laws, which had been ignored by the military during the war, now had to be followed.

The "Overcast" program was not prepared for the increasing public criticism of the employment of German military researchers. New rules had to be written for the transfer of scientists and technology. At the beginning of 1946 a way out was found with "Project Paperclip." As of 1948, around 500 German and Austrian top-level specialists and their families came to America through the program.[51] At the beginning of the 1950s the number of German families rose again; in the summer of 1953 the "Paperclip" lists included more than 700 specialists with around 1,500 dependents.[52] This second "wave" of immigrants with know-how also brought to the United States the engineer Tasso Proppe, whose opinions and experiences are covered in the following section.

The interest in using the scientific and technical abilities of the Germans was rekindled at the beginning of the 1950s, mainly because of the Cold War, which had been running at full steam since the Berlin Blockade (1948–9). Moreover, in August, 1949 the Soviet Union tested its first atom bomb, surprisingly quickly breaking the monopoly of the Americans and pulling itself into the atomic race. This shocked the American military and intelligence services, just as Sputnik would almost a decade later. In the course of the investigation of the causes of the Soviet success, it was remembered that at the end of the war the Soviets had recruited German atomic specialists. The obvious assumption was that these Germans had played an important role by the construction of the bomb.[53] Another unpleasant surprise came in the Korean War, in which Soviet weapons systems were used that German experts had helped to develop. The Americans feared that the "other side" had successfully used German know-how, which was vividly confirmed in the dogfights between Soviet-built and American jet planes in the sky over Korea.

The Army, Navy, and Air Force initiated a series of new military programs, which created a demand for more qualified personnel.[54] "Paperclip" experienced a renaissance: in November, 1950, a further program for the military use of German specialists began under the name "Project 63." Industry was also active, and under the program "National Interest Cases," attempted to find further experts in West Germany. These were supposed to help in the "defense of the Western world" by working in the defense branch of American industry.[55] In total, as of the middle of the 1950s, these three programs had brought around 1,000 German specialists with around 2,000 dependents to the United States.[56]

Tasso Proppe: a "Paperclip" engineer in the U.S.A.[57]

"Convictions are always interchangeable upon demand."[58] With this sentence, the German aeronautic engineer Tasso Proppe at the end of the 1980s described his work as an engineer-pilot in the Third Reich and subsequently as an experimental engineer in aircraft and rocket development in the U.S.A. Proppe thereby not only critically examined his own conduct, but also the pattern of conduct of those scientists and technicians who first of all under National Socialism, then after the end of the war worldwide, and finally by the middle of the 1950s once again in East and West Germany, were used in their areas of expertise without much interruption. The conduct thereby revealed for these individuals demonstrates a continuity of legitimation through use throughout the changing political and social system. Armaments, war, and Cold War were the central determinants of the collective mentality of German scientists and engineers in the first two-thirds of the twentieth century.

The notes on the impressions and experiences of Proppe and his family in the United States began with their arrival in America. The first letter back to Germany on March 11, 1953 was unusual in that it was a model for a circular letter which was to be reproduced in Germany[59] and then forwarded on to relatives, friends, acquaintances, and colleagues; by 1987, forty-two further circulars had followed. Proppe understood this medium as a communicative bridge between his old and new homes. First of all the letters came from the small coastal city of Coca in Florida, where Proppe worked at the nearby Patrick Air Force Base until the end of 1953. The engineer was then transferred to Hollomon Air Force Base in New Mexico in December 1953, and for around two and a half years the mail came from Alamogordo, a location 80 miles north of El Paso in Texas. In May 1956, Proppe finally transferred from the Air Force to private industry. He took a job with the aeronautic firm Convair in San Diego. There, at the southern tip of California, the Proppe family settled down. Their experiences and impressions were limited almost exclusively to the southern part of the United States, "where," Proppe perceptively noted, "not only is the climate different, rather the spirit also moves in different paths."[60]

In the first year the Proppes were still reporting on the noticeable differences between the old and new home. But the family very quickly got used to the different conditions. After a year the "American" appeared everyday and "normal." Reporting back on the basis of comparison became more and more difficult, for Proppe had more difficulty in recalling what conditions back home

174

had been like. Proppe saw a striking difference between Europe and America in the "tendency to blatant extremes," that is, the oscillations between the opposite poles in the society and nature.

The convictions and ideas mentioned by Proppe make especially clear that, for him, American society was influenced by regional, local, religious, and ethnic diversity. Families lived in their respective "neighborhood islands,"[61] whereby the national, linguistic, cultural, and religious differences did not merge together.[62] "Home" for the Proppes was always the location where they or the "offshoots" of the family lived. That was how they referred to the U.S.A. Proppe was disturbed by the strong influence of ideology on the social relationships in America during the Cold War, even when compared to his experiences in the Third Reich. It was precisely this ideological effect that he saw as the actual binding element of American society, essentially a mixture of politics, democratic rituals, naïve nationalism, superficial religiosity and moralism. "The freedom and democracy that is so loudly and over zealously praised," he wrote reflectively at the end of the 1980s, "is by closer inspection quite hollow."[63] However, he was already of this opinion in the middle of the 1950s. At that time he described at greater length why he found American freedom and democracy so unconvincing:

> Everything is certainly not rosy and wonderful. The superlatives in the dimensions of space, time, mass, standard of living, are always imposing. Compared to our earlier existence they give us the feeling of boundless distances and prosperity. This is often confused with the feeling of freedom. However, in the meantime we have found that freedom is a completely relative concept. Restrictions under which one grows up are not noticed, but it is hard to get used to new restrictions.
>
> If one discussed freedom with an American, then sooner or later the classic example of American freedom will come into the conversation: anyone can call the President of the United States a scoundrel, without anything happening to him. However, if one would dare to describe the democratic system in the U.S.A. as corrupt – and there was some justification for doing so – then one would be guilty of lèse-majesté, which could have the some consequences as listening to a foreign broadcast in Germany during the war. It can definitely happen that a person is accused of spreading communist propaganda by someone who is not involved but overhears him. From then on, this person has a file as a "security risk," is politically unreliable, and loses his position. Only someone who has a lot of money and can finance the long legal process of rehabilitation can hope to return to a job in the scientific profession. That is exactly happened to an acquaintance of ours, by the way a Jew. The feeling that our freedom is being robbed sneaks up on us Europeans at the beginning of the immigration process with the fingerprinting, which has happened to us hundreds of times. It

goes without saying that the Americans accept this, however, it would be an unbearable assault on their freedom, if when changing their apartment they had to register in the housing office; they shake their heads at what we let happen to us back home [in Germany].[64]

The Proppe family immigrated to the U.S.A. at a time when anticommunism was a central organizing principle of the American political system. The opposition of the system to the Soviet Union created the "military-industrial-academic complex," for which Proppe also worked. American politics oriented itself under these conditions both inwardly and outwardly substantially on its opponent the Soviet Union.[65] This fact was clear to Proppe early on, and he judged his professional conduct both in the Third Reich and in the U.S.A. self-critically and without illusions. In this he went further than the majority of his colleagues, who adopted the philosophy that, in all their professional conduct, they had served "only technology."[66] At the beginning of the 1970s he formulated his thoughts as follows: "We engineers, who have called ourselves the tools of technical progress, have, out of love of our work, prostituted ourselves to the politicians."[67] Almost thirty years later he became even clearer in his assessment of his professional conduct and behavior since the Third Reich:

> Five years after the immigration I received [American] citizenship (at my request, because the job more or less required it, and because I myself wanted it – in this connection, it simply went along with the job). One must repeat: 'I pledge allegiance to the flag...'...But it was also part of my job in Germany in aeronautic research that we sang the [National Socialist] Horst Wessel song. For the good of aeronautic science, we did exactly what later was required for supporting space flight (in the U.S.A.). In both cases it was also useful for defending the Fatherland. Technical progress was decisive: in the end that is good for everything – so it was said.
>
> We sang, or had to sing, "I devote myself, with heart and hand," under the German Emperor. For who? For the politicians? They are not worthy of it! For the Fatherland? This is basically a collection of families, who for economic and organizational reasons have come together. It was like this for the cave men. I have forged, hardened, and sharpened the knife for the band, in order to come up the wild boar that has been killed. When they began, to kill each other with it, I kept still and felt innocent, here as over there [in Germany]. One only puts on a tie, in order to stay as far out of the line of fire as possible. It does not matter much, which side one happens to be on. Those that remain will of course need people again, who can sharpen the knife.
>
> The national pride, which America has to cultivate artificially, because it is composed of many, often very different types of people,

has not swept us along. We had already learned that from the Nazis, who copied it, like so many other American things...The demands which are made on 'convictions' are minimal and bearable. Convictions are always interchangeable upon demand.[68]

This quotation is a summary of his perceptions and convictions as an engineer in Germany and the U.S.A. Proppe recognized and portrayed himself at once as an idealist, an opportunist, and a technocrat. He shared this self-image with many of his professional colleagues, who began their career in Germany, and ended it in the U.S.A.

Conclusion

Proppe was a wanderer between the systems, whose individual biography reflects the collective experience of the "century of catastrophes." Socialized in National Socialist Germany, he was wrenched out of his professional context by the German defeat. In the course of the "brain drain" from Germany into the victorious Allied powers, he shifted his professional and personal life to the U.S.A. and, in an American system of innovation deeply influenced by the Cold War, learned once again that scientific and technical elites in crisis periods of national security could legitimate their individual and collective pasts through their usefulness for the overriding goals of the political system in power.

In contrast to many of his colleagues, whose understanding of science as a politics-free space of social action was intensified still further by their experience under National Socialism, Proppe retained some insight in the inextricable connection between such scientific work and political applications. Of course, he ordered these under the residual category of "convictions," a type of superficial ideological varnish, which can be more or less arbitrarily accommodated to changing contexts. Here once again is the conception, deeply-rooted in the German educated middle class since the nineteenth century, of the superiority of German idealism over the pragmatism of western democracy, which the philosopher and socialist Ferdinand Tönnies expressed in 1887 as the oft-quoted contrast between "community and society."[69]

The counterpart to the interest on the part of the U.S.A., to offer a new home to German experts despite their involvement in the criminal politics of the NS regime, was their readiness to offer their services to their former enemy during the war. While the majority of the German scientists and engineers were forced to go to the USSR, the overwhelming majority of colleagues who went to work for the western Allies left their home voluntarily. A technocratic understanding of science and technology as an "arcanum" of the un-political, pained with the attractive prospect of shifting their work to much more attractive working conditions at the cutting edge of research, was the collective humus out of which the individual decision sprouted to legitimate oneself by being useful in a different system and surroundings.

The German specialists brought not only concrete scientific and technological knowledge which could be used in the Cold War. They also had experience with the organization of research in complex fields like aeronautics and rocket technology, which during the war years completed the quantitative and qualitative leap to Big Science. Even before the Manhattan Project, which through its spectacular nature has more obscured than illuminated our understanding of the development of Big Science as a new form of scientific production in the period between the wars, in Germany several Big Science centers were established in the course of rearmament and preparations for war.

This process, fed by the momentum inherent in science and influenced by social factors, has been demonstrated in this essay with the examples from the middle of the 1930s of the Centers for Aeronautic Research in Völkenrode near Braunschweig and for Rocket Research in Peenemünde, newly constructed on undeveloped land. On one hand, Völkenrode and Peenemünde, as well as their replicas in Tullahoma and Huntsville, stand in the continuity of the trans-Atlantic perception of research styles and forms, which since the second half of the nineteenth century had contributed considerably to the formation of the scientific and technological development in Germany and the U.S.A. On the other hand, the transfer of Big Science as an organizational model of science in the Third Reich to the U.S.A. in the postwar era by means of personnel illustrates the specific historical situation of the Second World War and the Cold War.

In this phase of the domination of the two scientific innovation systems by military and security goals, comparable problems led to remarkably similar results. Although the political and social systems were very different – and these differences should be emphasized explicitly – the patterns for connecting the scientific and military structures and goals were very similar. In both systems, for aeronautic as well as rocket research, the institutional location within a powerful military-industrial complex proved to be the necessary condition for their growth and potency for solving scientific problems. The polycratic structure of National Socialism corresponded in the U.S.A. to the differentiation of research along the lines of demarcation between the different arms of the armed forces (Army, Air Force and Navy).

If we finally examine once more the conduct of scientists and engineers, there are obvious parallels which result from the comparability of the historical contexts. Trust in the problem-solving ability of science and technology has been strengthened by the significance of science and technology for conducting war and national security and has pushed the debate over their roles in the extermination of whole peoples and endangering democratic conceptions of order to the limits of public discourse. Despite the differences in the political and social systems, experts legitimate themselves by their concrete usefulness, and this manifested itself in the era of war and Cold War above all through military-relevant research results and technical innovations. Vannevar Bush's programmatic book *Science: The Endless Frontier* appeared in the middle of July,

1945, and was the kernel of postwar science policy. In this climate, the German experts were not damaged by their NS past and were both potentially valuable and willing helpers in the quickly intensifying competition between the superpowers in the Cold War.[70]

Notes

1 Following Niklas Luhmann's concept of "legitimation through process," we use the concept of legitimation in a pragmatic sense. "Legitimation through use" refers to the process of knowledge transfer by means of individuals beyond the boundaries of national innovation systems, which largely masks the normative criteria of political conduct and recedes behind a *Vergegenständlichung* of the category of use. A thorough theoretical discussion of the concept "legitimation" would show that this is very controversial in sociology, political science, and legal studies, and difficult to comprehend. In political science, for example, legitimation is understood as a descriptive characteristic of the state, with which political conduct should be explained. Legal studies fuse the concept as identification for the validity of the law, i.e., the binding force of certain norms, decisions, or principles is assumed. In turn, among other meanings, sociologists speak of legitimation if binding decisions are taken over into its own decision-making structure. For Luhmann, legitimation signals the general willingness to "accept, within certain tolerances, decisions when the content is still undetermined"; Niklas Luhmann, *Legitimation durch Verfahren* (Neuwied am Rhein: Luchterhand, 1969), 28.
2 Steven Dickman, "Flucht vom sinkenden Schiff," *Die Zeit* (27 December 1991), 62.
3 Lany Weymouth, "A Russian Wernher von Braun," *The Washington Post* (January 23, 1992), A25.
4 Weymouth.
5 "Memorandum: Proposed Importation of German Scientists for U.S. Science and Industry Benefit," Secretary of Commerce Henry A. Wallace to Harry S. Truman (December 4, 1945), Official File Harry S. Truman Library.
6 "Speech given by Col. Putt at Dayton Country Club," (7 May 1946): Microfilm (MR), frame 0457, Historical Office, Bolling Air Force Base (BAFB).
7 John Gimbel, German Scientists, United States Denazification Policy, and the 'Paperclip Conspiracy,'" *The International History Review*, 7, 3 (August 1990): 455.
8 Gimbel, "German Scientists," 455. Michael J. Neufeld, *The Rocket and the Reich: Peenemünde and the Coming of the Ballistic Missile Era*(New York: Free Press, 1995), 270–1.
9 Burghard Ciesla, "Der Spezialistentransfer in die UdSSR und seine Auswirkungen in der SBZ und DDR," *Aus Politik und Zeitgeschichte* (3 December 1993), B49–50/93, 24–3; Christoph Mick, *Forschen für Stalin. Deutsche Wissenchaftler und Ingenieure in der Sowjetunion 1945–1955* (Munich: Oldenbourg 2000).
10 See Peter Galison and Bruce Hevly (eds), *Big Science: The Growth of Large Scale Research* (Stanford, CA: Stanford University Press, 1992), 1–17; Margit Szöllösi-Janze and Helmuth Trischler, "Entwicklungslinien der Großforschung in der Bundesrepublik Deutschland," in Margit Szöllösi-Janze and Helmuth Trischler (eds), *Großforschung in Deutschland* (Frankfurt am Main: Campus, 1990), 13–14; Helmuth Trischler, "Großforschung und Großforschungseinrichtungen," in Peter Frieß and Peter Steiner (eds), *Forschung und Technik in Deutschland nach 1945* (Munich: Deutscher Kunstverlag, 1995), 112–3; Derek de Solla Price, *Litte Science, Big Science*(New York: Columbia University Press, 1963); Peter Lundgreen *et al.*, *Staatliche Forschung in Deutschland 1870–1980*(Frankfurt am Main: Campus, 1986).

11 Helmuth Trischler, *Luft- und Raumfahrtforschung in Deutschland 1900–1970. Politische Geschichte einer Wissenschaft* (Frankfurt am Main: Campus, 1992); Helmuth Trischler, "Big Science or Small Science? Aeronautical Research in Germany 1900–1945," in Margit Szöllösi-Janze (ed.), *Science in Third Reich* (London: Berg, 2001, 79–110).
12 Gerhard A. Ritter, *Großforschung und Staat in Deutschland. Ein historischer Überblick* (Munich: Beck, 1992), 51–2.
13 Albert Betz, "Ziele, Wege und konstruktive Auswertung der Strömungsforschung," *Zeitschrift des VDI*, 91 (1949): 253; Albert Betz, "Entwicklungstendenzen der Forschung und ihre Gefahren," *Physikalische Blätter*, 13 (1957): 387.
14 Bettina Gundler, "Das 'Luftfahrtlehrzentrum': Luftfahrtlehre und Forschung an der TH Braunschweig im Dritten Reich," in Walter Kertz *et al.* (eds.), *Technische Universität Carolo-Wilhelmina Braunschweig 1745–1995: Vom Collegium Carolinum zur Technischen Universität* (Hildesheim: Olms, 1995), 509–31; Karl-Heinz Ludwig, *Technik und Ingenieure im Dritten Reich* (Düsseldorf: Droste, 1974), 271–83; Edward Homze, *Arming the Luftwaffe: The Reich Air Ministry and the German Aircraft Industry, 1919–39* (Lincoln, NB: University of Nebraska Press, 1976), 214.
15 See Burghard Ciesla, "Abschied von der 'reinen' Wissenschaft. 'Wehrtechnik' und Anwendungsforschung in der Preußischen Akademie nach 1933," in Wolfram Fischer *et al.* (eds), *Die Preußische Akademie der Wissenschaften zu Berlin 1914–1945* (Berlin: Akademie-Verlag, 2000), 483–513.
16 Heinrich Beauvais *et al.*, *Flugerprobungsstellen bis 1945* (Bonn: Bernhard & Graefe, 1998), 219–23.
17 See Helmuth Trischler, "Aeronautical Research in the Third Reich: Organization, Management and Efficiency during Rearmament and War," in Horst Boog (ed.), *The Conduct of the Air War in the Second World War: An International Comparison* (Oxford: Berg, 1992), 169–95; Ciesla, "Abschied," 499–513.
18 See Horst Boog, *Die deutsche Luftwaffenführung 1935–45. Führungsprobleme; Spitzengliederung; Generalstabsausbildung* (Stuttgart: Deutsche Verlags-Anstalt, 1982), 69; Ralf Schnabel, *Die Illusion der Wunderwaffen. Düsenflugzeuge u. Flugabwehrraketen in der Rüstungspolitik des Dritten Reiches* (Munich: Oldenbourg, 1994), 63–101. Also see Lutz Budraß, *Flugzeugindustrie und Luftrüstung in Deutschland 1918–1945* (Düsseldorf: Droste 1998) and Peter Fritzsche, *A Nation of Flyers. German Aviation and the Popular Imagination* (Cambridge, MA: Harvard University Press, 1992).
19 Michael Neufeld, "Weimar Culture and Futuristic Technology: The Rocketry and Spaceflight Fad in Germany, 1923–1933," *Technology and Culture*, 31 (October 1990): 725–52; see Frank Winter, *Prelude to the Space Age: The Rocket Societies: 1924–1940* (Washington, DC: Smithsonian Institution Press, 1983).
20 The connection between the use of poison gas and long-range rockets has not been given much attention. Rather the main question has been, why did the NS regime invest so massively in a technology which could hardly decide the course of the war. In fact, the resources used for rocket technology ripped great holes in other areas of the German armament industry. In contrast, the total amount of explosives in all of the A4/V2 long-range rockets launched was hardly larger than that of a single large British air attack on Germany. The number of victims – including the prison laborers who died producing the A4/V2, (around 16,000–20,000), as well as the civilians killed during the attacks (around 5,000) – were outweighed by the many hundreds of thousands who died during the Allied air attacks on Germany. Taking such relations into account, one is left with the "central paradox of Peenemünde." Yet the paradox is resolved, when one considers that the original intention might have been to use poison gas in the rockets. In such a case, the imprecision of the rockets would be of secondary importance and mass extermination would be possible. One can only speculate why Hitler did not agree to the use of rockets loaded with poison gas. It was

certainly of great importance that the long-distance rockets ready for mass production and the chemical weapons of mass destruction like Tabun suitable for use in them were only available at a time when the Allies were already in the position to react against Germany with massive air attacks. Hitler and the military stood before a sort of "prisoner's dilemma." If the German side had attacked with poison gas, then it would have had to expect a corresponding response. In principle it is astonishing that historians up until now have not paid enough attention to this dark side of the rocket project; see Ciesla, "Abschied," 496–7; Rainer Eisfeld, *Mondsüchtig. Wernher von Braun und die Geburt der Raumfahrt aus dem Geist der Barbarei* (Reinbek bei Hamburg: Rowohlt, 1996), 26; Neufeld, *Rocket*, 267–79.

21 Burghard Ciesla, "Spurensuche in Peenemünde," *Werkstatt Geschichte*, (15 December 1996): 119. See Neufeld, *Rocket*, 5–39; Hans Barth, *Hermann Oberth: "Vater der Raumfahrt"* (Munich: Bechtle,1991), 178 ff.; Volkard Bode and Gerhard Kaiser, *Raketenspuren. Peenemünde 1936–1996; eine historische Reportage* 2nd edn (Berlin: Links, 1996), 10 ff.; Jürgen Michels, *Peenemünde und seine Erben in Ost und West. Entwicklung und Weg deutscher Geheimwaffen* (Bonn: Bernhard & Graefe, 1997), 12 ff.; Heinz Dieter Hölsken, *Die V-Waffen. Entstehung – Propaganda – Kriegseinsatz* (Stuttgart: Deutsche Verlagsanstalt, 1984); Johannes Weyer, *Wernher von Braun* (Reinbek bei Hamburg: Rowohlt, 1999), 29–50.

22 Trischler, *Luftfahrtforschung*, 213–22; Adolf Baeumker, *Zur Geschichte der deutschen Luftfahrtforschung – Ein Beitrag*(Munich: Eigenverlag 1944), 31–44.

23 J.J. Green, R.D. Hiscocks, and J.L. Orr, "Wartime Aeronautical Research and Development in Germany, Part I," reprint from *The Engineering Journal*, (October 1948), 5; Leslie Simon, *German Research in World War II*, (New York: Wiley 1947), 12–24. See Hermann Blenk, "Die Luftfahrtforschungsanstalt Hermann Göring," in Karl Stuchtey and Walter Boje (eds), *Beiträge zur Geschichte der deutschen Luftfahrtwissenschaft und -technik*, vol. 1, (Berlin: Reichsdruckerei 1941), 463–561.

24 Margit Szöllösi-Janze, *Fritz Haber 1866–1934. Eine Biographie* (Munich: Beck, 1998), 15.

25 See Helmuth Trischler, "Im Spannungsfeld von Individuum und Gesellschaft. Aufgaben, Themenfelder und Probleme technikbiographischer Forschung," in Wilhelm Füßl and Stefan Ittner (eds), *Biographie und Technikgeschichte* (BIOS, 11 (1998), special issue), 42–58.

26 Thomas Mann, *Betrachtungen eines Unpolitischen* (Frankfurt am Main: Fischer, 1995), 13, 14.

27 Stefan Zweig, *Die Welt von Gestern. Erinnerungen eines Europäers* (Stockholm/Frankfurt am Main: Fischer Sonderausgabe 1944/1992), 14.

28 David C. Cassidy, *Werner Heisenberg. Leben und Werk* (Heidelberg: Spektrum Akademischer Verlag, 1995), 30. Also see the collective biography from Jonathan Harwood: "Mandarine" oder "Außenseiter"? Selbstverständnis deutscher Naturwissenschaftler, 1900–1933," in Jürgen Schriewer, Edwin Keiner and Christophe Charle (eds), *Sozialer Raum und akademische Kulturen. Studien zur europäischen Hochschul- und Wissenschaftsgeschichte im 19. und 20. Jahrhundert* (Frankfurt: Peter Lang, 1993), 183–212.

29 Lucie Varga, *Zeitenwende. Mentalitätshistorische Studien 1936–1939* (Frankfurt am Main: Suhrkamp, 1991), 120.

30 Notker Hammerstein, *Die Deutsche Forschungsgemeinschaft in der Weimarer Republik und im Dritten Reich. Wissenschaftspolitik in Republik und Diktatur 1920–1945* (Munich: Beck, 1999), 28.

31 See Cassidy, *Heisenberg*, 89–94, 121.

32 Ludwig Bölkow, *Erinnerungen* (Munich: Herbig Verlagsbuchhandlung, 1994), 17.

33 Neufeld, *Rocket*, 8.

34 Mark Walker, "Legenden um die deutsche Atombombe," Vierteljahrshefte für Zeitgeschichte, 38 (1990), 45–74; Herbert Mehrtens, "Mathematics in the Third Reich: Resistance, Adaption and Collaboration of a Scientific Discipline," in R.P.W. Visser et al. (eds), New Trends in the History of Science (Amsterdam: Rodopi, 1989), 151–66; several contributions to Monika Renneberg and Mark Walker (eds), Science, Technology and National Socialism (Cambridge: Cambridge University Press, 1994) and Szöllösi-Janze, Science.

35 Here see the contributions in Wolfgang Späte (ed.), Testpiloten (Planegg: Aviatic Verlag, 1993).

36 Gerhard Neumann, China, Jeep und Jetmotoren. Vom Autolehrling zum Topmanager. Die Abenteuer-Story von "Herman the German", eines ungewöhnlichen Deutschen, der in den U.S.A. Karriere machte (Planegg: Aviatic Verlag, 1989), 27.

37 See Helmuth Trischler, "Self-Mobilization or Resistance? Aeronautical Research and National Socialism," in Renneberg and Walker, 72–87.

38 Trischler, Luftfahrtforschung, 199–207.

39 Ian Kershaw, The Nazi Dictatorship: Problems and Perspectives of Interpretation, 3rd edn (London: Arnold, 1993).

40 Green et al., Wartime, Part IV (January 1949), 29.

41 "Report on Trip to Germany and Russia, September-October 1936," NACA, File 521, (NA) Record Group (RG) 255, National Archives.

42 Theodore von Kármán, Die Wirbelstrasse. Mein Leben für die Luftfahrt (Hamburg: Hoffmann und Campe, 1968), 327.

43 Theodore von Kármán, Where We Stand: A Report of the AAF Scientific Advisory Group, August 1945 (Ohio: Air Materiel Command, May 1946), 16.

44 The significance of Wernher von Braun and Huntsville has been often discussed in the literature. In contrast, the significance of the Arnold Engineering Development Center has hardly been noticed. For this reason, the following section will concentrate principally on the AEDC in Tullahoma/Tennessee and will handle Huntsville only briefly.

45 Die Tätigkeit deutscher Luftfahrtingenieure und –wissenschaftler im Ausland nach 1945 (Bonn-Bad Godesberg: Deutsche Gesellschaft für Luft- und Raumfahrt e.V., 1992), 22; "Australien warb Kriegsverbrecher an," Tagesspiegel (17 August 1999). For Huntsville see Harro Zimmer, Das NASA-Protokoll. Erfolge und Niederlagen (Stuttgart: Franckh-Kosmos-Verlag); Neufeld, Rocket, 270–1; Weyer, Braun, 88–134; Heinrich Schiemann, Erlebte Raumfahrt. Schauplätze und Begegnungen(Frankfurt am Main: Umschau Verlag, 1991), 76–81; Walter A. McDougall, The Heavens and the Earth: A Political History of the Space Age (New York: Basic Books, 1985), 119–23, 196–97 etc.

46 AEDC Staff and Office of Public Affairs (ed.), "Arnold Engineering Development Center. An Air Force Systems Command Test Facility" (Arnold Air Force Base, Tennessee: n.d.). It should be noted here that this Group sent out other important and consequential "memos" within the U.S.A. In May, 1945, Georg Schairer, aerodynamics expert and subsequently vice president at Boeing, sent a telegram letter to Seattle, in which he pushed for the project running at that time of a jet bomber from Boeing. This was the basis for the subsequent economic success of the company as the market leader in military and civilian large-scale jet planes; see John D. Anderson, A History of Aerodynamics (Cambridge: Cambridge University Press, 1998), 425.

47 Speech President Truman, Box 11.1, National Air and Space Museum (NASM).

48 AEDC Staff and Office of Public Affairs, "Chronological History Arnold Engineering Development Center," U.S.A.F., Box 88.1, NASM; Summary of the Report of the President's Air Policy Commission, Box 189, File 521, NA, RG 255; Abbendum to Short History of Unitary-Wind-Tunnel-Plan, Box 189, NA RG 255.

49 Burghard Ciesla, "German High Velocity Aerodynamics and their Significance for the U.S. Air Force 1945–1953," in Matthias Judt and Burghard Ciesla (eds), *Technology Transfer out of Germany after 1945* (Amsterdam: Harwood Academic Publishers, 1996), 93–106, here 100.

50 The first groups of German specialists were brought overseas in 1945/6 in the context of the secret operation "Overcast," which on September 3, 1946 was modified further. The designation "Paperclip" came from the marking of the personnel files of the selected German scientists and technicians with a paperclip. For more information on the origins and operations of this program and more background, see Clarence G. Lasby, *Project Paperclip: German Scientists and the Cold War* (New York: Atheneum, 1975); Tom Bower, *The Paperclip Conspiracy: The Battle for the Spoils and Secrets of Nazi Germany* (Boston: Little, Brown, 1987); John Gimbel, *Science, Technology, and Reparations. Exploitation and Plunder in Postwar Germany* (Stanford, CA: Stanford University Press, 1990); Linda Hunt, *Secret Agenda. The United States Government, Nazi Scientists, and Project Paperclip, 1945 to 1990*(New York: St. Martin's Press, 1991); Burghard Ciesla, "Das 'Project Paperclip' – deutsche Naturwissenschaftler und Techniker in den U.S.A. (1946 bis 1952)," in Jürgen Kocka (ed.), *Historische DDR-Forschung. Aufsätze und Studien* (Berlin: Akademie Verlag, 1993), 287–301; Burghard Ciesla, "'Intellektuelle Reparationen' der SBZ an die alliierten Siegermächte?" in Christoph Buchheim (ed.), *Wirtschaftliche Folgelasten des Krieges in der SBZ/DDR*, (Baden-Baden: Nomos Verlag, 1995), 97–109; Judt; Ciesla.

51 Up until January 31, 1946 129 specialists came to the U.S.A. under "Overcast." Around 80 percent of these belonged to the rocket researchers around von Braun, who were supervised by the U.S. Army. Two years later, the picture was very different. Of the 500 specialists, around 40 percent worked for the Air Force, 34 percent for the Army, 16 percent for the Navy, and 10 percent were the responsibility of the Department of Commerce. At this time, the responsible authorities also discussed sending the Germans back home, because their knowledge of new types of weapon technologies from the war had been "used up." Moreover, the Germans were merely "guests" of the American military who had gotten around the immigration laws of the U.S.A. However, the fact that the Germans had gained sensitive knowledge of the American circumstances spoke against sending them back. There was the concern that, upon their return, they might report on their experiences to the "other side." Thus in 1948 it was decided to make the approximately 500 German specialists and their approximately 1,200 dependents American citizens. A report from the Joint Intelligence Objective Agency (J10A) ran as follows: "The only way to prevent this experience and information from falling into the wrong hands is to encourage these Germans to become American citizens and to remain here the rest of their lives. Their supervisors feel they are learning to like America and the American way of life." By 1950 around 90 percent of the 500 experts had received the official "resident alien" status which allowed them to become American citizens after five years. Memorandum Joint Chiefs of Staff (JCS) for Director, JIOA, Subject: "The Kochel Wind Tunnel," (4 May 1948), File No. 1001, Record Group 319, Army–Intelligence, National Archives, Washington, D.C.; Ciesla, "Paperclip," 296–9.

52 "Statistical Report of aliens brought to the United States as "Paperclip", "Project 63" and "National Interest Cases", JCS, JIOA (1 July 1953). This document comes from the National Archives in Washington, D.C., and was generously provided to the authors by Jürgen Ast, director of an actor in the television documentary, "Aus der Hölle zu der Sternen" (Mitteldeutscher Rundfunk, 1993) on the rocket from Peenemünde.

53 The contribution of the Germans to the construction of the bomb was mainly limited to solving individual problems or answering questions in the area of basic research. In contrast, atomic espionage did play a large role in Los Alamos. For the actual contribution of the German atomic specialists and of atomic espionage, see Ulrich Albrecht, Andreas Heinemann-Grüder and Arnd Wellmann, *Die Spezialisten. Deutsche Naturwissenschaftler und Techniker in der Sowjetunion* (Berlin: Dietz Verlag, 1992) 48–82; Wladimir Tschikow and Gary Kern, *Perseus: Spionage in Los Alamos* (Berlin: Volk und Welt, 1996); David Holloway, *Stalin and the Bomb: The Soviet Union and Atomic Energy 1939–1956* (New Haven, CT: Yale University Press, 1994); and Pavel Oleynikov, "German Scientists in the Soviet Atomic Project," *The Nonproliferation Review* (Summer, 2000).

54 See in particular Stuart W. Leslie, *The Cold War and American Science: The Military-Industrial-Academic Complex at MIT and Stanford*(New York: Columbia University Press, 1993).

55 Letter from the American Ambassador to the Federal German Republic James Conant to the Secretary of State John Foster Dulles, American Embassy Bonn/Bad Godesberg (13 July 1956). The copy of this document comes from Jürgen Ast. This document is also cited by Hunt, *Secret Agenda*, 1 and 270.

56 However, most came with "Project Paperclip" to the U.S.A. For example, in July of 1953 the "Paperclip" experts made up more than 80 percent of the total number of Germans brought to the U.S.A. under "Paperclip," "Project 63," and "National Interest Cases." From the Statistical Report, 1 July 1953; Letter from James Conant (13 July 1956).

57 This example is based on the following: Burghard Ciesla, "'Tassos Rundbriefe' aus dem Land der Autos: *Auto*-mobile Kulturerfahrungen einer deutschen Ingenieurfamilie in der Neuen und Alten Welt," in Michael Wala and Ursula Leimkuhl (eds.), *Technologie und Kultur. Europas Blick auf Amerika vom 18. bis zum 20. Jahrhundert* (Cologne; Böhlan Verlag, 2000), 173–201).

58 Tasso Proppe, *Tassos Rundbriefe 1952–1987* (Munich 1990). The collection of the circular letters were printed for relatives and friends on the occasion of Proppe's eightieth birthday. We are grateful to Wilfried Ennenbach for sending us a copy.

59 Up until the beginning of the 1970s, the letters were first sent to Braunschweig, and then later to the German Society for Aeronautic Research. Beginning in 1963 the Proppes reproduced the letters themselves and sent them out.

60 *Tassos Rundbriefe*, b.

61 The time in New Mexico was exemplary in this regard, where the Proppes lived in a German community. Proppe wrote about this in February, 1954: "The Germans, around thirty scientists, work almost as a closed working group in one place. This does not make matters more pleasant, but cannot be changed. The Germans are very well respected, but, it seems to me, not particularly liked, because they are so refined and difficult to approach. The families have little contact with the outside world. *Tassos Rundbriefe*, No. 8, 6 February 1954, 75. The daughter, born in 1957 in Alamogordo, of the scientist Harald A. Melkus, who worked at the Holloman Air Force Base, described the community even more clearly: "Alamogordo was a town of barely 17,000 people. It was a typical southwest town with a large Spanish-speaking sector. As those earlier years of my childhood went by, more and more German families moved to the area. These German families were all very close friends and many of which knew one another from previous places that they had lived. With time, there became a very large German community living within a predominantly American-Mexican community. As for the children of these German families, we viewed life in Alamogordo as being almost totally surrounded by other Germans. My best friend throughout my childhood was German, and even though my brother and sister were almost eight and six years older than I; each of their best friends were also

German. Each of us had best friends from different families, yet each of those families were German. Therefore, all of us grew up with the same German traditions, customs and values in a very protected environment. This German community seemed to do everything together; we all went to the same church, same German delicatessen, same piano and ballet instructors, and on the same picnics together to White Sands and Cloudcroft. But perhaps what I remember most, were the many parties my parents had for their German friends."

62 Here also see Peter Lösche, *Die Vereinigten Staaten. Innenansichten. Ein Versuch, das Land der unbegrenzten Widersprüche zu begreifen* (Hannover: Fackelträger-Verlag, 1997), 11–18.
63 *Tassos Rundbriefe*, Im Rückblick, 433.
64 *Tassos Rundbriefe*, No. 11 (October 1955), 123–24.
65 See Lösche, *Innenansichten*, 13–14.
66 Here see Renneberg and Walker; Museum für Verkehr und Technik, *Ich diente nur der Technik. Sieben Karrieren zwischen 1940 und 1950* (Berlin: Nicolaische Verlagsbuchhandlung, 1995).
67 *Tassos Rundbriefe*, No. 19, December 1961, 263.
68 *Tassos Rundbriefe*, Im Rückblick, 434–35.
69 Ferdinand Tönnies, *Gemeinschaft und Gesellschaft. Grundbegriffe der reinen Soziologie* (Darmstadt: Wissenschaftliche Buchgesellschaft 1963).
70 See in particular Daniel L. Kleinman, *Politics on the Endless Frontier: Postwar Research Policy in the United States* (Durham, NC: Duke University Press, 1995); G. Pascal Zachary, *Endless Frontier: Vannevar Bush, Engineer of the American Century* (New York: Free Press, 1997); David M. Hart, *Forged Consensus: Science, Technology, and Economic Policy in the United States, 1921–1953* (Princeton, NJ: Princeton University Press, 1998).

8

WEAVING NETWORKS

The University of Jena in the Weimar Republic,
the Third Reich, and the postwar East German
state

Uwe Hoßfeld, Jürgen John, and Rüdiger Stutz

This chapter portrays a German university in the twentieth century whose
confused path of development and problematic circumstances during this
"Period of the Extremes"[1] are in many respects exemplary for the entire
university system in Germany. In particular, its history under the Swastika
and four-decades-long domination of the SED (*Sozialistische Einheitspartei*,
"Socialist Unity Party," or Communist Party in East Germany) touches the
nerve of their institutional self-understanding as a traditional educational
institution of European rank. After the upheavals of 1945 and 1989/90,
several generations of scientists at the Friedrich Schiller University were
confronted by critics questioning their life's work as university teachers,
although up until that time they had been esteemed in the scientific community. This resulted not least because of an abrupt paradigm shift in the
extremely sensitive value structure of science and politics. Only a minority of
the instructors and students thrown out of the universities in the Soviet
occupation zone had at their disposal what Mitchell Ash has described: the
ability of a few top researchers to build up new connective structures to politics in the postwar years, and to "construct continuities" in their career
paths.[2]

The dissolution of twelve departments at the University of Jena at the end
of 1990 did not come off much less dramatically, for this first phase of the East
German university renewal was accompanied by moral verdicts, political
barriers, but also painfully embarrassing attempts by the academics concerned
to justify themselves. This recent emotionally loaded process of "mastering
the past" at the University of Jena has up until now been handled only slug-
gishly and descriptively.[3] Thus the attempt will be made first of all to measure
the historical and semantic field in which the symbols and concepts of the
political foreign perception and institutional self-attribution of this time-
honored university in the twentieth century are "rooted."

186

With around 16,500 students, today the Friedrich Schiller University Jena belongs to the mid-sized universities in the Federal German Republic. As an educational institution[4] established in 1548/58, it embodies the traditional Thüringen State University.[5] The small federal state of Thüringen today also finances the Bauhaus-University Weimar, the Technical University Ilmenau, and the University of Erfurt[6] currently under (re)construction. Between 1945 and 1990, this region belonged to the eastern part of the four zones of Germany or to the German Democratic Republic (GDR). Like the other GDR universities,[7] after the collapse of this small German state in the course of 1990 the University of Jena was a part of the painful process of transformation into the Federal German university system. Since then the University has been the centerpiece of the regional capital, Jena, which has 100,000 inhabitants.

Alma Mater Jenensis or Salana sees itself today as an innovation-friendly traditional university which also does not want to hide the darker side of its long and, especially in the twentieth century, contradictory history.[8] It has in no way merely been drawn into the disputes and confusion of state politics and German international claims; during the two world wars, most of the university teachers and researchers placed themselves without reservation at the service of the state. However, just as chaotic conditions in general can produce new structures of order, after the two catastrophes of war this century the university also offered considerable opportunities for carrying through alternative plans for reform. Moreover, in this context not only were new university models discussed precisely in phases of upheaval and transition, but fundamental changes in social policy were also called for.

Thus in the following portrait of a "German university" during the twentieth century, three exemplary examples of such an extremely ambivalent conception of modernity flow together, closely connected with the names of the Jena rectors Abraham Esau, Karl Astel, and Franz Bolck. These university representatives and their respective political supporters understood themselves as making scientific contributions which extended far out into the society, towards "mastering the crisis" in the period of the second industrial revolution or for the "planned further development" of state socialism in East Germany as an alternative society to capitalist modernity. Through a view "from below," the perspective will be directed towards ambiguous connections between science and politics, which will be subsumed in what follows under the concept "network."

The usual administrative work contacts of the university to the state administration or party offices will be considered in this context, but so also will the hidden "connections" between science and politics. In this area, such informal networks can function as a transmitter of interests, but under certain conditions they can also act in the sense of a dampening "buffer." This methodological approach should avoid a schematic-dualistic consideration of university history during the "epochs of ideologies." This type of interpretation usually presupposes a political "external pressure," which "burdened" the university by treating it merely as a political object and led to all kinds of "deformations."

The politicization of science stands without question as an often described fundamental tendency of modern society of the twentieth century. But it appears worthwhile to examine more concretely the transformation unleashed among the students and especially among the faculty: how did the self-understanding of university teachers and students change inside the microcosm of the University of Jena? Which accommodation strategies developed inside the profession under the different political regimes? Which trend dominated, binding oneself in the new scientific-bureaucratic network, or detachment from politics?

This period of investigation, ranging between defeat in the world war and revolution in 1918–19, and the late 1960s, makes especially clear how explicitly barbaric and civilized tendencies paralleled each other at the University of Jena in the prototypical realization of this vision of modernity. At the end of the Second World War and after the military destruction of National Socialism, 1945 also represented the low point, where German academia stood before the shards of its actions and inactions and had to take new paths under different occupation regimes.

1945: The "Spirit of Weimar" and the "Myth of Jena"

As bands of U.S. troops liberated and occupied[9] Thüringen at the beginning of April, they moved into an area full of symbolic places extremely rich in contrasts for the history of German education, culture, and politics. Here lay Wartburg with its national and cultural symbolism, the double city Weimar-Jena[10] as the embodiment of the humanistic Germany of literary and philosophical classics; but also the Buchenwald concentration camp as the "gruesome negation of classical Weimar."[11] This incarnation of Nazi barbarism in the neighborhood of the classical places overshadowed the spirit of Weimar, which the protagonists of a spiritual new beginning referred to in 1945. In this regard they overlooked the fact that classical Weimar, as the would-be heart of German culture, had attracted Völkisch-nationalist movements since the later part of the German Empire.[12]

The National Socialist project of a "Third Weimar" was able to make a connection to these by claiming an alliance with the old cultural élites in order to build a new classical age based upon a politically, racially, socially, and cultur-ally "cleansed" "People's Community" (Volksgemeinschaft).[13] The symbol and guarantor of the established social discipline of the National Socialist society, secured by terror, was the Buchenwald concentration camp with its sarcastic motto *to each his own*. "Between us and Weimar lies Buchenwald," wrote the exiled germanist Richard Alewyn in 1949 about those German intellectuals and academics who, after the end of the NS regime, had been using Goethe as an alibi,[14] clearly separating Weimar and Buchenwald from each other, erasing the multiple references to this fatal proximity, and thus painting the legend of an unblemished classical city.[15]

In particular, in 1945 the humanistic intelligentsia invoked the "spirit of Weimar" in order to remove the universities, which supposedly had at their heart remained pure,[16] from the pressure for denazification by the occupying powers. They also sought to legitimize themselves as the intellectual elite of the new beginning by using the myth of the "spirit which had remained pure" of the other Germany of poets and thinkers.[17] These calls for the "spirit of Weimar" also referred to the situation after the First World War and the revolution of 1918/19. At that time Weimar was the place where the Weimar Republic was founded, and was represented by the Weimar Bauhaus and the reform-intensive state of Thüringen.

In 1945 people rightly argued that they could reconnect to this legacy, but they also repressed unpleasant things from their collective memory. After 1919/20 the system of democratic political parties, Bauhaus *avant-gard*, and the reform politics of the Thüringen state government found little support among the established social and intellectual-scientific élites. Instead, they supported the unholy alliance of a defensive front ranging from the honorable political middle to the radical right.[18] During the middle of the 1920s the state of Thüringen and its capital Weimar underwent a consequential political change of scene, degenerating into an experimental field for National Socialism.[19]

After 1933 the NSDAP covered their "exemplary region" of Thüringen[20] with a dense net of concentration and forced labor camps, armaments firms, and manufacturing plants of the most modern military technology. Because of them and Fritz Sauckel, regional leader of the NSDAP who rose in the ranks to become Plenipotentiary for Labor Mobilization and a taskmaster for Europe (*Fronvogt Europas*),[21] this region, respected for its cultural and educational intensity, was discredited worldwide. After the American bombing of February–March 1945, large portions of the classical and educational institutions in Weimar and Jena lay in soot and ashes. Only with difficulty could the coffins of the Weimar poets Goethe and Schiller, evacuated to Jena, be spared from the destruction ordered by the Police President.[22]

Because of this situation and its contradictions, the Americans and the Soviets, who followed after the change in occupation power at the beginning of 1945, could quickly get an understanding of conditions from the outside, but only with difficulty could they see the internal state of the Thüringen state university. "What did we know of the university of Jena?" In his memoirs, the administrative head of the Soviet military administration in Thüringen noted the honorable age, the legacy of Goethe, Schiller, Hegel, Fichte, Schelling, and Haeckel, the Ph.D. of Marx in Jena (1841), the reforms after 1918 – and that the University of Jena after 1933 was caught in an especially strong National Socialist grip.[23]

Measured by its tradition and reputation, the University of Jena was in no way merely one university among many, which in 1945 under Allied control overcame its NS past and had to make a new beginning. Jena was one of the older and best known German universities and enjoyed an international reputation as a center for classical German philosophy with a close connection to Weimar. Here stood the cradle of the student fraternities with their initial ambitions for freedom and democracy. The University of Jena carried the name of Friedrich Schiller, who functioned here as a history professor, and owed its classical heyday not least of all to the official efforts of Goethe. Thus its history was bound up with precisely these names which in 1945 stood for the other, better Germany of poets and thinkers. Thus the "Myth of Jena" inclined the officers of the Soviet occupation force to open the university as quickly as possible.[24]

Of course, the shadow of the National Socialist period and its prehistory darkened everything. The Thüringen state university did not receive Schiller's name in the reform period after 1945; rather the corresponding official papers were first handed over on November 10, 1934 by the National Socialist Minister of the Interior and Popular Education in Thüringen, Fritz Wächtler. The National Socialists thereby solidified their claim on the classical traditions.[25] Previously Thüringen had twice caused a national sensation: in the first "Thüringen University Conflict" during the years 1922/3, a majority of University of Jena faculty had repudiated the initiative-rich reform politics of a leftwing Social Democratic-led state government; and in 1930, against the expressed will of the rector and University Senate, the writer Hans F.K. Günther was appointed professor for social anthropology by the State Minister for Popular Education, Wilhelm Frick (NSDAP).

On the surface, both conflicts were over the autonomy of the university. In fact, this did reflect the intellectual situation of this and other German universities, and the change in environment and coordinates caused by the political culture of the Weimar Republic. As in the entire university system, only a small minority of the Jena professors and students supported the new republican democracy. The overwhelming majority either stood aloof from or opposed it, either still affected by the Wilhelmian Empire or fixated on a new "Third Reich." Students and faculty thereby politicized themselves in the pose of a supposed apolitical science, contributed to the destruction of this republic, and – willingly or unwillingly – prepared the ground for the National Socialists.

The question so pressing in 1945, where the intellectual Germany stood before and after 1933, therefore was directed mainly toward the intellectual élites in the universities and the hierarchical-oligarchicly organized universities. Examining the total development of German history since the Empire, the full professors (*Ordinarien*) appear to their critics to have been structurally incapable of democracy and to be just as responsible for the German disaster, as the "Spirit of Potsdam" superimposed over the "Spirit of Weimar."[26] This also cast doubt on the "Myth of Jena" and the metaphor with which the Thüringen State President Rudolf Paul reopened the University of Jena on October 15, 1945, "Only from the intellectuals can the rebirth come," thereby alluding to its history molded by crisis, upheaval, and new beginning.[27]

"Crisis and awakening": the structural genesis of the Thüringen state university

Among the older German universities established before 1800, the University of Jena belongs to the middle generation between the late middle ages[28] and those founded under the banner of the Enlightenment of the seventeenth and eighteenth centuries.[29] It came from the early modern period of humanism, the Reformation, the split into denominations. Territorial states now depended on the system of state churches, with their rapidly growing demand for academically trained civil servants, theologians, and lawyers. There was a close connection between theological and legal thought, as well as a general explosion in education[30] through printed books and publishing. After Marburg (1527) and Königsberg (1544), the University of Jena, which originated in 1548 as the high school of the small state Saxony-Weimar and received the Imperial University privilege in 1558, is considered a classical Reformation university. Even the conditions of its founding, following the war lost in 1547 by the Protestant Schmalkaldisch Federation, illustrate the connection between crisis, awakening, and new beginning so characteristic of its subsequent history.[31]

The new educational institution quickly became the structurally determining factor of the city, which had around 7,000 inhabitants at that time and which largely followed the lead of the university and became an important place for publishing houses. The University of Jena cast itself in the denominational political conflicts of the time first of all as a refuge of Lutheran orthodoxy theology. The early Enlightenment influence of the universal scholar Erhard Weigel (1653/99), began its rise to a philosophical university, which around 1800 brought it in close connection with classical Weimar, and to the pinnacle of the German universities and intellectual evolution[32] at that time.

Seen structurally, the University of Jena originated and developed as a university molded and supported by a small city in Saxony-Weimar and its four subsequent successor states (the last in 1918). This small and multiple-state affiliation had advantages and disadvantages for the University of Jena: on one hand there was a permanent financial need, but on the other hand there was considerable academic freedom, a policy of appointing controversial scholars who were taboo at other places such as Schiller, Fichte, or Hegel, and a certain liberalism lasting until the political ice age of the German Confederation, which made possible the student Wartburg Festival in 1817 as well as the fraternities and their intellectual mentors Heinrich Luden, Lorenz Oken, and Jakob F. Fries. The revolutionary situation after 1800, accompanying and forced by the Napoleonic expansion, and the reform founding of the University of Berlin in 1810, began the period of the modern type of university molded by Humboldt in the nineteenth and twentieth centuries, based on the neo-humanistic ideal of education, the unity of teaching and research, and the transformation to a state university.[33]

For the older German universities, this meant a deep transformation crisis. Many of them failed, including the University of Erfurt, which belonged to Prussia. Jena, which initially stood in the center of this process and was directly affected by the war in 1806, survived this university die-off and the contemporary process of concentration. In this context Jena embodies 1806 – the year of the Prussian defeat near Jena and Auerstedt, the collapse of the "Altfriderizianischen" state and the Prussian reforms – a renewed symbolic date for the connection between crisis and new beginning.[34] After a painful accommodation process, the University of Jena participated in the development trends. In the revolution of 1848, it stood in the center of German efforts at university reform.[35]

After 1850 it profiled itself, in the context of a still quite modest city of around 6,500 inhabitants, as a university suitable for the advancement of academic careers. A skillful promotion policy allowed it to bring significant scholars to Jena, whose names and research schools[36] once again helped Jena to an international reputation. As a traditional university with fraternities and with a Thüringen environment similar to Heidelberg, Jena attracted the increasingly elitist-nationalistic student corporations. After the turn of the century, a free student movement[37] took shape in opposition, but this movement first reached its goal of being a general student organization only after the revolution of 1918.[38]

A new phase in the structural development of the University of Jena began in conjunction with the development of big business in Jena in the 1870s and continued until the 1890s. The Carl Zeiss Foundation, created by the Zeiss and Schott Works and the industrial physicist Ernst Abbé in 1889–96, brought considerable resources to the financially strapped university.[39] This classic enterprise of the so-called science-intensive "new industries" acquired increasing significance for the University of Jena as its "fifth presentation state" (5. *Erhalterstaat*). This rather rare connection between university research and industry in this form and at this time gave the University of Jena its characteristic stamp from this point on, through to the end of the GDR.

In this climate of modernization a connective nexus developed between industry and the university, which was able to foster a specific work culture to attract new, important publishers (Gustav Fischer, Eugen Diederichs), and justifiably gave Jena the reputation of an *avant-garde* cultural and artistic city in the provinces.[40] In this regard, individuals at the university distinguished themselves, especially personalities from the Abbé circle like the lawyer Eduard Rosenthal and the physicist Felix Auerbach. This new milieu naturally demonstrated sharp contrasts with Jena's honorable-municipal, liberal-industrial, and conservative-university submilieus. Thus it resisted for a long time the political movements of "free students" (not in corporations) and assistant or associate professors, as well as the emancipation of women.[41]

The structural transformation quickly increased the number of inhabitants and students.[42] At the university, this transformation yielded a considerable expansion of subjects and process of differentiation, especially in the medical[43] and mathematical-scientific fields. The first courses were offered in new economic and social studies disciplines. Jena was thereby belatedly following the general trend in academia, but it also set off down new paths. Thus, for example, an associate professorship (*Extraordinariat*) for ear medicine (Eugen Weber-Liel) was established in 1884, an Institute for Psychiatry (Otto Binswanger) in 1891 and an Institute for Pediatric Medicine (Jussuf Ibrahim) in 1917 in the medical faculty. The Institutes of Technical Physics and Applied Mathematics (1902, Rudolf Ran), and Mathematics (1879, Johannes Thomar; 1898, August Gutzmer), and the associate professorships for scientific microscopy (1899, Hermann Ambrom) and theoretical physics (1889, Felix Auerbach) were established in the steadily expanding physics faculty. In 1865 zoology received a full professorship (*Ordinariat*) through Ernst Haeckel, and astronomy an associate professorship (Otto Knopf) in 1897.[44]

Paralleling the clear dominance of scientific subjects, around 1900 Jena began to define itself as a philosophical university with a strong scientific emphasis. Interdisciplinary circles were effective in this regard, like the "Lecture Evening" (*Referierabend*) founded in 1871 by the botanist Eduard Strasburger and the zoologist Ernst Haeckel and bound by the common umbrella of the philosophical faculty, which now became the "heart of the university."[45] In particular, individual personalities determined the climate of action of this Jena microcosm, which around 1900 sought philosophical answers to the crisis of modernity and thereby became the point of intersection for very different, in their concepts and effects very ambivalent intellectual currents.[46]

This was especially true for the zoologist Ernst Haeckel, thanks to whose efforts Jena became a Mecca "to where all zoologists make a pilgrimage"[47] and the *Genius loci Jenas* expressed in the following: "no other German university can credit itself with achieving so much and such great things with such skimpy means. Meanwhile exactly the material constraints of the empirical observation materials had their advantages, in which they stimulated the researching spirit to philosophical reflection."[48] Haeckel's reputation was founded, along with his later works on aesthetics in nature and on monism (*Art Forms in Nature*, 1899–1904; *The Riddle of the Universe*, 1900; *The Wonders of Life*, 1904), especially because of his service in popularizing Darwin's theory of evolution in Germany.[49] In this way, and with the German Monist League he helped found (1906), Haeckel proved to be a trailblazer of the Free Thinker Movement, and also of Social Darwinism and Racial Hygiene[50] – a tradition which was then carried on with an equally influential and consequential effect in Jena by his successor Ludwig Plate.[51]

The effect of Haeckel's intellectual adversary, the influential philosopher and Nobel Laureate for Literature (1908) Rudolf Eucken, was also very ambivalent. He was active as a philosopher in the media market, as a wandering preacher of the "Spirit of 1914" and finally after 1918 as the prophet and intellectual leader of the Eucken League (1920).[52] The other representatives of the philosophical climate of action in Jena – the pedagogue Wilhelm Rein, the mathematician and logician Gottlob Frege, or the pedagogue and philosopher Herman Nobel, at that time standing at the beginning of his career – proved to be excellent, pathbreaking scientists, whose penetrating thought produced very contradictory effects on the political culture of the later German Empire and the Weimar Republic. All of these processes still took place within the structural bounds of the University of Jena, supported by a small state, with the Weimar Grand Duke as nominal rector. The structural consequences first became clear after the "fundamental catastrophe" of the century, the First World War, the revolution which ended the monarchy, the creation of the Weimar Republic, and the establishment of the new state government in 1920.

The Weimar period

The First World War (1914–18) and the Revolution left deep marks on the university and scientific landscape at the University of Jena. Around 500 members of the university community, 27 percent of those present in 1914, were victims of the First World War. From the beginning there were different views of the First World War within the communities of students and professors, whereby the nationalist conservative forces held the upper hand. The historian Alexander Cartellieri, the zoologists Ernst Haeckel and Ludwig Plate, and the philosopher Rudolf Eucken, as well as such organizations and publications like the *Archive for Racial and Social Biology*, co-founded by Plate in 1904, the German Monist League, founded by Haeckel in 1906, the "Society >German State<" founded by Max Wundt, and the Eucken League founded in 1920 stand as prototypes of the political-ideological intellectual stance of the majority of the Jena professors *circa* 1918, who conceived the war as the "emergency defense" of Germany. In this tenor Plate remarked: "Thus it was self-evident that politically I had always stood on the right."[53]

Moreover, numerous Jena scientists signed the first "Professors' Manifesto" (1915), which included over 1,300 intellectuals, industrialists, bankers, and soldiers and was designed to justify Germany's actions in the First World War. In contrast, the Social Democrat and biologist Julius Schaxel[54] was sympathetic to the postwar revolution and years later vehemently fought against National Socialism and the ideas of the "Race-Günther."[55] Years earlier, the mathematician and physicist Karl Snell had emerged as a passionate opponent of Bismarck, but after the turn of the century these social-liberal traditions had faded more and more from of the University of Jena.[56]

After the armistice and the storm of revolution, the university faced grave problems in 1918/19. For one thing, the men returning home from war had to be reintegrated into the courses of study, so that the number in Jena swelled from an original 600 to 3,000. Moreover, individually they had their experiences in the trenches of a modern war to work through. Despite the Weimar Republic, most of them did not want to abandon the intellectual baggage of the fallen monarchy. In addition, during the postwar years the chauvinism of the war discharged in fashionable pseudo-scientific currents, which can be seen as a precursor to the subsequent racial-anthropological and eugenics movements. In particular, these currents circulated broadly among intellectuals and, thanks to the statements of Haeckel and his successor Plate, had a political significance which beyond Jena.[57]

Since 1919 the support of the social and economic sciences, as well as a reform of legal training, stood in the foreground of national efforts to push through an adaptation of the university administrative structures and curricula in order to catch up to international standards. These legislative initiatives from Prussia accelerated the efforts of the economic section at the University of Jena to establish a new profile, which was not least due to the economist Julius Pierstoff. The close connection of the economics curriculum to the practice he advocated corresponded to the founding of the Institute for Economic Law, which cooperated closely with the Zeiss Works and was led by Justus Wilhelm Hedeman, the first full professor for economic law in Germany. Hedeman also led a reform committee, which broke new ground in reforming the university constitution: the faculty reform of September 24, 1919, the Senate reform of July 25, 1920, and a decree on students' rights of August 23, 1920. By this last reform, the state of Thüringen legally recognized the general students committee which had been constituted in the revolutionary autumn of 1918.

However, these movements towards democratization were delayed by an overdue structural adaptation of the university. Only when the well-advanced project for establishing a Thüringen University for Economics, Administration and Technology in Eisenach became known – with the threat of competition thus before their eyes – did the restructuring of the University of Jena advance on October 20, 1923 with the formation of a legal and economics faculty. In the philosophical faculty, a majority of the full professors had resisted all suggestions of structural reform and had asserted the right of their faculty to stand "universally" above all the other divisions of the university. The spectrum of ideological stances in this large faculty ranged from Paul F. Finke, a philosopher who belonged for a time to the SPD, to his disciplinary colleague Max Wundt, a young conservative advocate of "Völkisch philosophy," to the founder of "Critical Pedagogy," Eberhard Grisebach, and finally to the German nationalist who succeeded Haeckel, the zoologist Plate.

Of course, the tone was set by those professors who either failed to understand or rejected the left-wing Social Democratic inspired university politics of the Thüringen Minister for Education, Max Greil. This repeatedly slandered education reformer was accused of trampling on the university's right to self-government. In fact, together with the development biologist Julius Schaxel, Greil wanted to find a place for applied social sciences in the philosophical faculty. Open conflict erupted between the leadership of the faculty and university and the state government when a psychological institute was created under the leadership of the Social Democrat Wilhelm Peters; it peaked in the appointments ordered by the government of a series of pedagogues to professorships, including Mathilde Vaerting, the second woman to hold a professorship in Germany. This "Thüringen University Conflict" gained national attention and ended with the dissolution of the Education Science Institute, founded just a few months before.

However, the University of Jena remained, both up until 1933 and in the second half of the 1940s, a very influential center for pedagogical reform and (social) psychology. The school model of Peter Petersen in particular gained international significance, although in the 1930s he began to support his pedagogical theories with *Völkische* concepts.

The structural renewal of the university did not come to a preliminary conclusion until after the political and economic crisis of the year 1923/4 with the founding of a Mathematical-Scientific Faculty on April 1, 1925. The sciences, beginning first of all with physics, as well as pediatrics and neurology, developed into the most important area of innovation at the University of Jena. Bolstered by an initially loose connection with the optical-precision mechanics industry and its technologies, this new faculty was molded by demands for a technical-scientific special education. This did contradict the university ideal of the German tradition of education, but was strongly supported by the national – liberal milieu around Zeiss.

In contrast, many professors in the philosophical faculty lost influence and reputation. Exceptional research results were connected with the influence of Max Wien, who demonstrated that the resistance of electrolytes at high field strengths was dependent on voltage. At the Physical Institute, Hans Busch developed the theoretical foundation of the modern electron microscope in 1926. Abraham Esau performed pioneering work in the area of shortwave and ultrashortwave technology at the Technical-Physical Institute. Esau's genius continued to encompass new fields of applied physics (materials testing, wireless telegraphy); however, beginning in 1939 he moved into the management of militarily relevant research technologies. In the first half of the 1940s, the theoretical physicist Georg Joos combined – like Abbé's successor Rudolf Straubel had before him – an Honorary Professorship for Physics with activity as the research chief in the central Zeiss combine.

As publications before and after 1933 demonstrate, the Jena physician and generally well-respected pediatrician Jussuf Ibrahim used the "social hygenic" categories of "healthy" and "degenerate," "sick and incurable", and a "society and population to be cleansed." Ibrahim worked at the University Children's Clinic, which was founded in 1917 as the Children's Hospital of the Carl Zeiss Foundation in Jena. There the pediatrician provided shelter and medical help to people threatened and persecuted during the Nazi period. His life's work was continued by Erich Hässler (1954–65) and Wolfgang Plevert (1965–86) as head of the clinic. Certainly the most significant scientific achievement came from Hans Berger, who discovered the bio-electric currents in the brain (EEG, electroencephalogram) and thereby entered the medical boundary area of neuropsychology.

The attractiveness of the latter divisions of the university ensured a constant stream of on average 2,000 to 2,400 students per semester in the period between the wars. By the end of the 1920s, a whole series of new institute buildings and social institutions were available to them. Around 400 students had a job on the side in order to finance their studies, which implies that the social structure of the student body was differentiated. However, a considerable part of the students in fraternities, verifiable for the Corps Thuringia and Frankonia, as well as the Burschenschaft Germania, could afford to unleash their contempt for the "polyps" of the hated Republic and thereby risk punishment. More importantly, a latent anti-Semitism coursed through the hazy circles of the student bars that from the outside seemed so inviting. To some degree, this explains why as early as the 1928 summer semester, there were hardly any Jewish students at the University of Jena: in the medical faculty, only 0.3 per cent compared to 6.1 per cent nationwide.

The German-*Völkisch*, Stahlhelm-related, and National Socialist right-wing extremism had gained a strong foothold in the milieu of the University of Jena.[58] This process took place early on in Jena, for this university, along with Leipzig, was one of the first universities in Germany to have a university branch of the NSDAP in 1925, the year Adolf Hitler reformed his party. In July of the same year the NSDAP proposed that the Thüringen State legislature "close the University of Jena to foreign Jews and Jewish instructors."[59] From then on the National Socialists considered Thüringen to be a "model state," "experimental field," or "model region."[60]

The NSDAP had its political breakthrough in the national parliamentary electrons of 1930. The National Socialists had a well-functioning party and election apparatus in Thüringen, like in the rest of the Reich, as early as 1928/9.[61] The consequence of this development finally came on January 23, 1930, when Wilhelm Frick was appointed as the first National Socialist state minister in Germany (for Education and the Interior). Only two years later, on August 26, 1932, a state government was formed dominated by the National Socialists under the leadership of the regional NSDAP leader and subsequent Reich Plenipotentiary for Thüringen, Fritz Sauckel. Thus when Hitler came to power on January 30, 1933, his party in Thüringen already had considerable experience in government and administration.

Along with this solid presence in state politics, the National Socialists were also able to gain a foothold in the Thüringen scientific landscape at a comparatively early stage. As early as 1930, Hitler, in one of his many visits to Weimar, suggested the establishment of a professorship for "racial questions and racial studies" at the University of Jena.[62]

Sauckel was happy to refer to this so-called "will of the Leader" and use science policy in order to demonstrate his independence of the Berlin authorities. Thus he considered this "task supported by the Leader" to make "the University of Jena more and more into a true National Socialist university" as "guarantee for its return to prominence" (February, 1937), building a "new type of faculty, living in and working from National Socialism" (March, 1937), and transforming the University of Jena into "a National Socialist scientific stronghold of the first rank" (March, 1943).[63]

The institutionalization of the subjects of racial studies and racial hygiene were the scientific emphasis of this network of science, ideology, and politics in Jena/Thüringen.[64] With the appointment of a non-habilitated philologist and writer Hans Friedrich Karl Günther as full professor for social anthropology at a traditional university, the National Socialists in Weimar and Berlin gave the recalcitrant senate a taste of their subsequent hiring policy.[65] At this time Günther and the subject of racial studies were given a high priority in the National Socialists' science policy. This was underscored by the demonstrative participation of high-ranking representatives of National Socialism at Günther's inaugural lecture on the subject, "On the Subject of the Racial Transformation of the Population of Germany since the Period of Wandering Peoples," on November 15, 1930 in the main auditorium of the university.

However, an amazing number of other prominent individuals from politics, science and the regional economy also attended. Thus a year before the formation of the Harzburger Front, the anti-republican front men in Jena held a rendezvous. Present alongside Hitler, Hermann Göring, and Frick were the well-known "homeland-style" architect Paul Schultze-Naumburg and Hanno Konopath, leader and founder of the "Nordic Ring." According to research by Helmut Heiber, this was the first and last time that Hitler was a guest at the German university.[66] Like all of Frick's politics, Hitler's presence underscored the will of the National Socialist leadership to use the coalition agreement with right-wing conservative parties at the state level in order to install Thüringen as a regional test-stage for the desired "legal seizure of power." Of course, this liaison was only of limited duration, but in this way "established facts" would be created in the personnel and education policies of the state. The Sauckel government could finally take advantage of this in the late summer of 1932, as did the national "coordination" in early 1933.

The Third Reich

Despite the distance kept from the "political quackery "of the NSDAP and the "sow herd tone" of their Storm Troopers (SA), the prehistory described above of National Socialism coming to power in Thüringen profoundly facilitated the building of bridges between the educated classes and the new regime after 1933.[67] As soon as the "Law for the Reinstatement of the Professional Civil Service" was passed on April 7, 1933, numerous politically and racially undesirable professors, instructors, and assistants could be fired without audible protests. Among those who lost their positions were the economist Paul von Hermberg, the psychologist Wilhelm Peters, the Jewish plant physiologist Leo Brauner, the pedagogues Mathilde Vaerting and Anna Siemsen, the economist

Berthold Josephy, the physicians Hans Simmel and Theodor Meyer-Steineg, the orientalist Julius Lewy, and the biologist Schaxel.

Moreover, in the further course of this year a whole series of economics professors left the university in order to forestall the dissolution of their section, which had already been decimated. Paul von Hermberg, Berthold Josephy, and Moritz Weyerman emigrated; Constantin von Dietze subsequently was counted among the trailblazers of the "Freiburg School" in economics. All Jewish scholars, co-workers, and assistant physicians were fired according to the law discussed above, together with a redirection of the professorship established by von Greil and occupied by socialists. Alongside this, educational institutions like the Home People's School Gera-Tinz or the Jena People's University Club were restructured fundamentally, because they had been especially hated as politically left-wing by the new people in power.[68]

As in the entire German university system, the structural "coordination" and the introduction of the "Leadership Principle" also significantly curtailed the academic self-administration, without completely eliminating it. This mandatory coordination of the leading committees in the university formally ended in the late autumn of 1933. The paramilitary "comradeships" formed in 1933/4 of the approximately 500 fraternity students and other student organizations took several intermediate steps before merging with the National Socialist Students Organization under the National Student Leadership of an SS colonel. Corresponding to this, the Regional Student Leader of Thüringen formulated in the second half of the 1930s the vigorous call for a "fighting science" and suggested that the Reich Governor create an "Institute for Political Science" at the university. Professors and students were forcibly incorporated into so-called "Instructor and Student Organizations."

However, this forced "coordination" was often effectively a "self-coordination." The established full professors still had significant freedom of movement, because National Socialism surrounded and penetrated the university structures rather than completely transforming them. Some things were new. For example, four Jena instructors taught the same subject (in the broadest sense) for different periods: racial studies. Along with Günther and his professorship for social anthropology (1930–5; from 1936, Bernhard Struck), the zoologist Victor Franz with his full professorship for "zoology, genetics, and history of zoology" (1936–45), Karl Astel with the professorship "for human breeding science and genetic research" (1934–35) and "human heredity research and race politics" (1935–45), and the zoologist Gerhard Heberer with the professorship for "general biology and anthropogeny" (1938–45) were all members of a type of "race quadrangle," a very exceptional academic constellation in the Third Reich.

Moreover, the National Socialists had learned a lesson from the protests against the appointment of Günther. In the following years the university committees were not involved in those appointments especially influenced by politics, including the case of the physician Astel in 1934, the appointment of the historian Johann von Leers, and of the linguist Bernhard Kummer in the winter semester of 1936–7. The "coordination" certainly was intended to exclude opponents and discipline members of the university staff, which not last of all was illustrated by the recruitment of faculty as "scientific workers" by the Security Service (SD) of the Reichsführer SS (like for example, the zoologist Victor Franz and the archivist Hermann Leutenberger.) In particular, the chemist Wolfgang Lintzel composed regular reports on the public political activities of his professional colleagues. These included assessments expressly made before eventual appointments by rectors. The great significance attributed by the SS to the infiltration of the University of Jena can be seen merely from the fact that two of its tenured faculty led the branch office of the SD in Jena until the beginning of 1938: the forensic scientist Gerhard Buhtz (leader), and Lintzel (his representative).[69]

On the other hand, the academic elites were fundamentally prepared to accommodate the generous offers of integration. Their large, silent majority showed itself around the middle of the 1930s to be more and more willing to following this lead. Physicists and mathematicians in particular saw the scientific standards in their disciplinary canon seriously threatened. Other university instructors worried about the abilities of the next academic generation. The physicist Abraham Esau was practically predestined to play the role of a middleman between scholarly intellect and political power, especially since it came during the period of his tenure as rector from 1932–5. He embodied the type of leading respected scientist who transformed himself during the second half of the 1930s into a technocratic-oriented scientific manager of the NSDAP.[70]

Esau was also responsible for the initiative to name the university after Schiller, his motivations for which remain unclear. Possibly in this way he wanted to make the university appear in a more favorable light to the Thüringen Education Minister. Esau had been at odds with the Minister in Weimar for a long time. Thanks to the mediation of Kurt Schmitt, Hitler's Reich Minister for Economic Affairs and Frick, the newly-appointed Reich Minister of Interior, Esau was named in May, 1934 to be Foundation Commissioner in the administration of the Carl Zeiss Foundation. Education Minister Wächtler could only comprehend this as a slap in the face, given his own regional and economic policy ambitions.

Wächtler used his own connections in Berlin. It was a complete surprise when the Reich Minister for Science and Education named the professor for political theology Wolf Meyer as the new rector, effective April 1, 1935. Meyer, who was called to Jena in 1933, was a Bavarian radio priest (*Rundfunkpfarrer*) who could preach like Luther and a fanatical supporter of the "Church Movement of German Christians," which, precisely because of its activism, eventually proved to be a failure. Moreover, he lacked any scholarly qualification for his professorship in practical theology. Meyer (later he called himself Meyer-Erlach) managed during his tenure in office to alienate almost all sides as well as most NS officials. His activism achieved precisely what appeared to have been overcome in Esau's first tenure as rector: a superficial politicization and corresponding rattling of the Jena faculty.[71]

Mark Walker in 1991 and more recently Benoit Massin have already drawn attention to the sometimes grave consequences of this, up until now, little noticed "apolitical ideology of science." According to this ideology, "good science or good scientists" and "bad science or bad scientists" exclude each other: "A 'real Nazi' could never be a good scientist, and conversely anti-National Socialist scientists were of course good scientists."[72]

The trained physician and specialist, fanatical race hygienist and politically astute public health politician Karl Astel embodied a completely different type of rector during the years from 1939 to the end of the war. Having held the professorship for human heredity and race politics since 1935, he radicalized day-to-day working conditions and appointments policy in many ways. Astel's name is connected with an especially thorough sterilization campaign in Thüringen, which according to his own information victimized between July 14, 1933 and the end of 1943 around 14,000 people, and in which the Jena University Clinics were significantly involved. Under these population policy considerations, in 1941 Astel called for giving priority to the construction of a Women's Clinic. The NSDAP local leader Paul Müller envisioned a "new university city" Jena: "the inner city, except for the clinics, the university institutes, and the botanical garden...must also disappear...so that on one hand the Firm Zeiss can extend its district to the current clinic and the pond, on the other hand the city could maintain the necessary breathing space by means of the undeveloped districts on the upper Philosophenweg."[73]

Thus he succeeded, with significant support from the Reichsführer SS, Heinrich Himmler, and the Reich Governor in Thüringen, F. Sauckel, to damage the trust held by scholars from all disciplines in newly conceived professorships, and to transform the University of Jena into a sort of National Socialist *universitas vitae*. This type of "transverse front" by SS professors cut across all faculties in order to be able to reach students in very different disciplines with their race ideological lectures.

In 1940 the lawyer Falk Ruttke was called to a "professorship for race and law" in Jena, and with Walter Weddingen there came in 1934 a supporter of "*Völkisch* economics." Moreover, in this regard one should also mention the pathologist Werner Gerlach, who belonged to Himmler's personal staff; Gerhard Buhtz, the successor to the forensic physician Ernst Giese; the agricultural scientist Konrad Meyer; who during the Second World War designed the "General Plan for the East" for the Germanic new settlement of the regions in Poland and the Soviet Union conquered by Germans; the historian Bernhard Kummer, head of the newly-established "Nordic Seminar"; Günther Franz[74], also known as "Peasant-Franz"; and the "Blood and Soil Ideologue" Johann von Leers, head of the "Seminar for Naval History and Prestige," newly established in 1942. Further, a lectureship for Music and Theater Studies was established, which was supposed to research the connections between race and art/music and was filled by the Weimar Chief Dramatist Otto zur Nedden. An associate professorship for technical chemistry was established in 1938, which was supposed to investigate new materials. According to the statements of Herbert Brintzinger, who received this position, he had been working on the synthesis of new chemical weapons since 1932.[75]

Whenever the personnel policies of Astel and the NSDAP regional leader Sauckel came into conflict with the Reich Education Ministry or other National Socialist entities, they referred to Hitler, who had empowered Sauckel to transform the University of Jena into a "truly National Socialist university," which was the "guarantee for its renaissance."[76] Astel's personnel policy had a disastrous effect on the Medical Faculty, where in October, 1938, he gave the professorship for psychiatry and the leadership of the psychiatric, mental health and outpatients clinic to the pediatrician Berthold Kihn, his close personal friend.

Among other things, Kihn participated in the so-called "children's euthanasia" and developed for this purpose the corresponding concepts and ideas that had existed before 1933.[77] The Stadtroda Clinic near Jena directed by Kloos actively participated in the so-called "children's euthanasia" and other murderous practices. Thus in 1939 the NS physicians journal *Ziel and Weg* (Goal and Path) wrote that: "The name Stadtroda has become a symbol of the definitive break with a weak, humanitarian past – not only in Germany, rather in the entire cultural world."[78] Ibrahim, the head of the Jena Children's Clinic, sent children classified as "incurable" there, in cooperation with Kihn and certainly also with Astel.

These "euthanasia" practices, in recent years first reported rather cautiously by local researchers, then subsequently publicized with great effect by Ernst Klee and Götz Aly, have since 1999 unleashed a lively debate over the role of Ibrahim and other scientifically recognized physicians in the Third Reich. This has been especially revealing for how the NS period has been handled by scholars and the public.[79] As a result, the University of Jena established its own investigative commission in order to examine Ibrahim's active and voluntary role in the "children's euthanasia."[80] This debate has become controversial, particularly because after 1945, in the general estimation of the DDR, Ibrahim was considered one of the "hereditary renewers" of the University and an honored citizen of the city, someone who had "earned" his place as "physician of the people," and "savior of the children."

A deliberate opposition to Astel's regime can be perceived only from his actual scientific opponent, the Catholic botanist Otto Renner.[81] In contrast to Renner, the director of the Women's Clinic (Walther M.H. Haupt) supported the radical anti-clericalism of the circle around Astel. In the medical faculty before 1939, only the surgeon Nikolai Guleke and the already retired forensic physician Ernst Giese were able to promote a profound humanism.

Like all science under National Socialism, the race myth and race idea formed the kernel of the Jena (Nazi) philosophy of science, which became especially clear in the lectures announced, the Ph.D. thesis themes given out, and the publications of individual authors like von Kummer, von Leers, and Stengel von Rutkowski. Unfortunately, this also caused the international community to distance itself from Jena's philosophic tradition (Schelling, Hegel, Fichte). The physician, SS Hauptsturmführer and "*Völkisch* philosopher" Lothar Stengel von Rutkowski, whose lectureship for race hygiene, cultural biology, and race hygienic philosophy was unique at the time, was thereby supposed to "build bridges for the understanding between hereditary biology and the cultural sciences..." Moreover, in 1944 Rutkowski founded the "Institute for Research into the Bolshevist World Danger" in Prague.[82]

In the area of medicine Astel stood out among others with his "research work." His office in Weimar (State Office for Racial Matters, founded in 1933) and his university institute (with a section for teaching and research, founded in 1934)[83] researched questions of population policy, hereditary and racial care and were also involved in the realization of race hygiene measures (sterilization, etc.). This "scientific chapter" counts among the blackest periods of Jena's history.[84] Subsequently the bio-statistician Erna Weber drew upon the "material" gathered together in Astel's office and institute. Her lectures were "influential for the national orientation of the university (under National Socialism).[85] Later, during the GDR period, Weber enjoyed an international reputation based upon her scientific work.[86]

In the natural sciences, especially the biological sciences, the scientific judgment of Jena scientists within the scientific community was ambivalent. Heberer's institute was founded in 1938 mainly because of National Socialist interests. However, alongside dubious investigations of the Indogermanic question and certain political tasks, it also achieved a respected international scientific success in the area of evolutionary biology. The collective work *The Evolution of Organisms* (1943), edited by Heberer during the middle of the Second World War, paralleled investigations in Russia and the Anglo-Saxon-speaking countries by making an important contribution to the establishment of the synthetic theory of evolution.[87] The botanists Heinz Brücher and Werner Zündorf, who also worked for a time at Heberer's institute, called into question and corrected theories of the hereditary propagated at the time (nuclear monopoly of heredity; cytoplasmic inheritance) with their cyto-genetic investigations. Their results complemented well the botanical investigations of the Renner School.[88]

In addition, in 1936 Brücher composed, with the encouragement of Astel and Stengel von Rutkowski, the first biography in the sense of hereditary biology of the zoologist Haeckel's family, entitled "Ernst Haeckel: Legacy of Blood and Intellect." In 1941 he was named by Himmler head of the SS Experimental Station in Lannach near Graz, and in 1943 he led an SS botanical-gathering commando to Russia. After the war he emigrated to South America, where he continued his work on the genetics of plants.[89]

The scientific influence of H.F.K. Günther remained marginal in the area of Jena science, for by 1935 he had already left for Berlin. It should merely be mentioned that through him the subject of anthropology was brought to the university and was then expanded by B. Struck through the Third Reich and into the GDR. The zoologist Victor Franz became director of the Ernst Haeckel House in 1935, entirely in the tradition of the patron who gave his name to the institute. Alongside imposing morphological and evolutionary biological investigations on the problem of the perfection or higher development in nature,[90] his efforts to expand the teaching presence, the expansion of the scientific and museum work, and his defense against attacks by leading National Socialist ideologues (Ernst Krieck, Bernhard Bavink) on the person of Ernst Haeckel should be emphasized.[91]

Since the attacks on Haeckel did not die away, in 1942 Franz decided to found an Ernst Haeckel Society under the auspices of Sauckel, with the support of Astel, having the approval of Alfred Rosenberg and others. This existed until 1945.[92] Of course, not all of Astel's "pipe dreams" were realized. Much of what he strove for remained more claim than reality. However, one cannot overlook the greater context, a sort of interdisciplinary scientific "transverse front" flanked by special lecture series, appointments, seminars, institutes, and faculty, which took advantage of the fact that joining the NSDAP helped one's career and leaned, in particular, on the intellectually and politically agile SS.

This all lay in the general trend of those years. Astel and his circle undoubtedly recognized this trend early on and tried energetically to utilize it in Jena. They relied on the appointment policy which had proven effective as early as 1930, whereby no consideration was shown to university committees. They used formal and informal channels, including, as was already common in professorial circles, using private meetings like the circle around Stengel von Rutkowski or the tea receptions Astel gave as rector.[93] This apparently "normal" strategy in no way reflected the intentions of the circle around Astel. Instead, they pursued a double strategy for achieving the *universitas vitae* they were striving for. The existing university system was supposed to be pierced "racially," at the same time ensuring a university capable of high scientific performance for the National Socialist system.

Astel left no doubt about this in his 1936 programmatic speech on "The Task" of the University of Jena.[94] Along with the "racial" foundation, he emphasized the university's scientific duties for the National Socialist "Four-Year Plan" Hitler had recently announced for quickly creating an army ready to be used and an economy prepared for war. From now on, the university would have to be organized in order to facilitate both goals. The usual practice, used since the precedent-setting case in 1930 and in Astel's own case, to appoint professors according to superficial political and ideological criteria, proved to be only of limited suitability. Often, this meant that ideologues who were interested in publicity and intrigue but without scientific substance found their way to professorships.

Individuals like von Leers, Kummer, and Richard Kolb were also judged failures by National Socialist officials. The actual type of researcher and teacher Astel strove for, capable of high scientific performance, socially active for the National Socialist *universitas vitae*, and closely connected to the SS, was embodied by the anthropologist Heberer or the historian Günther Franz, active in *Volkstum* research, who in 1938 initiated a special lecture series on "Germans and Jews" complementing the pogrom of the "Night of Broken Glass," and conceived Jena as a "Frontier University" of the "middle German nation" with its "face turned towards the East."[95]

In his tenure as rector of Jena during the war, Astel was responsible to a considerable degree for the scientifically creative service of the university in the war effort. Numerous institutes took up tasks for the armed forces. The traditionally close connections to the militarily important firms in Jena were intensified, but also to the Emergency Foundation for German Science (as of 1937, the German Research Foundation).[96] The universities, which were not closed in the first months of the war, experienced an enormous flood of students (male and female) and profound shift in the disciplinary and gender structure of the student body, especially around 1940–1. All of this had to be handled. And here it was not the politicized ideologues, rather the scientifically capable specialists who were needed. For this task, membership in the NSDAP was in no way the most important qualification.

However, Astel, like the regional NSDAP leader Sauckel, attempted, despite the stop placed on new party members, to bind capable and self-reliant individuals closer to the NS system. When in 1942 the regional party student leader complained about the hardly National Socialist and, in his opinion, unacceptable conditions in the law and economics faculty, Sauckel implied that Astel was responsible. And it was extremely characteristic of Astel that, in his article on the University of Jena in the *Brüsseler Zeitung* (March 13, 1941), along with a few of the NS activists he had helped appoint, he emphasized effective, respected scholars like the expert on criminal law Richard Lange, the economists Erich Preiser and Erich Gutenberg, or the reform pedagogue Peter Petersen.[97]

After the collapse of the NS regime and Astel's suicide, it was precisely this multiple social and scientific complexity that made it so difficulty to sort everything out, to separate the wheat from the chaff, and to determine true guilt or responsibility. The formal denazification process, with its questions of party membership and function in the NSDAP, largely missed the heart of the problem. Rivalries and conflicts over details were now enthusiastically repackaged as fundamental opposition or even an anti-Nazi position. All too quickly – and this holds to the present day – people reassured themselves with the thought that he who had remained scientifically capable and productive could scarcely have been truly caught up in the NS system.

The postwar period

The University of Jena was among the four German universities which were not closed during the first months of the war. For this reason the number of matriculating students temporarily rose to over 3,600, but between 1933 and 1938 it had been reduced from 2,771 to 1,104. Medical courses formed the focus of the curriculum, which had been restricted in many ways and which became more and more irregular as a result of the military drills and required service for the military activity in the Zeiss Works now demanded of the students. Seventy-five percent of the total building structures and 52 percent of all equipment had been damaged or destroyed by Allied air raids. However, the University of Jena suffered significantly less than the Universities in Berlin and Leipzig.

On October 15, 1945 the University of Jena was the first university to be reopened in the Soviet occupation zone. This decision by the occupying power had been preceded by early, if half-hearted initiatives for "self-purification" by university instructors and for cleaning away the rubble by students, but also by an insincere demand for a decisive new initiative in personnel and matriculation policies. The consensus for opening among ideologically opposed forces began a new phase of renewal, in which the first postwar rector Friedrich Zucker (1945–8; previously rector in 1928–9), and the president of the state of Thüringen, Rudolf Paul, strove to rebuild, denazify, and fill the professorships. This consensus broke completely apart during the "rector crisis" from 1948[98], in the course of which the Weimar occupation officials and the State Ministry of Education under Marie Torhorst succeeded in forcing on the university senate their favorite candidate for transforming the University of Jena into a "People's University," the biologist Otto Schwarz (rector 1948–51 and 1958–62).[99]

Among the older university instructors, these practices led to two waves of flight and emigration (at the end of the 1940s and end of the 1950s), including the physicist Friedrich Hund, the geologist Ludwig Rüger, the mathematician and philosopher Max Bense, the zoologists Jürgen W. Harms and G. Heberer, the botanist Otto Renner, and the agriculturist Friedrich Scheffer. Their dismay at the arbitrary acts, arrests and draconian persecution of nonconformist thinkers could not be overcome by the increased efforts by the SED after 1948 to provide better working and social conditions to the scientific intelligentsia. At the end of the 1940s and beginning of the 1950s teaching and research were governed by contradictory innovations, which aimed at a more practical experience for the students, but also encouraged a regression in the course of study to the school level. The ten-month school year with subsequent vocational training was introduced, along with the obligatory Russian and sport instruction, based upon a regular advisor system, and the very controversial "basic study of society from a scientific perspective" required of all courses of study.

After the Berlin Wall was erected on August 13, 1961, the University of Jena transformed itself into a mixture of regional medical center, productive scientific institution in the central mathematical and scientific disciplines, and an exhibition university of the "socialist university system" of the GDR, especially pampered by Walter Ulbricht and Kurt Hager. Since the beginning of the 1950s, the central officials in East Berlin had promoted the University of Jena for the education of desperately needed replacements in industry and the school system. This corresponded to the basic dogmatic "socially strategic" understanding of the communist planning elites in the "construction of the socialist basis" phase.

By 1958 at the latest, during the course of a long, drawn-out modernization program by the SED leadership, the scientific research capacity of the universities once again moved into the line of sight of the state planers. This practical "turning point" was expressed by Ulbricht especially in an ambitious "chemistry program," the realization of which would require peak scientific performance and new technologies for so-called key industries. Hager, Secretary for Science in the Central Committee (ZK) of the SED, also found the corresponding individual for the service function assigned in this connection to the University of Jena: Günther Drefahl, the full professor and director of the Institute for Organic Chemistry and Biochemistry, was appointed as rector in 1962.[100]

However, in the second half of the 1960s the leading party bureaucrats of the SED in East Berlin ran up against the "fragmentation of the scientific potential" in the Academies of Science and universities in the GDR. Given the fact that there were eighty-nine institutes and clinics, as well as a whole series of independent facilities placed directly under the rector, twenty courses of study for a diploma (roughly analogous to a master's degree) and twenty-two confirmations of subjects in teacher training, as well as the shortage of funds, a structural reorganization now made sense to many people at the University of Jena.[101] However, by the end of the 1960s this had led to the complete orientation of the University towards the highly specialized needs of the large industrial firms Zeiss and Schott. In the medium term, the removal of the already meager funds for equipment and personnel from the basic research typical of universities was threatened. The technology relevant potential of the university had been concentrated since 1966 in the main research areas of mathematics, physics, economic cybernetics/technology of scientific apparatus, and the chemistry of glass.

This was accompanied by a renewed stiffening of the state bureaucratic centralization: on July 13, 1967 the Ministry for Higher and Technical Education was formed in order to be able to carry through at the universities the science, research, and industrial policies announced by Ulbricht at the Seventh Party Congress of the SED. Officially this so-called Third Reform of Higher Education had already begun with the 1965 "Law for the Unified Socialist Education System." However, at the University of Jena this "reform from above" was actually prepared by rector Drefahl (1962–8) and to an even greater degree by his successor Franz Bolck (1968–83). Moreover, important impulses for the conceptual elaboration of this so-called structural policy of Ulbricht came from the large industrial firms at the basis of the GDR "People's

Economy," which were hungry for modernization. This was documented by the intervention of the General Director of the VEB (*Volkseigener Betrieb*, firm owned by the state) Carl Zeiss Jena at the beginning of February 1967 by Politbüro member and Secretary of the Economics Section of the Central Committee of the SED, Günther Mittag.

Without warning, the Zeiss Works had received a huge order for the final processing of high-precision equipment for photo-lithographic processes. The Soviet Union needed this equipment for the expansion of their molecular electronics industry. The General Director of Zeiss, Ernst Gallerach, took advantage of the opportunity and confronted the leading economics policy maker in the GDR with far-reaching demands. If these were not fulfilled, then Zeiss could not satisfy the standards of high quality specified by the Soviets. Along with allowing Zeiss to seize selected firms, the reallocation of labor, and the new construction of a fully climate-controlled "clean hall," Gallerach demanded that around 500 scientists and technicians from universities be reassigned to Zeiss.

This formed the starting point for an expansion, planned in just a few weeks and approved in just a few months, of the capacity of this large Jena firm to levels set by official decrees in 1967 and 1968. Between 1969 and 1975, around 6,000 university graduates were committed to this site. According to these central plans, a large proportion of them were supposed to be trained at institutes from the University of Jena working closely with industry. According to a prognosis from the SED local leadership in Gera, by the year 2000 Jena should have expanded to around 183,000 inhabitants, of course with an exceptionally high percentage of intelligentsia.[102]

In this period of revolutionary change at universities, the introduction of a new rector at Jena was not coincidental: *Magnifizenz* Prof. Dr. sc.med. Dr. h.c. Franz Bolck. His inaugural address thus had special significance; in large part, it was like a dirge sang for the *Universitas litterarum*. The new rector spoke of the fundamental ideas of the university being superceded by the "conception of a socialist university." That meant, to preserve the core of the university, but to transform it. He continued:

> thus we have a responsibility that should be taken very seriously. For it is up to us to shape the university so that the fragmentation comes to an end, the concentration on the crucial points of emphasis is completed, however, the elements of a true science and fundamental research are retained, indeed, enhanced and further developed. The methods are the socialist communal work with its organizational forms and socialist competition. The areas of science and teaching which we want to nurture here, must also yield the harmony of the whole like the instruments of an orchestra.[103]

However, traditional "performers," institutes and disciplines, were to be expelled from this "ensemble." According to the directive issued by the Rector's office on April 4, 1968, "The Development of the Friedrich Schiller University up until 1980," the training of agriculturists, agrarian pedagogues, economic lawyers, pharmacologists, of teachers in the geo-sciences and literature, as well as students of the science of antiquity, prehistoric archaeology, and music would be stopped at the University of Jena.[104] Inside the university, there was talk of sacrificing the *Universitas litterarum* on the alter of economic policy. Still the historian and classicist Friedmar Kühnert demonstrated how effective wrangling outside the so-called structurally-determined branches could be. After unavoidable compromises with the administration of the university, he brought together the previous disciplines with their valuable collections in the Institute for the Science of Antiquity (1969/77) and thereby preserved them from the dissolution already proscribed.

In any case, the "model example" of the new structures of the University of Jena, which had been defended by Bolck at the Berlin Rector Conference on April 17, 1968, was in sight. Up to three-quarters of the entire educational and research capacity of this university was to be dedicated to the construction of scientific apparatus in the VEB Carl Zeiss. This "thinking along structural lines" also included closing the productive agricultural faculty, although in 1966–7 it had divided into several institutes and sections in order to take into account the trend at the time towards an "agricultural factory." An essential point on the way to a "remodeled" university with regard to research strategy was marked by the visit of Walter Ulbricht to Jena at the end of April, 1968. In his meeting with the rector on April 26th, which had other participants, including the General Director of Carl Zeiss, the head of the GDR state emphasized the model function of the reform at the University of Jena for the entire university system of the East German State. Jena should not be a "special case," but rather an "exemplary model."[105]

Ulbricht expressly admonished Bolck with regard to his speech. It must not appear to the public that, in the future, fundamental research and university education would be determined exclusively by industrial and economic considerations. According to Ulbricht's remarks, the rector "had made too many concessions with regard to the concept of the scientific-technical revolution." It must have been painful for those present, that the leader of the state and party had to slow down the rector of the University of Jena and his zeal for modernization. Ulbricht urged him to legitimize more cleverly the dramatic change in the university structure. Corresponding to this, the "essence of the scientific-technical revolution under the conditions of the socialist revolution" should first of all be emphasized, which in the future would also be "embodied" architecturally in the center of the university and industrial city Jena.

In his very confidential discussions with the leadership of Zeiss the day before Ulbricht had already brought this up:

> One cannot leave a city in such a condition when it has a firm that achieves such a masterly performance...Jena is very built up, one began

to build up Lobeda. I drove by it slowly; the devil should come and take this architect away. We can hardly slap together the houses like it is usually done under capitalism. That does not correspond to the level of Zeiss, that must not be. The architects also have to rise to the world level. I do not know how the state officials could have approved that. We are improving this construction and adding a few high-rise buildings in order to turn this village into a city. Now I have learned that the architects have not dared to come to grips with the city center...My suggestion is: build further in Lobeda, but draw up a general plan for the heart of the city. Determine which historic buildings have to remain. The Academy for Construction and the Ministry for Construction Engineering must provide the corresponding people. There is too great of a contrast between Zeiss and the city, which leaves a provincial impression.[106]

However, the First Secretary of the Central Committee of the SED hardly considered the old buildings of the University of Jena worth preserving. It was a sign of the hypertrophic modernity consciousness of the time that he wanted to tear down the original site of the University of Jena, similar to what happened to the Leipzig University Church.[107] Over the course of 1969, demolition teams gave an entirely new meaning to Ulbricht's main theme, determining "breakthrough points" for scientific-technical progress. The original buildings of the Friedrich Schiller University had come under the crossfire of the planners of a spacious scientific complex, the middle of which would include a new research high-rise for the VEB Carl Zeiss. Alarmed by the threatened destruction of the old university, a small group of courageous scientists centered around the microbiologist Hans Knöll turned to well-known individuals in the circle around Ulbricht for support. These individuals included the politically astute chairman of the Research Council of the GDR, the plasma physicist Max Steenbeck.

At the end of the 1960s, Steenbeck, along with Manfred von Ardenne, was one of the most famous East German scientists and, moreover, enjoyed the special confidence of Ulbricht, so that he could go further than other university institute heads without getting into trouble. Steenbeck had in fact asked Ulbricht on March 7, 1969 for more information on the fate of the historic university buildings in Jena. But even this close confident was merely informed by the highest state official that, "assuming the planned development and formation of the city Jena, the entire ensemble of the Jena Collegium cannot be left in the city center." Ulbricht merely left it up to the Central Institute for Microbiology and Experimental Therapy and the Institute for Magnetic Hydrodynamics to set up a university museum "in one of the new buildings at the Old Market in Jena" and thereby calm the anger at the University of Jena. Caught up in their modernization intoxication, only a few days later F. Bolck and E. Gallerach told Hager they also approved of the total demolition of the time-honored original university building, indeed "with emphasis."[108]

Ulbricht had already mentioned in his speech at the House of the People in Jena that it was difficult to reshape the city center destroyed by bombs, if a few "old teeth" are left standing. Obviously these had to be "removed, in order that a 'healthy bite' can be fashioned." This comparison corresponded to his fundamental conviction, almost as if the new concept of a city center could use architectural means to anticipate the "organic" correction and reciprocal penetration of the basic function work-housing-recreation in the future socialist society. In this holistic-organic way of thinking, the "body of society" corresponded to the construction of cities through its hierarchically ordered coordination of its individual "components."

In the leading conceptions of the city planners and architects, this image opposed that of the "capitalist city" of the past, which had produced an unbridgeable contrast between the "industry" and "city" areas. For this reason, a prognosis group at the Gera regional office of the SED considered such "qualitatively new interaction between the university, the VEB Carl Zeiss, and the development of the city as a whole" – as well as drastic solutions to the problem of traffic in the inner city by means of a monorail train. This was supposed to connect the places where production, research, and education took place directly with the most important housing areas.[109] In conclusion, the secretary of the Gera regional office, Herbert Ziegenhahn, quoted to Ulbricht in August 17th, 1970 the goal of a total investment on the scale of more than 1.5 billion GDR marks in order to create "a real model for the socialist city."[110]

The microbiologist Knöll had tried and failed to mobilize the President of the National Council of the National Front, Erich Correns, against this demand. Along with Steenbeck's reserved mediation, Knöll only found support among the meritorious curator of the University of Jena, Günter Steiger, and backing from the university teachers Herbert Bach (biology and anthropology), Georg Uschmann (historian of science), and Bernhard Wächter (artist). The birthplace of the Friedrich Schiller University was spared from demolition, but was intentionally "pressured" by the university high-rise building erected between 1968 and 1972 on the border of the city center. The round building, which was not very user-friendly, had twenty-six spacious floors and, after an intervention by the economic-political section of Gera regional office of the SED with the Ministerial Council of the GDR in early 1971, was more or less signed over to the university.

The original conception planned to use it as a new research high-rise of the VEB Carl Zeiss. However, Bolck had readily accepted this offer to take over the new building in October 1971, for the "alternative" was to use ruined work halls in the middle of the main Zeiss works in Jena. He had previously protested against such an unreasonable suggestion in an unusually sharp form to the administrative positions in Berlin. In a corresponding letter the rector again pointed out the precarious space shortage at the university. According to his information, 47 per cent of its buildings had been erected before 1900. The areas for research and teaching were scattered among 104 buildings (without

211

housing or cafeterias) throughout the entire city, of which around 20 per cent were housed in spaces under 500 square meters. Over 100 rooms had to be rented, some of which were only 10 meters square.

This calamitous shortage of space, which had existed for decades, had an especially disadvantageous effect on the medicine and the university clinics, so that this sober assessment could also be read as a renewed warning call by the scientific elites for "long overdue modernization." In the medium term, the working conditions in the area of medicine continued to worsen noticeably. Thus it was no coincidence that the flood of applications for emigration to the German Federal Republic during the late 1980s included many health care personnel from the clinics of the University of Jena.[111]

Of course, when the university took over the high-rise in the center of the city, the expansion of the five "structure-determining sections" researching the basic technologies for the construction of scientific apparatus was cut out of the budget, even though it had been approved by the Ministerial Council in early 1971. On March 12, 1970 Rector Bolck had deliberately assured the head of state Ulbricht that, through his far-sighted planning, all preconditions had been fulfilled for a disciplinary appropriate use of the new building complex. Ulbricht then instructed his personal secretary and the Chairman of the State Planning Commission to add the required additional investments to the perspective plan for the 1970s. But precisely these arbitrary practices of Ulbricht led to the state of affairs that 240 so-called decisive structural construction projects had to be financed "extraordinarily" by the end of 1969.

Credit inflation caused the amount of money circulating in the GDR to rise by 149 percent between 1965 and 1971. A hidden inflation was the result. Shortages of supplies unavoidably appeared. Of course the finance system malfunctioned, and this could only partially be compensated for with loans drawn from banks in the Federal German Republic and elsewhere in the West. This was the background to the latent power struggle between Ulbricht and Erich Honecker, which came dramatically to a head in the years 1970/1.[112] The removal from the budget of the new university construction in Jena-Lobeda by the Chairman of the Ministerial Council of the GDR was already evidence of a reversal away from the modernization process from the late Ulbricht era, which could no longer be financed by the state, which had been pushed through to the detriment of consumer goods, and which, as far as the rebels around Honecker were concerned, was endangering the SED's hold on power.[113]

However, the regimented leadership structure of the "ordered" university reform of 1968 continued to mold the universities in the GDR up until the end. Since the faculties had been removed as an intermediate level of self-administration, the new section directors were directly responsible to the rectors of their university. In this way, the regionally dominant industrial combines and the universities were supposed to intervene with each other in a symmetric double axle on the rope of command. More realistically, this reform adopted the principle of an authoritarian individual cell leader and took away the competent advice of the Academic Senate and the deans from the rector, while the scientific and social councils created to replace them atrophied into accessory organs of legitimization. Pressed between the always bolder behavior of the General Director of the VEB Carl Zeiss and the personnel-political "suggestions" of the Secretary of the University party leadership, the freedom of movement, of the rectors of the University of Jena from then on became ever narrower.[114]

The dogmatic party indoctrination remained a constant companion of the working process for these years, perhaps most obvious with the blatant expulsion of nonconformist students from the social studies faculty in 1976–7. Beyond this, the State Security Service operated at the universities.[115] Their concealed surveillance was not limited to the military-relevant areas of university contract research carried out for Zeiss, but also spied on instructors and students in specific scientific and medical facilities. Everyday life was influenced by an insidious remilitarization of students' education, which reached from the combat team in the industrial firms, over obligatory courses of study for civil defense, to the so-called military sport festivals each year. Only in the second half of the 1980s did this militarization diminish.

The total number of matriculating students moved back and forth by around 5,000 between 1960 and 1980, with highest level, produced by the Third University Reform, reaching 7,318 in 1972. Women formed the majority of the direct students for the first time in 1970, far overrepresented in teacher training. Young families in the student dormitories and women students with small children hinted at the self-determined lifestyles existing since the 1970s, but the housing shortage caused considerable problems for the university administration and its attempts to find housing for them.[116]

When the long-term effects of the Third University Reform are judged, a broad spectrum is opened, ranging from noble claims to shattered reality.[117] The "dialectical transformation" of Humboldt's university into a "model" of a socialist *Universitas*, painted so eloquently by Rector Bolck in his inaugural speech, proved to be a phantom. The young scientific disciplines in the newly established sections were plagued with all kinds of teething pains or were clearly stillborn; for example, the "socialist science of organization." In any case, at the beginning they were hardly able to stand out within the entire university, as their zealous advocates of the new structural model had hoped and propagated.

On the contrary, their educational profile, which was supposed to transcend traditional disciplinary boundaries, threatened to narrow to that of a technical community college or polytechnic. The responsible political functionaries in Berlin, in conjunction with the local industrial and scientific managers, tried to justify the forced dissolution of the smaller disciplines and/or their transfer to other educational institutions in the GDR as the unavoidable costs of modernization. More thoughtful voices among the faculty and in the scientific councils of the humanities, cultural sciences, and social sciences were subjected to unceasing suspicions; they were denounced as "cripplers of technical progress."[118] This anti-culture of empty phrases created wounds in individual university teachers that would not heal, and reopened during the political turmoil of the years 1989/91.

Conclusion

In its several centuries of history, the University of Jena has enjoyed great prestige and scientific recognition from around the world. Moreover, the city and its large-scale industry were often described as a "laboratory of modernity." However, the history of the Friedrich Schiller University is much more complicated than this common depiction implies. This university not only produced widely recognized genius; it also fundamentally supported the infrastructure of terror in the nearby Buchenwald concentration camp. This "shadow side of modernity" also casts an ambivalent light on the entire history of this university, which from the beginning was molded by extreme transitions, including periods of crisis, glory, defeat, and the expectation of a new beginning.

Founded in the early modern period in the age of humanism, reformation, and confessional orthodoxy, the University of Jena developed during the following centuries into an institution of moderate liberalism and tolerance with regard to scientific methodology. Its highpoint came around 1800, when the close personal and institutional networks of the "double city Weimar-Jena" rightly gave it the reputation of a "philosophical university."

Thanks to the steady support provided by the Carl Zeiss Foundation since the nineteenth century, this university, located at the "new" industrial site of Jena, was able to avoid the fate of the many universities that had been closed within the German Federation. The intellectual intensity and creative force radiating out of Jena in turn created the preconditions for successfully overcoming the structural transformation into a modern university. As early as the 1860s, humanists were confronted by the both challenging and stimulating proximity of scientific "materialistic" traditions of thought, which built upon the influential work of the zoologist and natural philosopher Ernst Haeckel.

The First World War marked a deep divide in the history of the University of Jena, facilitating the hitherto unknown politicization of its faculty and students who did not accept the republic and suffered under the oppressive spiritual and mental burdens caused by the war and subsequent defeat. For this generation of students and many university faculty members, a mythical, heroicized war experience remained the focal point of their lives and worldviews. This nationalistic, chauvinistic processing of the so-called war experience intensified among them during the 1930s and led some of them to call for innovations in basic research "significant for state policies" or national defense "in the interest of the Fatherland." This promoted a "decivilization" of scientific work, teaching, and study[119] and accelerated the retreat from the Humboldtian principle of fundamental, disinterested university research already in process.

If one surveys the relationship between science and ideology in the twentieth century, two contradictory fundamental trends emerge from our case study. On one hand, in the area of medicine, but also in the scientific and humanistic faculties, disciplines not well established at the University pushed for their inclusion in the "great network"[120] of science and politics. They forced this networking in order to mobilize resources in an ideologically laden context and complete the consolidation of their young disciplines in the university system. On the other hand, the central disciplines of the University and methodically conservative full professors and their colleagues sought to preserve the professional autonomy of the University operation and to defend their disciplinary autonomy against the implied trend towards ideologically-determined sciences.

The self-mobilization of the Jena scientific and academic elites for the "service to the People and the Fatherland" definitely followed in the context of the dramatic social-political upheavals of 1932–3 and 1945, and therefore should be interpreted as an accommodation and survival strategy of the University. But it went along with an intensified orientation towards applications and practice by scientists across faculty boundaries, and especially in disciplines which took on the functions of ideological legitimation or military-industrial service during the National Socialist period and the GDR.

As a result of tight budgets and the demands placed on an university in Thüringen, beginning in the 1920s the University of Jena took on more and more service functions for the regional economy, state administration, culture, and religion. This transformation into a service university was in no way accompanied by rapidly rising numbers of students. It was much more connected to the system-specific requirements of the respective power elites which developed Jena into a National Socialist, and after 1968 state socialist, "model university." During the NS period, this excessive "progressionism in the provinces" took on threatening characteristics, because the leading medical group from the University milieu supported and radicalized it.

In the medical faculty, on the dividing line between hereditary biology, human genetics, and anthropology, young disciplines and ambitious scientists established themselves. These used scientific methodology to support the myth of "National Socialist progress."[121] A disproportionately high portion of Jena physicians were members of the NSDAP and its subsidiary organizations, and a few of the heads of clinics sought to support racism publicly with scientifically based arguments about "culling" and "sorting." Possibly as a result of this politicization, in the decades after 1945 many physicians kept a noticeable distance from state socialism in East Germany.

After the harsh denazification phase of 1945–6, scientists, physicians, and technically oriented university faculty could often count on toleration from the new people in power, since in the short run the SED regime was dependent upon these functional elites. However, in the 1950s, precisely these areas of the University formed the centers of resistance to the system. For this reason, the University mainly repressed and severely punished students of medicine and younger faculty members in mathematics and the sciences. At the same time, the reform socialist opposition at the University of Jena found its greatest support here, especially among the SED members of the chemical institutes. In contrast, by the end of the 1950s, the humanities and social sciences were already entirely dominated by half-educated party-line agitators.

After the construction of the Berlin Wall on August 13, 1961, most university students understandably sought to accommodate themselves to the new conditions. It was clear that, for the foreseeable future, two German states would exist. Their conduct fell in a broad spectrum that swung from belief in socialist reforms, imperturbable daily routine, and conscious detachment from the Ulbricht regime. Nevertheless, the overwhelming majority of university staff remained true to their unbroken loyalty to the state and – year-by-year – they more or less demonstrated this. Of course, only a minority of younger university faculty from the rapidly expanding disciplines connected to the industrial technologies of Carl Zeiss and Schott Jena still expected something from the "technocratic reforms"[122] of the 1960s. During the later Ulbricht years, these progressive visions challenged the main ideology of the "triumphal march of the scientific-technological revolution" in state socialism, but in the end were unable to outmaneuver the failure of this attempted modernization.

Jena and its university are often characterized as places of "pure spirits" and fundamental renovation, at least in the post-1945 legends of rebuilding and self-renewal after the end of National Socialism. However, this myth existed only in the desires and values of the educated, middle-class university elite. Indeed, the recent history of the Friedrich Schiller University makes clear that in science the "noble" and "good" cannot be cleanly distinguished from the pathologies of the twentieth century.

Notes

1 See Eric J. Hobsbawm, *The Age of Extremes: A History of the World, 1914–1991* (New York: Vintage, 1996).
2 See Mitchell Ash, "Wissenschaftswandel in Zeiten politischer Umwälzungen: Entwicklungen, Verwicklungen, Abwicklungen," NTM, 3, (1995): 1–21, and Mitchell Ash, "Scientific Changes in Germany 1933, 1945 and 1990: Towards a Comparison," *Minerva*, 37 (1999): 329–35.
3 See Jürgen John, "Wissenschaft und Politik – die Jenaer Universität im 20. Jahrhundert," in Herbert Gottwald (ed.), *Universität im Aufbruch. Die Alma mater Jenensis zwischen Ost und West. Völkerverbindende Vergangenheit und europäische Zukunft einer deutschen Universität* (Jena: Druckhaus Mayer-Verlag, 1992), 239–61; *Vergangenheitsklärung an der Friedrich-Schiller-Universität* (Leipzig: Evangelische Verlagsanstalt, 1994).
4 See Helmuth G. Walther, et al., *Aufbrüche. 450 Jahre Hohe Schule Jena. Ausstellungskatalog* (Jena: Selbstverlag der Friedrich-Schiller-Universität, 1998); for the entire history, see Max Steinmetz (ed.), *Geschichte der Universität Jena 1548/58–1958*, 2 vols, (Jena: VEB Gustav Fischer Verlag, 1958/62); Erich Maschke, *Universität Jena* (Cologne: Böhlau Verlag, 1969); Siegfried Schmidt (ed.), *Alma mater Jenensis. Geschichte der Universität Jena* (Weimar: Hermann Böhlaus Nachfolger,

1983); Gottwald, *Universität im Aufbruch*; Thomas Pester, *Im Schutze der Minerva. Kleine illustrierte Geschichte der Universität Jena* (Jena: Verlag Dr. Bussert and Partner, 1996), as well as the short sketch from U. Köpf, "Friedrich-Schiller-Universität Jena," in L. Boehm and Rainer A. Müller (eds), *Universitäten und Hochschulen in Deutschland, Österreich und der Schweiz. Eine Universitätsgeschichte in Einzeldarstellungen* (Düsseldorf: Econ Taschenbuchverlag, 1983), 211–15.

5 Considering the time period handled here, Thüringen refers to the state formed in 1920 from seven small states, in 1945 reestablished and founded again in 1990. In the NS period this overlapped with the structure of the NSDAP regional section, and in the GDR from 1952 through to 1990 was replaced by the districts Erfurt, Gera, and Suhl. See Jürgen John, *Grundzüge der Landesverfassungsgeschichte Thüringens 1918 bis 1952*, in *Thüringische Verfassungsgeschichte im 19. und 20. Jahrhundert* (*Schriften zur Geschichte des Parlamentarismus in Thüringen*, 3) (Jena: Wartburg-Verlag, 1993), 49–113; Beate Häupel, *Die Gründung des Landes Thüringen. Staatsbildung und Reformpolitik 1918–1923* (Weimar: Böhlau-Verlag, 1995); Karl Schmitt (ed.), *Die Verfassung des Freistaats Thüringen* (Weimar: Böhlau Verlag, 1995); Karl Schmitt (ed.), *Thüringen. Eine politische Landeskunde* (Weimar: Böhlau Verlag, 1996).

6 Founded 1379/92, dissolved in 1816 after integration into the Prussian state, refounded in 1993.

7 For example, the Universities of Leipzig (1409), Rostock (1419), Greifswald (1456), Halle (1694) and Berlin (1810), the Bergakademie Freiberg (1765) as well as the Technical University Dresden (1961).

8 The history of the University of Jena in the twentieth century has only been partially researched, and the period after 1945 is burdened by holes in our knowledge and one-sided interpretations. For this reason, the Academic Senate formed a special historical commission at the beginning of 1999; H.R. Böttcher, *Vergangenheitsklärung an der Friedrich-Schiller-Universität Jena* (Leipzig: Evangelische Verlagsanstalt, 1994).

9 See Klaus-Dietmar Henke, *Die amerikanische Besetzung Deutschlands* (Munich: Oldenbourg, 1996), 657–776.

10 See Jürgen John and Volker Wahl (eds), *Zwischen Konvention und Avantgarde. Doppelstadt Jena-Weimar* (Weimar: Böhlau Verlag, 1995).

11 See notes from April, 1945, on the status of Weimar, reprinted in *Thüringische Landeszeitung* (4 March 1995) Supplement, 3; for the history of the concentration camp at Buchenwald, see Gedenkstätte Buchenwald (ed.), *Konzentrationslager Buchenwald 1937–1945. Begleitband zur ständigen historischen Ausstellung* (Göttingen: Wallstein, 1999).

12 Among the many titles on Weimar as "Cultural City of Europe, 1999," see Justus H. Ulbricht, "Wo liegt Weimar? Nationalistische Entwürfe kultureller Identität," in Ursula Härtl, Burkhard Stenzel, and Justus H. Ulbricht (eds), *Hier, hier ist Deutschland...Von nationalen Kulturkonzepten zur nationalsozialistischen Kulturpolitik* (Göttingen: Wallstein Verlag, 1997) 11–44; Peter Merseburger, *Mythos Weimar. Zwischen Geist und Macht* (Stuttgart, Deutsche Verlags-Anstalt, 1998); Volker Mauersberger, *Hitler in Weimar. Der Fall einer deutschen Kulturstadt* (Berlin: Rowohlt, 1999).

13 "Deutscher Festtag unter leuchtender Maiensonne. Im Gau Thüringen wachsen die Fundamente einer neuen Klassik," *Thüringer Gauzeitung. Der Nationalsozialist*, 101 (3 May 1937), reprinted in Thomas Neumann (ed.), *"Wir aber müssen eine Welt zum Tönen bringen...". Kultur in Thüringen 1919–1949* (Erfurt: Landeszentrale für politische Bildung, 1998) 198–201 (Rudolf Heß used the term "new classic" in his speech commemorating the founding of the Weimar NS "Regional Forum"); also see the programmatic speech by Sauckel on "New Weimar" (24 August 1939 and 18

October 1940), *Reichsstatthalter in Thüringen*, 186, 272–5; 187, 269–78, in Thüringisches Hauptstaatsarchiv Weimar (ThHStAW).

14 See Richard Alewyn, "Goethe als Alibi," in Karl Robert Mandelkow (ed.), *Goethe im Urteil seiner Kritiker. Dokumente zur Wirkungsgeschichte Goethes in Deutschland, Volume IV: 1918–1982* (Munich: Verlag C.H. Beck, 1984), 333–5.

15 See Peter Krahulec *et al.* (eds), *Buchenwald-Weimar. April 1945. Wann lernt der Mensch?* (Münster: LIT, 1994); Jens Schley, *Nachbar Buchenwald. Die Stadt Weimar und ihr Konzentrationslager 1937–1945* (Cologne: Böhlau Verlag, 1999).

16 See Axel Schildt, "Im Kern gesund? Die deutschen Hochschulen 1945," in H. König *et al.* (eds), *Der Fall Schwerte und die NS-Vergangenheit der deutschen Hochschulen* (Munich: 1997), 223–40; for Jena see Jürgen John, "Die Jenaer Universität im Jahre 1945," in Jürgen John, Volker Wahl, and Leni Arnold (eds), *Die Wiedereröffnung der Friedrich-Schiller-Universität Jena 1945. Dokumente und Festschrift* (Rudolstadt: Hain Verlag, 1998), 12–74.

17 For Jena, see the memorandum by the lawyers and economists Richard Lange, Erich Preiser, Erich Gutenberg and Hermann Schultze von Lasaulx, "Zur Haltung der deutschen Intelligenz gegenüber dem Nationalsozialismus und ihre Bedeutung für die Zukunft Deutschlands," from September 1945, partially reprinted in John, Wahl and Arnold, *Die Wiedereröffnung*, 212–20.

18 See the recent publication by Häupel (1995); Jürgen John, "'Weimar' als regionales intellektuelles Reform- und Experimentierfeld," in Wolfgang Bialas and Burkhard Stenzel (eds), *Die Weimarer Republik zwischen Metropole und Provinz. Intellektuellendiskurse zur politischen Kultur* (Weimar: Böhlau Verlag, 1996), 11–21; Justus H. Ulbricht, "Willkomm und Abschied des Bauhauses. Eine Rekonstruktion," *Zeitschrift für Geschichtswissenschaft*, 46 (1998): 5–27; Hans Wilderotter and Michael Dorrmann (eds), *Wege nach Weimar. Auf der Suche nach der Einheit von Kunst und Politik. Ausstellungskatalog* (Berlin: Jovis Verlagsbüro, 1999).

19 See Lothar Ehrlich and Jürgen John (eds), *Weimar 1930. Politik und Kultur im Vorfeld der NS-Diktatur* (Weimar: Böhlau Verlag, 1998).

20 See Detlev Heiden and Gunter Mai (eds), *Nationalsozialismus in Thüringen* (Weimar: Böhlau Verlag, 1995).

21 There is still no biography of Sauckel; see first of all Peter Hüttenberger, *Die Gauleiter. Studie zum Wandel des Machtgefüges in der NSDAP* (Stuttgart, Deutsche Verlags-Anstalt, 1969); Dietrich Eichholtz, "Die Vorgeschichte des General-bevollmächtigten für den Arbeitseinsatz," *Jahrbuch für Geschichte*, 9 (1973): 339–83; Peter W. Becker, "Fritz Sauckel – Generalbevollmächtigter für den Arbeitseinsatz," in Ronald Smelser and Rainer Zitelmann (eds), *Die braune Elite I. 22 biographische Skizzen* (Darmstadt: Wissenschaftliche Buchgesellschaft, 1993), 236–45; Manfred Weißbecker, "Fritz Sauckel," in Kurt Pätzold and Manfred Weißbecker (eds), *Stufen zum Galgen. Lebenswege vor den Nürnberger Urteilen* (Leipzig: Militzke Verlag, 1996), 297–331.

22 See Volker Wahl, *Die Rettung der Dichtersärge. Das Schicksal der Sarkophage Goethes und Schillers bei Kriegsende 1945* (Weimar, 1991).

23 See I.S. Kolesnitschenko, "Der Neubeginn der Friedrich-Schiller-Universität Jena nach dem zweiten Weltkrieg," in *Neubeginn. Die Hilfe der Sowjetunion bei der Neueröffnung der Friedrich-Schiller-Universität Jena* (Jena: Selbstverlag der Universität, 1977), 14 ff.; I.S. Kolesnitschenko, *Bitwa posle wojny* (Moskwa: Woenisdat, 1987), 75 ff.

24 The University of Jena was ceremoniously reopened as the first university in the Soviet occupation zone (SBZ) on October 15, 1945, a short time after Göttingen (British zone) and Marburg (American zone) and at the same time as Tübingen (French zone), and began instruction in the first days of December, 1945. Thus by the standards of all occupation zones it was one of the first universities again capable of instruction, and the first among the universities destroyed by the war. For the

details as well as the reopening of all German universities, see John, Wahl and Arnold, *Die Wiedereröffnung*, 449–51.

25 See Hans Herz, "Zur Namensverleihung "Friedrich-Schiller-Universität Jena" am 10. 11. 1934," *Alma Mater Jenensis. Universitätszeitung*, 14 (1990/91). On the morning of the day named above, the official 1934 Reich Schiller Festival in Weimar took place in Hitler's presence. Reich Minister and Reich Leader of the NSDAP Josef Goebbels gave the main address. He applauded the naming of the University of Jena.

26 A classic text in this sense on the "Unity of Intellect and Act" as a fundamental principle of "German essence" came from the Jena neo-Kantian Bruno Bauch: see Bruno Bauch, *Der Geist von Potsdam und der Geist von Weimar. Eine Rede bei der von der Universität Jena veranstalteten Feier des Jahrestages der Gründung des Deutschen Reiches, gehalten am 18. Januar 1926* (Jena: Verlag Gustav Fischer 1926); for a criticism from the perspective of the time, see the speech made by Wilhelm Girnus on November 6, 1945, reprinted in John, Wahl and Arnold, *Die Wiedereröffnung*, 334–6.

27 "Opening Remarks (15 October 1945)," reprinted in John, Wahl and Arnold, *Die Wiedereröffnung*, 263–8; also see Rüdiger Stutz, "Hochschulerneuerung unter Besatzungsherrschaft. Der Landespräsident von Thüringen als 'politischer Rektor' der Universität Jena (Sommer 1945 bis Frühjahr 1946)," in John, Wahl and Arnold, *Die Wiedereröffnung*, 75–102.

28 The universities founded in the fourteenth or fifteenth centuries include Prague (1348), Vienna (1365), Heidelberg (1386), Cologne (1388), Erfurt (1379/92), Würzburg (1402), Leipzig (1409), Rostock (1419), Greifswald (1456), Freiburg (1457), Trier (1473), Mainz (1476) and Tübingen (1477). Those established at the beginning of the modern period include the universities of Wittenberg (1502) and Frankfurt an der Oder (1506), from which Wittenberg became the starting point and model University of the Lutheran Reformation.

29 Including Halle (1694) and Göttingen (1737).

30 See Rüdiger vom Bruch, *Die deutschen Universitäten 1734–1980* (Frankfurt am Main: 1985); Thomas Ellwein, *Die deutschen Universitäten vom Mittelalter bis zur Gegenwart* (Königstein: Hain, 1985); Rainer A. Müller, *Geschichte der Universität. Von der mittelalterlichen Universität zur deutschen Hochschule* (Munich: Callwey, 1990).

31 The Ernestine princely house, as head of the Schmalkaldischen Union, lost Kurwürde, Kurlande, and the University of Wittenberg in 1547 as a result of the "Wittenberg capitulation"; the remaining duchy of Saxony-Weimar founded the University of Jena as a replacement. See Walther *Aufbrüche*.

32 See Gerhard Müller, Klaus Ries, and Paul Ziche (eds.), Die Univesrsität Jena. Tradition und Innovation um 1800 (Stuttgart: Franz Steiner Verlag, 2001)

33 See Thomas Pester, *Zwischen Autonomie und Staatsräson. Studien und Beiträge zur allgemeinen deutschen und Jenaer Universitätsgeschichte im Übergang vom 18. zum 19. Jahrhundert* (Jena: Academica and Studentica Jenensia e.V. 1992).

34 See Jürgen John, "'Jena 1806' – Symboldatum der Geschichte des 19./20. Jahrhunderts," in Gerd Fesser and Reinhard Jonscher (eds), *Umbruch im Schatten Napoleons. Die Schlachten von Jena und Auerstedt und ihre Folgen* (Jena: Verlag Dr. Bussert and Partner, 1998), 177–95.

35 See Frank Wogawa, "Universität und Revolution: Jena und die "hochschulpolitischen" Reformbestrebungen 1848," in Hans-Werner Hahn and Werner Greiling (eds), *Die Revolution von 1848/49 in Thüringen. Aktionsräume. Handlungsebenen. Wirkungen* (Rudolstadt: Hain Verlag, 1998), 445–74.

36 See Ernst Schmutzer (ed.), *Wissenschaft und Schulenbildung* (Jena: Selbstverlag der Universität, 1991).

37 See Meike G. Werner, "Die akademische Jugend und Modernität. Zur Konstituierung der Jenaer Freien Studentenschaft 1908," in Jürgen John and Volker

Wahl (eds), *Zwischen Konvention und Avantgarde. Doppelstadt Jena-Weimar* (Weimar: Böhlau, 1995), 289–309.

38 The trend was set during the world war at the Jena Conference of German Student Organizations. See the *Verhandlungsschrift der Tagung deutscher Studentenausschüsse in Jena am 19. und 20. Januar 1918 zur Vorbereitung eines deutschen Studententages.*

39 See the edition of sources with commentary, Friedrich Schomerus, *Werden und Wesen der Carl Zeiss-Stiftung an der Hand von Briefen und Dokumenten aus der Gründungszeit (1886–1896) dargestellt* (Jena: Verlag Gustav Fischer, 1940); see Rüdiger Stolz and Joachim Wittig (eds), *Carl Zeiss und Ernst Abbe. Leben – Wirken – Bedeutung. Wissenschaftshistorische Abhandlung* (Jena: Universitätsverlag, 1993).

40 See Detlef Ignasiak, *Das literarische Jena. Von den Anfängen bis in die ersten Jahrzehnte unseres Jahrhunderts* (Jena: Jena Information, 1985); Volker Wahl, *Jena als Kunststadt. Begegnungen mit der modernen Kunst in der kleinthüringischen Universitätsstadt zwischen 1900 und 1933* (Leipzig: Seemann Verlag, 1988).

41 See Gisela Horn (ed.), *Die Töchter der Alma mater Jenensis. Neunzig Jahre Frauenstudium an der Universität Jena* (Rudolstadt: Hain Verlag, 1999).

42 1890: 13,400 inhabitants, 640 students; 1900: 20,700 inhabitants, 759 students; 1910: 38,000 inhabitants, 1,859 students.

43 See Bernd Wilhelmi (ed.), *Jenaer Hochschullehrer der Medizin. Beiträge zur Geschichte des Medizinstudiums* (Jena: Selbstverlag der Universität, 1987).

44 Moreover, the financial resources of the Zeiss Foundation were used during these years to found two institutes, which except for Göttingen, were not to be found at any other university: technical physics and technical chemistry, see Steinmetz, *Geschichte der Universität Jena*, 52. Thus the funds of the foundation first and foremost benefited the aspiring scientific disciplines.

45 See Schmidt, *Alma mater Jenensis*, 237; for philosophy in the narrow sense see Max Wundt, *Die Philosophie an der Universität Jena in ihrem geschichtlichen Verlauf dargestellt* (Jena: Gustav Fischer, 1932).

46 Also see the conference organized in 1999 by the Philosophical Institute of the University of Jena, under the title "Die Angst vor der Moderne. Philosophische Antworten auf Krisenerfahrungen. Der Mikrokosmos Jena um 1900".

47 Susann Hensel and Rosemarie Nöthlich, *Jena und seine Universität. Wissenschaftshistorische Streifzüge durch fünf Jahrhunderte* (Jena: Wartburg Verlag, 1991), 39.

48 See Ernst Haeckel, *Über die Biologie in Jena während des 19. Jahrhunderts* (Jena: Verlag Gustav Fischer, 1905), 15. This is the printed version of Haeckel's lecture before the Medical-Scientific Society of Jena (17 June 1904).

49 See Georg Uschmann, *Geschichte der Zoologie und der zoologischen Anstalten in Jena 1779–1919* (Jena: Gustav Fischer, 1959); Heinz Penzlin, *Geschichte der Zoologie in Jena nach Haeckel (1909–1974)* (Jena: Gustav Fischer, 1994); Erna Aescht et al. (eds), *Welträtsel und Lebenswunder. Ernst Haeckel – Werk, Wirkung und Folgen* (Linz: Stapfia 56, 1998).

50 See Horst Groschopp, "Den Deutschen eine neue Kultur. Forderungen und Tätigkeiten des 'Weimarer Kartells' von 1907 bis 1914," in Jürgen John and Volker Wahl, *Zwischen Konvention und Avantgarde*, 257–88; Horst Groschopp, *Dissidenten. Freidenkerei und Kultur in Deutschland* (Berlin: Dietz, 1997); J. Reulecke "Rassenhygiene, Sozialhygiene, Eugenik," in Diethart Kerbs and J. Reulecke (eds), *Handbuch der deutschen Reformbewegungen 1880–1933* (Wuppertal: Peter Hammer Verlag, 1998), 197–210.

51 For the biography of Plate, see Penzlin (1994), 13ff.

52 See Friedrich W. Graf, "Die Positivität des Geistigen. Rudolf Euckens Programm neoidealistischer Universalintegration," in Rüdiger vom Bruch, Friedrich Wilhelm

Graf, and G. Hübinger (eds), *Kultur und Kulturwissenschaften um 1900. Teil II: Idealismus und Positivismus* (Stuttgart: Steiner, 1997), 53–85.

53 See Ludwig Plate, "Kurze Selbstbiographie," *Archiv für Rassen- und Gesellschaftsbiologie*, 29, 1 (1935): 86–7. For example, also see Alexander Cartellieris, "Der Krieg und die wissenschaftliche Arbeit," as well as Haeckel's arguments: *Vernunft und Krieg* ("Raison et Guerre. La Réconciliation,") *Deutsch-französische Revue*, 1 (October 1913); *Englands Blutschuld am Weltkriege* (Eisenach: Jacobis Buchhandlung, 1914); *Ewigkeit. Weltkriegsgedanken über Leben und Tod, Religion und Entwickelungslehre* (Berlin: Reimer, 1915), etc. In 1915 the university library in Jena established a war archives, see Steinmetz (1958), 510. "The Weimar Republic," *History Workshop Journal*, 41 (1996): 117–53.

54 See Nick Hopwood, "Producing a Socialist Popular Science"; Nick Hopwood, "Biology Between University and Proletariat: The Making of a Red Professor," *History of Science*, 35, 4 (1997): 367–424.

55 See Uwe Hoßfeld, "Die Jenaer Jahre des 'Rasse-Günther'. Zur Gründung des Lehrstuhls für Sozialanthropologie an der Universität Jena," *Medizinhistorisches Journal*, 34, 1 (1999): 47–103.

56 Schmidt, *Alma mater Jenensis*, 224.

57 See Daniel Gasman, *The Scientific Origin of National Socialism: Social Darwinism in Ernst Haeckel and the German Monist League* (New York: American Elsevier, 1971); Daniel Gasman, *Haeckel's Monism and the birth of facist ideology* (Frankfurt am Main: Peter Lang, 1998); Jürgen Sandmann, *Der Bruch mit der humanitären Tradition. Die Biologisierung der Ethik bei Ernst Haeckel und anderen Darwinisten seiner Zeit* (Stuttgart: Gustav Fischer, 1990); Pat Shipman, *The Evolution of Racism: Human Differences and the Use and Abuse of Science* (New York: Simon and Schuster, 1994); Klaus Thomann, Dieter Thomann and Werner Friedrich Kümmel, "Naturwissenschaft, Kapital und Weltanschauung. Das Kruppsche Preisausschreiben und der Sozialdarwinismus," 3 parts, *Medizinhistorisches Journal*, 30, 2 – 4 (1995): 99–143, 205–43, and 315–52; Eve-Marie Engels (ed.), *Die Rezeption von Evolutionstheorien im 19. Jahrhundert* (Frankfurt am Main: Suhrkamp, 1995); Uwe Hoßfeld, "Menschliche Erblehre, Rassenpolitik und Rassenkunde (-biologie) an den Universitäten Jena und Tübingen von 1934–45: Ein Vergleich," *Verhandlungen zur Geschichte und Theorie der Biologie*, 1 (1998): 361–92; Heidrun Kaupen-Haas and Christian Saller (eds), *Wissenschaftlicher Rassismus. Analysen einer Kontinuität in den Human- und Naturwissenschaften* (Frankfurt am Main: Campus, 1999); Uwe Hoßfeld, "Haeckelrezeption im Spannungsfeld von Monismus, Sozialdarwinismus und Nationalsozialismus," *History and Philosophy of the Life Sciences*, 21 (1999): 195–213.

58 See Susanne Zimmermann, *Die Medizinische Fakultät der Universität Jena während der Zeit des Nationalsozialismus* (Berlin: VWB-Verlag, 2000), 80.

59 Schmidt, *Alma mater Jenensis*, 224.

60 See Helmut Heiber, *Universität unterm Hakenkreuz. Der Professor im Dritten Reich. Teil 1* (Munich: KG Saur, 1991); Uwe Hoßfeld and Jürgen John "Die Universität Jena im 'Dritten Reich,'" *Uni-Journal Jena*, 11 (1998): 20–1; Jürgen John et al., *Geschichte in Daten. Thüringen* (Munich: Koehler and Amelang, 1995); Paul Weindling, *Health, Race and German Politics between National Unification and Nazism, 1870–1945* (Cambridge: Cambridge University Press, 1989); Paul Weindling, "'Mustergau' Thüringen. Rassenhygiene zwischen Ideologie und Machtpolitik," in Norbert Frei (ed.), *Medizin und Gesundheitspolitik in der NS-Zeit* (Munich: Oldenbourg, 1991), 81–97.

61 See Donald R. Tracey, "The Development of the National Socialist Party in Thuringia, 1924–1930," *Central European History*, 8 (1975): 23–50; Rüdiger Stutz, "Im Schatten von Zeiss. Die NSDAP in Jena," in Heiden and Mai, *Nationalsozialismus in Thüringen*, 119–42.

62 Thüringen was without exception the first in all of Germany to institutionalize racial science. See Fritz Dickmann "Die Regierungsbildung in Thüringen als Modell der Machtergreifung," *Vierteljahrshefte für Zeitgeschichte*, 14 (1966): 454–64; Andreas Dornheim et al., *Thüringen 1933–1945. Aspekte nationalsozialistischer Herrschaft* (Erfurt: Landeszentrale für politische Bildung Thüringen, 1997); Lothar Ehrlich and Jürgen John, *Weimar 1930. Politik und Kultur im Vorfeld der NS-Diktatur* (Cologne: Böhlau, 1998); Heiden and Mai, *Nationalsozialismus in Thüringen*; Günter Neliba, *Wilhelm Frick. Der Legalist des Unrechtstaates. Eine politische Biographie* (Paderborn: Schoeningh, 1992); Günter Neliba, "Wilhelm Frick und Thüringen als Experimentierfeld für die nationalsozialistische Machtergreifung," in Heiden and Mai, *Nationalsozialismus in Thüringen*, 75–96; Merseburger, *Mythos Weimar* (1998); Mauersberger, *Hitler in Weimar* (1999).

63 Karl Astel to Heinrich Himmler (16 November 1937); Fritz Sauckel to Bernhard Rust (8 March 1943); Bestand U, Abt. IV, No. 16, Universitätsarchiv Jena [UAJ].

64 See Robert Proctor, *Racial Hygiene: Medicine under the Nazis* (Cambridge: MA, Harvard University Press, 1988); Robert Proctor, *The Nazi War on Cancer* (Princeton, NJ: Princeton University Press, 1999); Weindling, *Health, Race and German Politics*; Peter Emil Becker, *Sozialdarwinismus, Rassismus, Antisemitismus und Völkischer Gedanke. Wege ins Dritte Reich, Teil II* (Stuttgart: Georg Thieme, 1990); Peter Weingart et al. (eds), *Rasse, Blut und Gene* (Frankfurt am Main: Suhrkamp, 1992); Benôit Massin, "From Virchow to Fischer: Physical Anthropology and Modern Race Theories in Wilhelmine Germany," *History of Anthropology*, 8 (1996): 79–154; Stefan Kühl, *Die Internationale der Rassisten* (Frankfurt am Main: Campus, 1997); Uwe Hoßfeld, *Gerhard Heberer (1901–1973) – Sein Beitrag zur Biologie im 20. Jahrhundert* (Berlin: VWB-Verlag, 1997).

65 This appointment tarnished not only the University of Jena, but also the German universities as a whole, where the subject anthropology had, up until that time, had a good reputation. Hoßfeld, "Die Jenaer Jahre".

66 Heiber, *Universität unterm Hakenkreuz*, 73.

67 Schmidt, *Alma mater Jenensis*.

68 For the entire time period: Jürgen John, "Die Weimarer Republik, das Land Thüringen und die Universität Jena 1918/19–1923/24," *Jahrbuch für Regionalgeschichte*, 10 (1983): 177–207; Jürgen John, *Thüringen und Jena in den 1920er Jahren. Zum landes- und kommunalpolitischen Wirkungsmilieu Adolf Reichweins*, in Martha Friedenthal-Haase (ed.), *Adolf Reichwein. Widerstandskämpfer und Pädagoge* (Erlangen: Palm and Enke, 1999), 23–49.

69 See ZR 770, Nr. 8, Bll. 61–74, Bundesarchiv, Zwischenarchiv Dahlwitz-Hoppegarten.

70 ZA I 12243, Nr. 6, Bundesarchiv, Zwischenarchiv Dahlwitz-Hoppegarten; short, handwritten cv, in which Esau writes that he is in charge of radio in the regional NSDAP office in Thüringen.

71 See Hans-Joachim Sonne, "Die politische Theologie der deutschen Christen," *Göttinger Theologische Arbeiten*, 21 (1982): 232–49; also see the contribution from Tobias Schüfer and Ralf Meister-Karanikas in Thomas A. Seidel (ed.), *Thüringer Gratwanderungen. Beiträge zur fünfundsiebzigjährigen Geschichte der evangelischen Landeskirche Thüringens* (Leipzig: Evangelische Verlagsanstalt, 1998).

72 See Benôit Massin, "Anthropologie und Humangenetik im Nationalsozialismus oder: Wie schreiben deutsche Wissenschaftler ihre eigene Wissenschaftsgeschichte?" in Heidrun Kaupen-Haas and Christian Saller (eds), (1999), 13; Mark Walker, "Legends Surrounding the German Atomic Bomb," in Teresa Meade and Mark Walker (eds), *Science, Medicine and Cultural Imperialism* (New York: St. Martin's Press, 1991), 179–80, 184–8.

73 Nr. 13097, Bestand BACZ, Unternehmensarchiv der Carl Zeiss GmbH, Jena; also see Nr. R 43 II/ 938b Abt. R, Best. R 43, Bundesarchiv.

74 See Wolfgang Behringer, Bauern-Franz und Rassen-Günther, "Die politische Geschichte des Agrarhistorikers Günther Franz (1902–1992)," in Winfried Schulze and Otto G. Oexle (eds), *Deutsche Historiker im Nationalsozialismus*, 2nd edn, (Frankfurt am Main: Fischer Taschenbuch Verlag, 1999), 114–41.

75 For the development of chemistry in Jena, see H. Schmigalla (ed.), "Chymia Jenensis. Chymisten, Chemisten und Chemiker in Jena," *Studien zur Hochschul- und Wissenschaftsgeschichte*, 6 (1989).

76 ThHSTAW, Thüringisches Volksbildungsministerium, Best. C 155, Bl. 262 (Erlaß Sauckels vom 2 February 1937).

77 See Berthold Kihn, "Die Ausschaltung der Minderwertigen aus der Gesellschaft," *Allgemeine Zeitschrift für Psychiatrie*, 98 (1932): 387–404.

78 See Götz Aly, "Tuberkulose und 'Euthanasie,'" in Jürgen Pfeiffer (ed.), *Menschenverachtung und Opportunismus. Zur Medizin im Dritten Reich* (Tübingen: Attempto Verlag, 1992), 137.

79 *Tabuisierte Vergangenheit? Aspekte der Jenaer Medizingeschichte im Nationalsozialismus. Tagung am 24 January 2000; Tabuisierte Vergangenheit – part II: Zur Aufarbeitung der NS-Euthanasieverbrechen in der DDR. Tagung vom 13 March 2000.* Also see Susanne Zimmermann, *Die Medizinische Fakultät der Universität Jena während der Zeit des Nationalsozialismus* (Berlin: VWB, 2000).

80 See "Bericht der Kommission der Friedrich-Schiller-Universität Jena zur Untersuchung der Beteiligung Prof. Dr. Jussuf Ibrahims an der Vernichtung 'lebensunwerten Lebens' während der NS-Zeit, Jena," (Jena: Selbstverlag, 2000).

81 See Jost S. Casper and Manfred Eichhorn, "Otto Renner (1883–1960)," *Sonderschriften der Akademie gemeinnütziger Wissenschaften zu Erfurt*, 31 (1997): 93–142.

82 See Monika Leske, *Philosophen im Dritten Reich* (Berlin: Dietz Verlag, 1990); A. Hamann, "Männer der kämpfenden Wissenschaften," (Jena: unpublished Hausarbeit, 1999); Uwe Hoßfeld, "Staatsbiologie, Rassenkunde und Moderne Synthese in Deutschland während der NS-Zeit," in Rainer Brömer, Uwe Hoßfeld and Nicolas A. Rupke (eds), *Evolutionsbiologie von Darwin bis heute* (Berlin: VWB-Verlag, 2000).

83 The university institute Astels was supported from the middle of the 1930s to the end of the war with 5,000 RM per year by the Carl Zeiss Foundation. See Nr. 550, Bl. 22, Abt. C, Thüringer Volksbildungsministerium, ThHStAW.

84 See Zimmermann, *Die Medizinische Fakultät* and Hoßfeld, "Menschliche Erblehre," "Die Jenaer Jahre."

85 See Hoßfeld, "Menschliche Erblehre."

86 During the NS period she wrote a book with the title *Einführung in die Variations- und Erblichkeitsstatistik* (Munich: J. F. Lehmanns Verlag, 1935), which, with altered contents, served in its third edition in the GDR as a textbook, *Grundriß der biologischen Statistik für Naturwissenschaftler, Landwirte und Mediziner* (Jena: Gustav Fischer, 1957). Also see her articles "Neue Ergebnisse der Zwillingsforschung," *Volk und Rasse*, 12 (1937): 145–6; "Neue Ergebnisse der Zwillingsforschung bei verschiedenen Krankheitsgruppen," *Volk und Rasse*, 12, (1937): 358–61; "Die rassenhygienischen Gesetze und Maßnahmen in Deutschland," *Neuland*, 7/8 (1941): 96–100 etc.

87 See Hoßfeld, *Gerhard Heberer*; several authors in Thomas Junker and Eve-Marie Engels (eds), *Die Entstehung der Synthetischen Theorie* (Berlin: VWB-Verlag, 1999); Wolf-Ernst Reif, Thomas Junker and Uwe Hoßfeld, "The Synthetic Theory of Evolution: General Problems and the German Contribution to the Synthesis," *Theory in Biosciences*, 119 (2000): 41–91.

88 See Jonathan Harwood, *Styles of Scientific Thought: The German Genetics Community 1900–1930* (Chicago: University of Chicago Press, 1993); Jost S. Casper and Manfred Eichhorn, "Otto Renner".

89 See Ute Deichmann, *Biologists under Hitler* (Cambridge, MA: Harvard University Press, 1996); Hoßfeld, *Gerhard Heberer*, Uwe Hoßfeld, "Das botanische Sammelkommando der SS nach Rußland 1943 oder: Ein Botaniker auf Abwegen," *Verhandlungen zur Geschichte und Theorie der Biologie*, 3 (1999): 291–312.

90 See Viktor Franz, *Die Vervollkommnung in der lebenden Natur* (Jena: Gustav Fischer, 1920); Viktor Franz, *Geschichte der Organismen* (Jena: Gustav Fischer, 1924); Viktor Franz, *Der biologische Fortschritt* (Jena: Gustav Fischer, 1935).

91 See Erika Krause and Uwe Hoßfeld, "Das Ernst-Haeckel-Haus in Jena. Von der privaten Stiftung zum Universitätsinstitut," *Verhandlungen zur Geschichte und Theorie der Biologie*, 3 (1999): 203–32.

92 See Uwe Hoßfeld, "Evolutionsbiologie im Werk von Victor Franz – Voraussetzungen, Bedingtheiten und Ergebnisse," (Jena: unpublished Magisterarbeit, 1993). Also see the two yearbooks of the society published in 1943 and 1944 with the title "Ernst Haeckel eine Schriftenfolge".

93 Hoßfeld, *Gerhard Heberer*, 81–3.

94 Held in the main auditorium of the university at the opening of the winter semester 1936/1937 on November 6, 1936 in the presence of F. Sauckel; published in the Jena Gustav Fischer Verlag as volume 24 of the "Jenaer Akademischen Reden".

95 See Günther Franz (ed.), "Die geschichtlichen Grundlagen des mitteldeutschen Volkstums. Eine Vortragsreihe," in *Das Thüringer Fähnlein. Monatshefte für die mitteldeutsche Heimat*, 7 (1938): 361–405.

96 Thus the DFG gave financial support to research projects in Astel's institute; see Hoßfeld, "Menschliche Erblehre," 375.

97 See Hein Retter (ed.), *Peter Petersen und der Jenaplan. Von der Weimarer Republik bis zur Nachkriegszeit. Berichte – Briefe – Dokumente* (Weinheim: Deutscher Studien Verlag, 1996); Hein Retter (ed.), *Reformpädagogik zwischen Rekonstruktion, Kritik und Verständigung. Beiträge zur Pädagogik Peter Petersens* (Weinheim: Deutscher Studien Verlag, 1996); Hein Retter, "Peter Petersens Identitätsbalancen vor und nach 1933," in Tobias Rülcker and Jürgen Oelkers (eds), *Politische Reformpädagogik* (Frankfurt am Main: Verlag Peter Lang, 1998), 563–89; Andreas von Prondczynsky (ed.), *Lokale Wissenschaftskulturen in der Erziehungswissenschaft* (Weinheim: Deutscher Studien Verlag, 1999), 75–187.

98 See several articles in Rüdiger Stutz (ed.), *Macht und Milieu. Jena zwischen Kriegsende und Mauerbau* (Rudolstadt: Hain-Verlag, 2000).

99 John et al., *Die Wiedereröffnung.*

100 See Jörg Roeseler, "Wende in der Wirtschaftsstrategie. Krisensituation und Krisenmanagement 1960–1962," in Jochen Czerny (ed.), *Brüche, Krisen, Wendepunkte* (Leipzig: Urania Verlag, 1990), 171–84.

101 See Ralph Jessen, *Akademische Eliten und kommunistische Diktatur: die ostdeutsche Hochschullehrerschaft in der Ulbricht-Ära* (Göttingen: Vandenhoeck and Ruprecht, 1999); Herbert Gottwald, "Die Hochschulreform unter besonderer Berücksichtigung der Entwicklungen an der Universität Jena," in Karl Strobel and Gisela Schmirber (eds), *Drei Jahrzehnte Umbruch der deutschen Universitäten. Die Folgen von Revolte und Reform 1968–1974* (Vierow bei Greifswald: SH-Verlag, 1996), 161–169; Rüdiger Stutz, "Die verordnete Reform. Die Friedrich- Schiller-Universität und die Hochschulpolitik in der späten Ulbricht-Ära," *Uni-Journal*, 6 (1998): 20–21.

102 IV B–2/6/459, Bl. 253, Bezirksparteiarchiv, Bezirksleitung SED Gera, Thüringisches Staatsarchiv Rudolstadt [Th Rud]; for a comparison, in the year 1968 84,140 inhabitants lived in Jena.

103 Franz Bolck, "Aufsätze und Reden als Rektor der Friedrich-Schiller-Universität Jena," *Jenaer Reden und Schriften* (1983): 16.

104 See Nr. 2744, Bl. 45–83, Bestand VA, UAJ.

105 See Nr. 2744, Bl. 84ff., Bestand VA, UAJ.

106 Zentrales Parteiarchiv der SED [ZPA], Abt. SAPMO, Best. Büro Mittag, DY 30/IV A 2/2.021/577, Bl. 137, Bundesarchiv.

107 See Christian Winter, *Gewalt gegen Geschichte. Der Weg zur Sprengung der Universitätskirche Leipzig* (Leipzig: Evangelische Verlagsanstalt, 1998).

108 Quoted from Christoph Graudenz, "Das Jenaer Universitätshochhaus 1968–1972. Ein Beitrag zur Stadtgeschichte," (Jena: unpublished Magisterarbeit, 1996), 93; as well as the solidly researched chapter "Die städtische Neubebauung und die Jenaer Öffentlichkeit – Der Kampf Jenaer Wissenschaftler für den Erhalt des Collegium Jenense" in Graudenz, 28–43, and 89–92.

109 IV B–2/6/460, Bll. 60–65, BPA, BL SED Gera, Th Rud; also see the traffic planning in the design competition "Neugestaltung Stadtzentrum Jena 1968" of the urban construction division of the Technical University Dresden from Prof. János Brenner and Dr. Bernhard Gräfe which won the first prize according to this, the research high-rise was supposed to be built "in the form of two triangular prisms"; *Deutsche Architektur*, 19 (1970): 1, 24ff.

110 IV B–2/2/241, Bll. 80 and 86, BPA, BL SED Gera, Th Rud; see the paragraphs from Günter Mittag and Walter Ulbricht from the report by Herbert Ziegenhahns in July 1970 in DY 30/IV A 2/2.021/577, Bll. 104–111, ZPA, Abt. SAPMO, Best. Büro Mittag, Bundesarchiv.

111 See Rüdiger Stutz, "'Durchbruchstellen' des technischen Fortschritts – Walter Ulbricht und die Umgestaltung der Jenaer Innenstadt (1967–1971)," Supplement to von Michael, Diers *et al.* (eds), "Der Turm von Jena. Architektur und Zeichen," *Minerva. Jenaer Schriften zur Kunstgeschichte*, 9 (1999).

112 See Monika Kaiser, *Machtwechsel von Ulbricht zu Honecker. Funktionsmechanismen der SED-Diktatur in Konfliktsituationen 1962 bis 1972* (Berlin: Akademie Verlag, 1997).

113 See Stutz, "'Durchbruchstellen,'" and the article by John Conelly in Mitchell Ash (ed.), *Mythos Humboldt: Vergangenheit und Zukunft der deutschen Universitäten* (Vienna: Böhlau, 1999).

114 NY 4182, Nr. 942, Bl. 35–37, ZPA, Abt. SAPMO, Nachlaß Ulbricht, Bundesarchiv.

115 See Gerhard Kluge and Reinhard Meinel, *MfS und FSU* (Erfurt: Wellendorf, 1997); Gerhard Kluge, *Der "NATO-Professor" Walter Brödel. Dokumentation* (Erfurt: Landesbeauftragter des Freistaates Thüringen, 1999).

116 Schmidt, *Alma mater Jenensis*, 500ff.

117 See Helmut Metzler, "Wechselverhältnis von Industrie und Universität. Zeiss und die dritte Hochschulreform," in *Universität im Zwiespalt von Geist und Macht* (Jena: Schriftenreihe des Jenaer Forums für Bildung und Wissenschaft e.V., 1996), 96–111.

118 Thus spoke the rector in a programatic "collective contribution" with the General Director of the Zeiss Works on the dynamic relationship between science and technology. Both addressed the question, what would it mean to comprehend science as a main productive force. They came to the conclusion "that up until our present time, it has been difficult for the intellectually-creative person to fully comprehend these facts...and to act accordingly." This was mainly due to the "retrospective

a-technical orientation of the humanistic tradition in education," whereby they *expressis verbis* dismissed Wilhelm von Humboldt's main thoughts on the disinterestedness of university education as out of date. Franz Bolck und Ernst Gallerach: "Wissenschaft – Industrie – Hochschulreform," in *Sozialismus und Universität. Walter Ulbricht zum 75. Geburtstag gewidmet* (Jena: Selbstverlag der Friedrich-Schiller-Universität, 1968), 10–23, here 13; see also the article by Rüdiger vom Bruch in Ash (ed.), *Mythos* (Vienna: Bühlau, 1999).

119 Herbert Mehrtens, "Kollaborationsverhältnisse: Natur- und Technikwissenschaften im NS-Staat und ihre Historie," in Christoph Meinel and Peter Voswinckel (eds), *Medizin, Naturwissenschaft, Technik und Nationalsozialismus. Kontinuitäten und Diskontinuitäten* (Stuttgart, GNT Verlag, 1994), 13–32, here 25, "Entzivilisierung".

120 Peter Schöttler, "Von der rheinischen Landesgeschichte zur nazistischen Volksgeschichte oder Die 'unerhörbare Stimme des Blutes,'" in Winfried Schulze and Otto Gerhard Oexle (eds), *Deutsche Historiker im Nationalsozialismus* (Frankfurt am Main: Fischer Taschenbuch, 1999), 89–113, here 105.

121 "Der Reichsstatthalter in Thüringen," Bd. 186, Bl. 9 Thüringisches Hauptstaatsarchiv Weimar.

122 Jörg Roesler, "Das kritische Jahr der Reform," *Neues Deutschland*, (14 July 1998), 12.

9

FRIEDRICH MÖGLICH

A scientist's journey from fascism to communism

Dieter Hoffmann[1] and Mark Walker[2]

"He was arrested under Fascism because he spoke out against the Nazis. After the collapse [of the Third Reich] he was one of the first to contribute unconditionally to the new start [made by the Communist East German state]."[3] This is a quote from an obituary for the physicist Friedrich Möglich. However, a closer examination of his life history makes clear that this statement served a myth. The circumstances that led to his conflict with the National Socialists and his arrest were much more complex than implied by the statement that he spoke out against the Nazis. Indeed, there was a lot more to "Nazi physics" than Johannes Stark and the other so-called "Aryan physicists." Friedrich Möglich is a compelling example of a very different type of "Nazi physicist": young, professionally talented and respected, the student and protégé of Stark's nemesis Max von Laue, and an ideological opponent of "Aryan physics."

The "unconditional" support for the "new start" also suggests conduct by Möglich which should be examined closely. In this case, as in general, the official German Democratic Republic (GDR) historiography and propaganda emphasized the discontinuities between the GDR and National Socialism and encouraged a one-sided anti-fascist myth. The life history of the physicist Friedrich Möglich is thus an example of the kind of continuities which molded the development of the GDR and, along with anti-fascism, the technocratic elite of the "reconstruction generation" in particular. An examination of Möglich's life also offers the possibility of comparing east and west German apologia when dealing with the National Socialist past.

Friedrich Möglich was born on October 12, 1902 in Berlin. His father was a civil servant and second lieutenant in the reserve; his mother was born in England. He went to secondary school (*Gymnasium*) in his hometown and finished his final school exams (*Abitur*) in 1920. Thereafter he studied physics and mathematics at the University of Berlin. In January of 1927 he received his doctorate under Max von Laue with work on "Diffraction Phenomena from Bodies of Ellipsoidal Shape,"[4] which received the unusual designation of "excellent." The work generalized Mie's theory of light dispersion from the sphere to the ellipsoid and closed a physical and mathematical gap in Maxwell's theory.

What sticks out in this work is the skill Möglich demonstrates with the methods of mathematical physics. The Ph.D. thesis was so good and comprehensive that it enjoyed the undivided attention and recognition of Möglich's professional colleagues. Thus it was not only published in the *Annalen der Physik*,[5] but rather before Möglich's dissertation was finished, Ernst Gehrcke, editor of the *Handbuch der physikalischen Optik* (*Handbook for Physical Optics*), invited him to write the articles on the ponderomotive effects of radiation and on the mathematical theory of deformation for the famous compendium.[6]

After his dissertation was finished, Möglich went to Max Born in Göttingen with funds from the Emergency Foundation for German Science in order to study the modern developments of the quantum theory there. As early as 1928 he returned to Berlin in order to continue his scientific education and take an assistant position with Richard Becker at the Technical University in Charlottenberg and, beginning in 1929, at the Institute for Theoretical Physics at the University of Berlin under Max von Laue and Erwin Schrödinger. In February of 1930 he completed his *Habilitation* on "Quantum Effects in Vibrating Continua"; at the same time he received the *venia legendi*, i.e. the right to teach physics at the university.[7] The published version of his Habilitation[8] fit into a very pressing problem at the time, the investigation of the reciprocal effect between matter and an electromagnetic field, and it was part of early efforts to create a quantum electrodynamics.

Thanks to this work and other publications from this period,[9] Möglich established himself as an expert in quantum mechanics. His investigations attempted to develop further Paul Dirac's proposals, for example on the phenomenon of spin. Möglich thereby demonstrated exceptional skills in the mathematical handling of problems, but little physical originality;[10] indeed, a characterization that fits for Möglich's entire scientific production.[11] Nevertheless, Möglich's contribution was so important to the critical physicist Wolfgang Pauli that he cited it in his famous Handbook article on wave mechanics.[12] Shortly before Möglich turned thirty, he had finished his academic education and demonstrated professional competence and ambition. Not least of all, thanks to influential teachers, he was ideally prepared for a scientific career, even if Schrödinger's report on Möglich's *Habilitation* referred to the already mentioned lack of true physical "originality" and further speculated that Möglich perhaps would only "reach Privatdozent and no further."[13]

Of course, something else makes Möglich's life history to this point interesting and distinguishes him from most of his scientific contemporaries. On October 1, 1932 he joined the National Socialist German Workers Party (NSDAP). This came at a time when Hitler had achieved his first significant political victories, but, as Möglich remembered,[14] National Socialism "was still out of the question among German academics and one received nothing but ridicule and scorn." Möglich's early allegiance to Hitler thus cannot be explained only by opportunism, for in the fall of 1932 the National Socialist seizure of power appeared anything but probable. On one hand, Möglich's allegiance had its roots in growing up in a home molded by national pride.[15] On the other hand, he kept a critical distance from the "parliamentary swindle and haggling over political office of the Weimar Republic and especially to the immeasurable corruption, which reached into the organization of the working class."[16]

Möglich committed himself to the party and the tasks that came with it, not as a brawling SA man, but rather within the framework of the Cultural Department of the Berlin's Regional Leadership of the NSDAP.[17] The need for such a committed and competent "fighter" was certainly greater here than in the street, for there were not all that many established scientists who had publicly pledged allegiance to the NSDAP before Adolf Hitler came to power on 30 January 1933.[18] Naturally their career opportunities improved significantly after the NS "seizure of power."

Thus for example in 1934, Möglich became a local official in the quasi-mandatory organization for university teachers.[19] His professional position was strengthened in parallel to his political rise, for in the fall of 1933 he was paid to lecture on the mathematical principles of physics along with his activities as assistant and Privatdozent. A little later the faculty asked the Ministry to name him a non-tenured adjunct professor because of his success in teaching and research.[20] Although this application was turned down, by the middle of the 1930s Möglich had moved into a position in the academic hierarchy which qualified him for a professorship. As a colleague noted, Möglich was "extremely ambitious."[21]

Möglich's political activities apparently did not tarnish his relationship and collaboration with his teacher Max von Laue, although Laue belonged to the few prominent and active opponents of the NS regime among scientists. Laue, and probably other physicists critical of the NS regime as well, obviously made a distinction between idealistic-opportunistic NS protagonists like Möglich and those others who, like Johannes Stark or Philipp Lenard, not only followed Hitler, rather also advocated abstruse scientific ideas, tried to exploit their political influences by altering modernity in physics in particular and science in general, and sought to dominate their profession.

Thus on July 6, 1933 Laue presented a work by his student Möglich at a meeting of the Physical-Mathematical Section of the Prussian Academy of Sciences and proposed it for publication in the renowned publications of the Academy.[22] A few weeks later, however, Laue's courageous stand in the Academy meeting of December 14 hindered Johannes Stark's election to the Academy.[23] Both of these actions took place in the name of a supposed "apolitical science," even though, at that time, Möglich's NS worldview was not only known, but he also had a reputation as a "Big Nazi."[24]

However, Möglich's days as a "Big Nazi" were numbered. He had already come into conflict with the NS University Teachers League at the end of 1934 and on January 1, 1935 was fired from his position as a local party official "because he has no way fulfilled the expectations which one had for an 'old'

Party comrade."[25] Little is known about the reason for this conflict; Rompe's obituary was clearly apologia when it spoke of the "open rejection of Hitler's politics,"[26] subsequently even of the open rejection of "the anti-Semitism and of the anti-Jewish laws and refusal to carry out directives of the Propaganda Ministry."[27]

The roots of the conflict should probably be sought in a certain naïveté and political idealism. Möglich probably sympathized with the left wing of the NSDAP gathered around Gregor Strasser and Ernst Röhm, and may have been so alienated by the brutal purge of Röhm and his followers, as well as the general political opportunism and political practice of the NSDAP after they came to power, that he revolted.[28] The subsequent developments show, however, that this can in no way be seen as a break in principle with National Socialism.

His conflict with National Socialism escalated in the following months, but this had little to do with political differences; rather it was the result of a love affair. At the beginning of 1934 Möglich made the acquaintance of Countess Else von Bubna-Littitz-Gerisch, who at that time was still married to a Dr. G. Lechler, a colleague in the NSDAP Cultural Department. After this marriage ended in divorce in the fall of 1934, Möglich became the lover of the Countess, who at this time lived both in Berlin and Budapest. When Möglich returned to Berlin from a trip to Budapest in August of 1935, he learned that the Gestapo was investigating him for illegal currency transfer and "racial defilement" (*Rassenschande*), and threatened him with arrest. Since Möglich suspected political intrigue behind these charges and, as he himself wrote,[29] "in these times one" could not "hope for objectivity and justice from the Gestapo," he decided to flee Germany. Max von Laue was also helpful when it came to Möglich's flight, by driving him late at night to the German-Czech border,[30] something that Laue later remembered "at that time was easier to do than one would think possible today."[31]

In the following months Möglich stayed in London and Paris, sometimes together with Bubna. While in Paris he heard the news from his mother that he had received the call to a professorship at Heidelberg. At this time he could not bring himself to return to Germany and take the position. Thus he informed the Ministry of Education that he could not take the job because of an illness, with the result that he was given leave for the winter semester 1935/6. In March of 1936 the Ministry approached him again and asked him to take over the vacant professorship for theoretical physics at the University of Heidelberg for the coming summer semester.[32]

This appointment not only fulfilled his professional hopes and personal desire to secure a certain standard of living (not least with regard to his planned marriage to Bubna), it was also a very prestigious offer for Möglich. Particularly in physics, Heidelberg belonged to one of the leading German universities, and moreover it was a place where until recently Philipp Lenard, representative of the so-called "Aryan Physics" and notorious opponent of modern theoretical physics, had worked and still held great influence. Apparently, Möglich had not fallen so low in the favor of the powerful. Despite a few misgivings, Möglich decided in May of 1936 to return to Germany, although not before he had checked with the German Embassy in Paris and received the assurance that he and Bubna were not threatened with any kind of prosecution.

Möglich was mistaken. After his return he first of all tried to explain the reasons for the time he spent abroad to his superiors in the Reich Ministry of Education and to portray these as a purely private trip and not as flight or even preparation for emigration. That was not easy, for in the fall of 1935 he had sent a confidential letter to a friend and colleague in the Ministry, Professor Leyhausen, only to have his "friend" forward it on to the Security Service of the SS. In this letter Möglich described his relationship with Bubna and at the same time criticized the "excessive anti-Semitism" of the summer of 1935. In this context he declared that: "there is only the Bubna-Möglich case. If I return, then the countess must also be able to return and finally the Ministry must grant permission (for us) to marry. If this is not granted, then that would be an incentive for me once again to leave Germany."[33]

Even if Möglich apparently did have friends and patrons in the Ministry and Party apparatus, after his return in the summer of 1936 the charges had not gone away; instead the Gestapo and Security Service of the SS reopened their investigation of him. For a short period of time Möglich returned to his activities at the university[34] and also worked especially hard to legalize his relationship with Bubna and, through expert reports, for example from the Reich Office for Hereditary Research, to verify her Aryan ancestry.[35] He did this for love, even if it was also clear to him that the professorship he desired, as a job in the civil service, required that "my wife also be Aryan."[36]

The basis for the criminal investigation were the laws "for the Protection of German Blood and German Honor" which Hitler had announced in September, 1935 at the Nuremberg Party Conference of the NSDAP. These so-called "Nuremberg Laws" provided legitimization for the National Socialists' anti-Semitism. Henceforth not only marriage, but even "extra-marital intercourse between Jews and citizens of German or related blood" was forbidden and could be "punished by jail or prison."[37] Möglich was bound by these laws in particular as a long-standing Party and SA member and especially because of the significant Party functions he had exercised, with the result that his failure to honor them was investigated especially thoroughly. Not only was the legal and police apparatus involved, but the corresponding organs of the Party were as well.[38]

Möglich had long hoped that the matter would be cleared up to his benefit, but, to his own surprise, he was arrested by the Gestapo on January 15, 1937 and placed in "protective custody," a status whereby he could be held indefinitely without any rights to see a lawyer, etc. In the meantime, the investigation by the Reich Office for Hereditary Research had yielded the fact the Bubna had been listed as a member of a Jewish community in Berlin, so there could be no more doubt about her Jewish ancestry. The Nuremberg Laws could now be applied. Together with Möglich, Bubna was now also taken into protective custody, but after a thorough interrogation was released. After another subsequent interrogation by the Gestapo in February of 1937, she committed suicide.

As the indictment of the General State Attorney detailed before the Berlin State Court on February 9, 1937, Möglich was accused specifically of "continuing in Berlin and other places, from the middle of September 1935 to 1937, extramarital sexual relations as a citizen of German blood with the countess von Bubna-Littitzt-Gerisch, born Hirschberg, a Jew, on 5 May 1895 in Berlin, a crime against article 2, 5th paragraph of the Law for the Protection of [German] Blood."[39] On July 6, 1937 this case came before the 8th Greater Criminal Court in Berlin, which finally ended with Möglich's acquittal. This acquittal was justified because, although the objective fact of racial defilement had been proven, the State could not prove that Möglich had known at that time that she was a member of the Jewish community and thereby had willfully committed the crime.

Because of this verdict Möglich was immediately released, but the matter was not thereby ended. His professional rehabilitation made no progress, although immediately after his acquittal his mentor Max von Lane contacted the Dean and Curator of the University and urged them to "give back his assistant position and teaching contract."[40] Möglich himself appealed to the Prorector of the University and attempted to justify his behavior.[41] This remained unsuccessful, and neither the University nor the Reich Education Ministry even considered reopening Möglich's future as an academic teacher after his acquittal.

Since Möglich (as the judgment of July 1937 stated) "had behaved carelessly as a Party member and German comrade" and had damaged the prestige of the Party, he had also become unacceptable as a university instructor, i.e., as model and teacher of young students from the standpoint of National Socialist ideology. However, the Education Ministry wanted to wait with the definitive legal decision on his dismissal from university service until a proceeding to expel him from the party had been carried out. Countless letters from the Education Ministry to the various Party officers during the fall of 1937 and early part of 1938 requested again and again information on the status of the Party proceedings against Möglich and pressed for a quick completion of the proceedings.[42]

Möglich himself was informed in January of 1938 that proceedings to expel him from the party because of "sexual intercourse with a Jewess" had begun at the district court IVA of the NSDAP in Berlin.[43] These ended in June 1938 with the expected guilty verdict, that is, with Möglich's expulsion, since "the accused had acted contrary to the efforts of NSDAP, in that he had sexual intercourse with a carrier of Jewish blood."[44]

The "Möglich Affair" could have ended here. Instead, it preoccupied various offices in the NSDAP and Friedrich Möglich himself for almost three years more. Indeed, Möglich became the driving force behind the proceedings because he was unwilling to accept expulsion from the Party. The motives for his obstinate perseverance are not entirely clear. Undoubtedly at that time Party membership was a necessary condition for Möglich to restart and continue his activity at the university. In a letter to the NSDAP Regional Court from July 17, 1939 he himself noted that "Professorship means a civil servant career. Setting aside all ideological considerations, separation from the Party would make this impossible."[45] Taking into account the reality of life in National Socialist Germany, with the immense increase of anti-Semitic propaganda and harassment as well as increasing political ideological intolerance and persecution by the Party and state repression apparatus, Möglich must have understood, at least at some level, that his hopes for rehabilitation were unrealistic and that it might be better to cut his losses.

Why did he need to rescue his honor? Most probably because he wanted to come to some kind of arrangement with the NS regime. This was certainly not merely due to opportunism, but rather also because of the ideological conviction which had led him into the ranks of the NSDAP even before the NS "seizure of power" and to which, at this point in time, he still fundamentally adhered. Moreover, he was at that time undoubtedly convinced that National Socialist Germany was not only to be accepted as a whole, but also to be supported to the best of his ability. Naturally he was aware of the shortcomings and brutality of this society, for they had affected him personally, but he nevertheless assumed that these were exaggerations of the NS revolution, of secondary importance when compared to Hitler's great successes and sooner or later correctable through reforms. This was a stance which a large majority of Germans shared and which found additional support through Hitler's domestic and foreign policy successes. At that time (1938–40) Hitler was at the pinnacle of his power and popularity.

Like so many of his countrymen, Möglich wanted to participate in these successes, all the more so since, as an "old fighter," he believed that he had earned preferential treatment; had he not already supported Hitler when the latter enjoyed anything but fame and honor? Should this fail because the "misfortune of his love" for Countess Bubna had come between Möglich and the Party and he had become the victim of "boundless anti-Semitism"? Given this background, the "harshest and most disgraceful Party punishment...[should not be the] only possible solution"[46] and he therefore, despite all the consequences, fought for his rehabilitation.

Moreover, in the course of this effort he not only employed the usual National Socialist rhetoric. In an emphatic objection to the verdict and his expulsion from the Party, on the night before the so-called "Night of Broken Glass," he wrote:

> However, the harshest judgement for me would be expulsion from the Party...because an eventual expulsion would be an incomparably harder blow for me than for many others. My work was devoted to a scientific research and education in the sense of the Leader [Hitler].

The possibility of exact scientific work, the content and task of my life, is exterminated if the motive power of the Party is taken away. I know that I have made a mistake. I believe, however, that the generosity of the Leader does not destroy where not wickedness, rather human weakness committed a mistake. Please examine it once more. In the meantime I will demonstrate even more strongly through work and conduct that I am what I formerly was: Adolf Hitler's soldier.[47]

Möglich could still be seen as "Adolf Hitler's Soldier" in 1938–9, when he visited a laboratory in the Siemens works wearing a SA uniform.[48] In the course of his attempted appeal against expulsion from the Party, he also did not simply deny his guilt, rather he attempted to justify or to downplay it and asked in particular for "a mild verdict and to leave (him) in the Party."[49] In this sense the district court noted that he had admitted "to have made a mistake...but [appeared] not to have comprehended the consequences of his actions in the National Socialist sense."[50] He informed the regional court that "for me expulsion from the Party means the most grievous insult to my honor which could be done to me."[51]

After the regional court in Berlin upheld the decision of the district court in November, 1939, Möglich was granted the right to appeal to the Highest Party Court because of the "early profession of allegiance by the accused to the Party and his active work for it" as well as "in consideration of the circumstance, that this case represents a fundamental decision."[52] In the summer of 1940 this court then subsequently and definitively upheld his expulsion from the Party, whereby the opinion declared that Möglich "had not understood the racial foundation of National Socialism and, as his actions demonstrate, rejected it. The accused thereby violated the efforts of the Party and forfeited the right to continue to belong to it. In such a case, milder punishment cannot be granted to him for fundamental reasons, even taking into consideration his long-standing membership in the Party and service to it."[53]

The judgment of this Highest Party Court was much harsher in tone and content than the opinions of the other courts. This was probably due to the fact that, in the meantime along with racial defilement, the political aspect of the Möglich case came to the attention of the judges. Two weeks after the regional court had met, the leader of the Berlin NS University Teachers League called attention to the letter which Möglich had written in November 1935 from Paris to his supposed friend Leyhausen in the Reich Education Ministry and which contained "insults to National Socialism."[54] As already mentioned, Möglich had, among other remarks, spoken of "boundless anti-Semitism" and "an all too great intellectual narrowness" and committed himself without reservation to his lover.[55] Once the letter was requested by the Highest Party Court from the Reich Security Main Office and, with some delay and thanks to the direct involvement of Rudolf Hess, transferred over,[56] Möglich was called upon to respond to it.

This was certainly done in part in the hope that the compromising letter would convince Möglich that his case was hopeless and thereby avoid bringing the case before the Highest Party Court. The letter from the Court threatened Möglich very clearly and noted that his 1935 letter could be used against him.[57] But this did not dissuade Möglich. He merely argued that the letter contained nothing new and indeed corroborated the position he had taken in previous court proceedings. However, he did admit that a "few sentences from the letter to Professor Leyhausen [are] today incomprehensible to me and I regret them very much. They can only be explained by an exceptional embitterment and the feeling, that I had suffered injustice. Today I recognize that my panic at the time has made my position much worse."[58] Once the Highest Party Court had definitively upheld Möglich's expulsion from the Party, the Reich Education Ministry hastened to ban him permanently from the university and deny the application he had submitted the year before to be named "Dozent of the new order"[59] by revoking the "right to use the designation Dozent."[60] Möglich's academic career in Hitler's Germany was thereby terminated once and for all.

Thanks to friends and connections, which had also helped him survive the previous years, a blacklisting did not mean total exclusion from his profession. In particular, it was once again his mentor Max von Laue, who helped him find a position in industry. Thanks to Laue's mediation, in the autumn of 1937, thus immediately after his release from protective custody, Möglich received short-term positions at different Berlin industrial firms and the Observatory in Babelsberg.[61] In this context the most promising position, which probably also came through Laue's mediation, was a contract as advisor for Osram, one of the largest firms in the German electrotechnical industry.

As Rompe writes,[62] Laue knew two members of the Board of Directors, A. Meier and K. Mey, well; perhaps connections to the Siemens Family could also have played a role, for Möglich had known Hildegard Siemens since the autumn of 1937, and married her the next year.[63] This activity as an industrial advisor guaranteed Möglich more than his mere survival; it made him financially independent. From his own information,[64] his monthly income was around 1,700 RM, triple his previous income as an assistant (600RM) and significantly higher than the average income at the time of a managerial employee in the metal industry (around 1,000 RM).[65] Despite the tainted expulsion from the Party, it was possible to live well in the Third Reich.

At Osram, Möglich got to know Robert Rompe, who worked there as a physicist and whose commitment to communism meant that an academic career in the Third Reich was closed to him.[66] By his own statement, Rompe was asked by Laue "to keep an eye on Möglich,"[67] advice he then made full use of. In the following years Möglich and Rompe not only developed an intensive scientific collaboration between them, rather also became close friends. This close friendship became very valuable for Möglich. In the Third Reich, especially during the years 1939–40, it certainly played a decisive role in turning Möglich increasingly away from National Socialism, both ideologically and politically. Thus Möglich did not learn to hate the National Socialists only because of his concrete personal experiences. By the end of the war he had become a consequential "anti-Nazi."[68] As such he came once again in contact with the Gestapo in 1942–3, who interrogated him with regard to listening to foreign radio broadcasts, specifically the BBC.[69] However, this denunciation apparently had no adverse consequences for Möglich.

The relationship between Möglich and Rompe was also scientifically productive. Although up until that time Möglich had devoted himself almost exclusively to questions of quantum theory and problems of superconductivity, solid state physics now increasingly found its way into the area of his investigations. This research was not restricted to the area of theoretical physics, rather, as Rompe wrote in his obituary,[70] included "the meeting with productive practice," "the intervention of 'pure' research in the development of technology" and expanded Möglich's scientific perspective. These perspectives and experiences, which at first were certainly seen as a necessary evil, then as a virtue, became an advantage for Möglich's career in the postwar period and for the scientific reconstruction in the subsequent GDR. The socialist model of science and science policy was orientated towards the direct connection between pure and applied research as well as a close reference to technology and practice.

However, even in the context of his activity as an advisor, Möglich always returned to fundamental questions of physics. Thus in 1938 he published a short original article in Die Naturwissenschaften on the main relationship between electrons and neutrons,[71] and in 1939 worked together with Rompe on a publication which dealt with the physical and cognitive consequences of the existence of a specific shortest length.[72] Another "metaphysical" problem kept him busy at this time, for his competency in matters of quantum mechanics led him into the area of biophysics. He began to take up questions concerning the physical or quantum theoretical basis of life or heredity. The spiritus rector of the biophysical interest was certainly once again Rompe, who led Möglich at the beginning of the 1940s to the discussion circle in Berlin-Buch around the biophysicist Nikolas W. Timoféeff-Ressovsky.

A special colloquium discussed questions of modern genetics at the Kaiser Wilhelm Institute for Brain Research, whose radio-biological department was headed by Timoféeff. Along with Rompe, Möglich, Timoféeff-Ressovsky, K.G. Zimmer, and G. Born, participants included H. Stubbe, P. Jordan, N. Riehl, W. Weizel, and, up until his emigration, M. Delbrück.[73] Two articles by Möglich and Rompe for Die Naturwissenschaften emerged from this discussion[74] and examined the question of what modern physics, that is, quantum mechanics, and not least the so-called "Hit Theory" could contribute theoretically to the creation of "biophysical models." In this context, the mechanisms of energy propagation and transformation for radio-biological processes, and to what degree they could cause mutations, were a special interest. These publications played a role in the development of modern molecular biology and contributed in particular to contemporary discussions on the relationship between genetics and molecular biophysics.

Another problem area for Möglich was superconductivity, which he had worked on for many years and which led him directly to solid state physical research. He had already come into contact with it as Laue's assistant. In the context of Laue's efforts to set up a (phenomenological) theory of superconductivity, Möglich and his teachers investigated the destruction of the superconductive state by a magnetic field and the question of the current distribution in superconductors. This work, which Laue presented on 18 May 1933 to the meeting of the Physical-Mathematical Section of the Berlin Academy,[75] provided an important stimulus for the discovery of the so-called displacement effect by W. Meissner and R. Ochsefeld a few months later in the low-temperature laboratory of the Imperial Physical-Technical Institute in Berlin.[76]

In the following years, Möglich took up questions of the theory of superconductivity again, and it was certainly this activity that preserved his scientific contact with Laue over the years. Möglich tried in this context to further the design of Laue's theory of superconductivity, which deviated from the phenomenological theory of London in a few points. In the tradition of his teachers, Möglich tried to develop an electrodynamic theory of superconductivity which avoided the application of London's *additional terms*. Immediately after the war, Möglich and Rompe attempted in a series of works to explain superconductivity on the basis of the special interaction of electrons, the so-called plasma oscillations.[77] This attempt took its place in the worldwide efforts after the war to find a microscopic theory of superconductivity.[78]

Möglich and Rompe followed the idea that superconductivity was a collective quantum phenomenon and sought the concrete explanation for the phenomena in the clarification or quantification of the Coulomb interaction of the electrons. Similar ideas were developed in this connection not only by Möglich and Rompe, but also by other contemporary physicists. However, in 1950 the discovery of the so-called isotope effect made it clear that the electron–electron interaction did not underlay the phenomenon of plasma oscillation; instead the interaction with the crystal lattice, i.e., the electron–photon coupling, was central.[79] A direct path lay from this recognition to the so-called BSC theory of the American physicists John Bardeen, John R. Schrieffer, and Leon N. Cooper. Möglich did not take this path, but continued to attempt to make the model of plasma oscillations more precise, for example by the incorporation of hydrodynamic considerations.

Möglich's stubbornness was in part a sign of the scientific isolation of postwar Germany and especially the GDR. Financial reasons alone made it impossible to travel to the important conferences on low temperature physics (for example, the MIT conference in 1949). Moreover, scientists like Möglich were only a peripheral part of the scientific community of the time. Along with those objective reasons came subjective ones, which had to do with the Cold War and ideological blockades. Thus participants at a solid state physics conference (Dresden 1954) remembered how Möglich not only grappled with the ideas of the American physicists theory of superconductivity in a scientific way, but also based his critical stance on ideology and denounced their theory as a "product of American imperialism."[80] Möglich thereby certainly did not so much follow his own convictions as serve the ideological spirit of the time and the rhetoric of the Cold War. For example, at the same time there were continuous discussions of Pauling's resonance theory of chemical bonding, which the Communist Party philosophy then also condemned in the name of dialectical materialism.

However, beginning in the 1940s, it was not the theory of superconductivity that stood in the center of Möglich's scientific production, but rather problems of solid state physics. In his research activity at Osram, he was especially concerned with questions of materials along with the physics of gas discharge. This led him even further into the area of solid state physics, which then became his main domain for the remainder of his life. Solid state physics at that time was an area which was still in its infancy with regard to its physical basis. It was only with the development of quantum mechanics that a theoretical approval was found which could explain the phenomena of solid states, namely their electrical, optical, and thermal properties.[81]

A turning point in the application of quantum mechanics to solid states came with the path-breaking work of Felix Bloch on the behavior of electrons in crystal lattices in 1928. At the beginning of the 1930s, the concept of energy bands was developed in order to describe the electronic spectrum in solid states, which became the standard model for modern solid state physics. This so-called band model provided a general theoretical approach for handling solid state properties, but little was known about the concrete band structure of the individual matter and materials, and the model had to be brought into agreement with the large number of observations available and tested in detail against the new effects.

The consolidation of the theoretical and experimental basis of this field in the context of quantum mechanical concepts stood at the center of solid state physical research in the 1930s and 1940s. Möglich's work should be seen in this context. He was mainly interested in the theory of semiconductors and especially in using the tools of quantum mechanics to penetrate to the mechanism of energy transformation and energy conversion in solid state for radiant processes and related processes. The choice of such a research emphasis was not coincidental, for Osram was especially interested in the development of new light sources and radiation detectors, whereby not least a better understanding of the physical basis of the processes connected with it could be beneficial. Thus Möglich carried out investigations of the theory of solid isolators, of the mechanism of illumination and of the energy transformation in crystal phosphorus as well as of the internal photo effect of illuminants.[82] He also participated in a secret military research project on the development of new types of photographic conductors, so-called photographic metals.[83]

Möglich's and Rompe's theoretical investigation of a new type of interaction mechanism for the radiation-free energy release of stimulated electrons in crystals was an especially valuable part of this project.[84] The concept of the so-called multiple collisions or the multiphonon recombination is today generally recognized in semiconductor physics and is applied in the modern semiconductor theory for handling of the electron–photon interaction and the description of the effects connected to it. When compared to the international research standard, this research was perhaps no scientific breakthrough worthy of a Nobel Prize, but they were solid research achievements which established Möglich's reputation as one of the best "authorities on solid state theory in Germany."[85]

After the war, this reputation was responsible for Möglich being charged with the reconstruction of solid state research in the Soviet Occupation Zone (SBZ)/GDR. However, it is clear that these administrative and organizational functions caused his scientific productivity and the originality of his publications to suffer. In this context, it is still worth mentioning that it was Möglich and his colleagues who oriented the research in the institute towards investigations of the CdS crystal and thereby took an important international development in semiconductor research into account, for the CdS advanced during the 1950s and early 1960s to become a sort of model substance for this research area.

But Möglich still had to survive the war years and the downfall of the Third Reich before he could participate in the development of a research center or institute for solid state research. He did this surprisingly well. Once again he was able to draw on the help of friends and colleagues. When it appeared that he might be drafted into the armed forces, Heinz Schmellenmeier provided help. Schmellenmeier earned his doctorate in physics in 1935 and was a member of the Communist Party since 1932. He had problems with the National Socialists at almost the same time as Möglich, however, for clear political reasons. He was arrested in connection with the destruction of a Berlin resistance group, but due to lack of evidence he was released by the Gestapo after several months of protective custody. After his release, he was first of all employed as an industrial physicist. In early 1941 he founded his own company, the Development Laboratory Schmellenmeier, which was devoted to the development and production of "devices and processes which were technically or scientifically interesting." This small firm, at first practically a one-man company, essentially financed itself through research contracts from Army Ordinance.

This connection offered the possibility of signing a consultant's contract in the autumn of 1941 with Möglich (and Rompe) on the further development and realization of a patent application which described the "Possibilities of Producing Decimeter and Centimeter Waves."[86] The obvious military-technical relevance of this research (development of radar) not only emphasized the military importance of Schmellenmeier's Development Laboratory, it also facilitated a U.K. classification ("indispensible," exempting the individual from other military service) for Möglich and Rompe; moreover, Schmellenmeier was able to make similar arrangements in a short period of time for Richard Gans, a well-known Jewish professor, and thereby save him from deportation to Auschwitz.[87] The contact between the three scientists even intensified during the last years of the war. In 1943 Schmellenmeier's Development Laboratory, as well as parts of Osram, were evacuated to Oberlausitz because of the increasing bombing attacks on Berlin. Here they held common scientific colloquia, discussed what would be done after the war and, as Schmellenmeier recalled,[88] did not live at all badly.

Möglich experienced the end of the war and the entrance of the Red Army at the beginning of May, 1945 in Olbersdorf. The transportation system and communications network had totally collapsed in the first postwar months, so that insurmountable obstacles stood in the way of a fast and orderly return to his hometown, Berlin. Möglich had therefore decided to stick it out with his family in Olbersdorf, since life was much easier to organize there than in bombed-out Berlin. Apparently Möglich first returned to Berlin in the summer of 1945, at a time when the Potsdam Conference had already made Berlin into a four-sector city of the Allies. Möglich's apartment lay in the western part of Berlin, which was now administered by the American occupation forces. When he returned he learned that the Americans were interested in him and had inquired in the Friedenau neighborhood about the "atomic physicist" Möglich; according to a family story,[89] a military policeman was even posted in front of the door.

The Allies' hunt for German "specialists" was running at full speed at that time.[90] A group of German atomic physicists had already been brought to the Soviet Union in the early summer of 1945,[91] his mentor Max von Laue and members of the German "uranium club" were interned in England,[92] and the American project "Paperclip" had gathered together a group of German rocket specialists around Wernher von Braun.[93] Although all of these actions were secret, they nevertheless were noticed by the German civilian population. The wildest rumors and speculations found their way to Möglich's ears. As far as he was concerned, he wanted neither a "dollar contract" nor to be required to go to the Soviet Union.[94]

For this reason he turned to his old friend and colleague Robert Rompe, who had experienced the end of the war in Berlin-Buch. Part of Osram was also evacuated during the last months of the war to the northern perimeter of Berlin. There they enjoyed the hospitality of Timoféeff and used the empty rooms of the Kaiser Wilhelm Institute for Brain Research, which had been evacuated to Göttingen away from the approaching Red Army. After the arrival of the Red Army the institute complex was placed under Soviet protection and thus remained mostly free from the encroachment, plunder, and other unpleasantness of the first weeks of the postwar. However, Timoféeff was soon arrested by the Soviet KGB and deported to a forced labor camp[95] and Riehl was "won over" for research work on the Soviet atomic bomb project.[96]

In contrast, thanks to his communist past and good contacts with the Soviet occupation officials, Rompe belonged to the "untouchables." Since he was moreover immediately enlisted for the political reconstruction of the Soviet zone, he was therefore an ideal contact for Möglich. Rompe supposedly assured Möglich that he would be able to remain in Berlin and would have nothing to fear from the Russians if he came to Buch and continued his research in the "Institute of Medicine and Biology" there, which in the mean time had been placed directly under the Soviet authorities.[97] Thus in the summer of 1948 Möglich and his family did not return to their apartment in Berlin-Friedenau, rather chose the lesser evil and settled down under the protective and certainly also more than bearable umbrella of the Soviet research complex in Buch.

Möglich's return and continuation of the research institute in Buch must be seen as part of the efforts of the Soviet occupation authorities and their German communist allies to bring the research potential of the KWI institutes in the Soviet zone under their control and encourage as many evacuated scientists as possible to return. One hoped in this way to be able to install a direct successor organization to the prestigious Kaiser Wilhelm Society in the Soviet occupation zone, since thanks to directives from the Allied Control Council, the organization itself appeared to stand shortly before on the brink of dissolution.[98] Moreover, such research organizations, directly subordinate to the ministry in Moscow, fit into the general concept of the Soviet occupation authorities in the immediate postwar period.[99] In this way they were able not only to offer the scientists something and hinder their departure to the western zone, but at the same time they also wanted optimally to exhaust the scientific and industrial potential of Germany and use this instead for the reconstruction of the Soviet Union.

Only in 1946 did this policy fundamentally change. The Soviet officials now dismantled on a large scale, even entire scientific institutes, and transported them together with part of the institute personnel to the Soviet Union[100]; parallel to this development, the remaining institutes, as well as other research institutions outside the universities, were increasingly subordinated to the newly-founded German Academy of Sciences, which was to take over science policy role played by the KWG in the Soviet zone and subsequent GDR.[101] Möglich also intervened in the internal discussions on the future of the KWG institutes. In a letter to Max von Laue, who had just returned from English detainment, Möglich asked his mentor to use his influence to ensure that nothing was done in Göttingen which might further damage the already stained relationship between the general administration in Berlin and the Berlin "counter-government" under Haveman and "wake sleeping dogs."[102] However, in this regard Möglich's teacher turned a deaf ear.[103]

Along with his activity in Buch, as of October, 1945 Möglich worked as an advisor in the German Administration for Public Education (DVV) in the SBZ. Here he worked especially closely with Rompe,[104] who since September, 1945 held the powerful post of Main Departmental Official for Higher Education and Science in this agency; moreover, in the meantime Rompe had also moved up into the Party Executive of the KPD (German Communist Party) and its successor SED (Socialist Unity Party) and thus was among the politically most influential scientists in the SBZ and subsequent in the GDR. Indeed, one can go further and claim that he was the "cadre chief" of the entire higher education field and that without his support, no appointment could be successful. The activity of the DVV was concentrated at this time especially on the reconstruction of higher education and the fastest possible reopening of the universities in the SBZ.

Reconstruction was necessary here in every aspect, for the greater part of the scientific research institutions and universities were not only materially destroyed, but the National Socialist personnel policies had left deep ideological traces and dislocation. This situation was summed up by Rompe in an internal memorandum in the following point: "one can distinguish four groups among the university faculty: 1. The truly clean scholars, whose number was never that large; 2. Then a very large number of usable specialist scholars, who lacked only civil coverage and willingness to take a stand; 3. Pronounced opportunists, whose number was not small; 4. Active fascists, most of completely insufficient professional quality, of whom there apparently were not that many."[105]

Thus in the institutes and universities of all occupation zones, not only was the rubble created by war and bombing cleaned away, but there was also an extensive denazification of the personnel. However, in this connection it quickly became clear that a tension existed between the "elevated political goals of cleansing and the need for specialists for the reconstruction of an economy capable of functioning."[106] This was especially true for the Soviet occupation zone, in which the political and ideological boundary conditions appeared to favor a rigorous policy of firing, but in which on the other hand the shortage of skilled workers and qualified scientists was especially grave – not least because of the extensive evacuations to the west during the last months of the war. Moreover, denazification in the SBZ followed a double strategy: on one hand to remove the protagonists of the National Socialist regime from their posts and on the other hand to use these – in some cases severe[107] – interventions in the personnel of the universities in order to clear the way for a new, if politically conformist elite.[108]

Friedrich Möglich's life history can be seen as a concrete example for this "double strategy" of the cadre policies in the university systems of the SBZ/GDR. When the University of Berlin was reopened with a ceremony on 29 January 1946 and resumed teaching operations for the summer semester, Möglich belonged to its faculty, almost exactly ten years after his suspension by the National Socialist authorities. Given the fact that around three-fourths of the faculty had to quit because of their National Socialist past, this was in no way a foregone conclusion; it is also difficult to see Möglich's appointment as an art of official rehabilitation. The circumstances and background of his appoint-ment speak against such an interpretation.

As early as October, 1945, Christian Gerthsen, the only full professor of physics remaining in Berlin, had pointed out the shortage of staff for the teaching of physics at the University of Berlin in a letter to the Central Administration for Public Education and submitted the suggestion that the faculty, "after its restoration to suggest Prof. Dr. Möglich and Dr. Rompe for the theoretical professorships."[109] A short time later a "characterization" of Möglich was submitted, which not only pointed out his scientific merit but also discussed his political past. The designated rector, the philologist Johannes Stroux, thereby noted the "Möglich originally belonged to the Nazi Party, but protested, as already mentioned, in writing against the Nazi government, after he had become concerned of their perniciousness for the German people. The conse-quences of this step were arrest, loss of his position, expulsion from the Party because of demonstrated rejection of Party goals. Either the arrest, or also the expulsion from the Party because of rejection of Party goals would suffice according to the report of the four antifascist parties to annul his membership in the Nazi Party. Moreover, during the war Möglich confirmed himself as an antifascist by acting illegally. He certainly should not be counted among the former NSDAP."[110]

This argument appeared sufficient for those responsible individuals in the Central Administration to appoint Möglich on 12 January 1946 "Full Professor with Chair for Theoretical Physics in the Faculty for Mathematics and Sciences"[111] But they did not appear to be certain about the matter, for in the weeks between his appointment and the beginning of the semester the DVV, and in particular Rompe, asked for evaluations of Möglich from respected but personally well-known colleagues. Given the details of this case, this should be seen as an act of insurance.

The letters from Max von Laue, the Bonn physicist Walter Weizel, and Rudolf Seeliger from Greifswald all praised the quality, versatility, and produc-tivity of Möglich's scientific achievement and noted that his appointment in the Third Reich "failed due to difficulties which had nothing to do with scien-tific or academic activity."[112] The vice president of the DVV, the Berlin physician Theodor Brugsch, was relieved and could report to the president of this agency, Paul Wandel, that "the evaluations were more successful than we could have expected, we have made no mistake with Möglich's appointment, rather we have placed the right person in the right place." In order to oppose future, if perhaps also current rumors circulating about Möglich, Brugsch suggested that "the evaluations of Möglich also be circulated among the faculty."[113]

In retrospect it is not possible to determine precisely to what degree Möglich actually was the subject of rumors and speculation during the postwar years. Inquiries and interviews conducted by the authors all demonstrated that Möglich's National Socialist past was largely unknown among students and his younger colleagues. Among older colleagues this appears to have been a taboo subject – certainly often thanks to their own "brown" past. Apparently Möglich's National Socialist part only became the subject of public discussion on one occasion, in the shadow of the Cold War.

Political conflicts at the University of Berlin led in 1948 to the founding of the Free University in the western part of the city. In early 1948, politically undesirable students were expelled from the University in East Berlin and sharp public conflict arose between the Student Council and the Central Administration for Public Education or University Administration, in which Möglich was involved as Chairman of the University Disciplinary Committee. When Möglich obligatorily approved the disciplinary measures, one of the people affected, the head of the SPD University Group, Otto Stolz, publicly denounced Möglich's National Socialist past, naming not only the precise dates of his NSDAP membership but also his activity in the Cultural Department of the Regional Party Leadership of Berlin. It was also mentioned that Möglich's personnel files were extremely incomplete and merely a "recommendation signed by Professor Rompe" was there.

Those attacks led to inquiries and an explanatory correspondence within the DVV which glossed over the facts and argued that although Möglich had been a member of the NSDAP, "as early as 1933 [he] repeatedly got into trouble with this Party...which ended in 1938 with his expulsion from the Party."[114] Naturally no references were made to his earlier activities, the concrete circumstances of his expulsion, and especially to his determined resistance to it. However, more than an internal explanation appeared necessary, for Rompe passed this on to the rector of the university since "perhaps not all the members of the senate...know these facts." Furthermore he urged the rector "to take the necessary disciplinary measures" if the author of the accusatory piece had unacceptably "received access to the personnel files of the professors maintained by the University."[115]

Instead of confronting the past, in Möglich's case as with others, the past was repressed. Moreover, the way Möglich's personal questionnaire from the Academic was handled in 1948 fits into this picture. Möglich said no to the question of "membership in NSDAP organizations," but left the column "membership in NSDAP" blank and instead referred to an attached certificate in which one of the highest officials, Vice President Bruguh of the DVV, certified that "in agreement with the Highest Soviet Military Administration in Karlshorst, he is to be judged as politically unburdened. His continual antifascist activity during the Nazi regime is known to this administration."[116]

The general denazification was therefore applied in an extremely flexible way. Möglich's membership in the NSDAP, which was in no way merely formal, was removed from discussion in the GDR. Subsequent discussions of this question are not known. Even if such discussions did happen, then they were handled internally and held "under the covers," or, as in the case of the physicist and re-emigrant Martin Strauss, who blamed his difficulties with integration into academic life in the GDR to the career of the "Old Nazi" Möglich,[117] they were dismissed as purely personal conflicts.

Such gentle treatment of Möglich and his National Socialist part was undoubtedly due to the influence that his friend and colleague Robert Rompe enjoyed in the scientific and political life of the early GDR and his ability to take care of certain things. However, a central role was played by the fact that the officials in the SBZ were working intensively to retain the technocratic elite, to integrate them in a controlled fashion and thereby win them over for the reconstruction of the destroyed country. In this sense the chairmen of the SED, Wilhelm Pieck and Otto Grotewohl, in a memorandum from 1952 urged the regional leaders of the Party to "win over the former nominal members of the Nazi Party where possible and make them active participants in the democratic reconstruction."[118]

Especially in the area of the natural sciences, there had been a strong exodus of highly qualified specialists. This began with the expulsion of Jewish scholars during the National Socialist period, and was continued in the last months of the war with the evacuation or flight of entire institutions and research teams to the West, and continued in the postwar period almost without interruption through the conscription by the Allies as well as a continually increasing migration of scientists and other technically trained people to the Western zones of occupation. Symptomatic of this was the example of the Kaiser Wilhelm Society, the largest and most important research institution outside of the universities in Germany. Not only was the general administration evacuated from Berlin to Göttingen, but most of its institutes were also evacuated and deserted by the end of the war.[119] The membership of the Prussian Academy of Sciences had a similar fate, for only 25 percent of its members, mainly representatives of the philosophical-historical section, were still in Berlin by the end of the war.[120] The situation was better at the universities with regard to the personnel, but here the material and intellectual destruction was disastrous. A large portion of the buildings and laboratories lay in ruins and the majority of the professors had to be fired because of their involvement with National Socialism.

Although the destruction and shortages of the first postwar years were grave, for those who remained they also offered diverse and profound opportunities for career development. This was certainly the case for Friedrich Möglich, whose appointment at the newly opened University of Berlin was not yet the high point of his postwar career. As director of the prestigious Institute for Theoretical Physics, Möglich was expected to bring research and education back into gear at the University and revive the old traditions of this prestigious institute. Given the grave postwar problems, this was no easy task. Laue wrote to him when he began his job: "may you be able to preserve the old spirit in the unavoidably new form."[121] The institute grew quickly with respect to its personnel, space, and equipment beyond its prewar level: the institute used more rooms in the mainly undestroyed west wing of the university, and at the end of the 1940s the director had four paid assistants.

This expansion was an expression of the increasing significance which physics and especially theoretical physics played in the natural sciences during the postwar period. In it was reflected the need to make up for past shortcomings in the training of qualified physicists, for both the war as well as the restrictive National Socialist university policies had torn great holes in the physicist population. Therefore it was hardly a surprise when the number of physics students even in the first postwar years exceeded the prewar level. Since theoretical physics had traditionally enjoyed a privileged place in physics education,[122] Möglich was quickly overwhelmed by his increasing extra-university responsibilities and the scope of his teaching duties. This led to the situation where Möglich not only made the assistants responsible for the review sessions, but they often had to stand in for his lectures. This did not necessarily strengthen Möglich's professorial authority, but on the other hand, one did not have to put up with much under Möglich. The institute felt like a "bourgeois enclave,"[123] and was largely free from ideological indoctrination and only oriented toward professional competency.

Even in the immediate postwar period this was nothing to be taken for granted, especially not in a divided Berlin dominated by the Cold War. For example, Möglich did not denounce his students when they visited events at the Free or Technical Universities in West Berlin. There is even evidence for a case where he publicly urged his students to visit a guest lecture by Lise Meitner at the FU instead of his lecture, which resulted in the few students in the communist youth organization reporting the matter immediately to the rector of the University.[124] It was also not typical for this period that Möglich actively helped a few of his students to find jobs at West German universities. Möglich was thus thinking and acting in terms of Germany as a whole, not just the GDR. In this way he was similar to his mentor Laue, who from the other side of the Iron Curtain also tried to keep open and preserve communication with the East. Moreover, this "Germany as a whole" attitude did not fundamentally contradict the official GDR policies of the time, for the certainty of eventual German unity, naturally under socialism, belonged to the essentials of their official politics and ritual language.

In the context of his university activity, Möglich also should be credited with reviving the so-called Laue colloquium. Möglich's colloquium, which continued the tradition of the famous Thursday Colloquium of Berlin physicists, quickly developed into an all-Berlin physics forum. This could not shine as brightly as it had in the Weimar period, with its many Nobel laureates. Nevertheless, the Berlin physics community was once again united in common discussions as Möglich tried to cultivate the spirit of an undivided and apolitical physics, until the Cold War increasing undermined this claim and finally called it completely into question. The Colloquium must also be seen as one of the early seeds of a Physical Society in the GDR. Möglich, as always together with Rompe, was one of the leading participants in this society, which was formally founded in 1952.[125] There are even indications that Möglich was the desired candidate for the president of the society, but his election could not be pushed through. Along with generational problems between older and younger members of the managing board, his National Socialist past may also have played a role. Up until his death Möglich participated actively in the managing board, where among other things he concerned himself with contacts with the regional Physical Societies in West Germany. This was a task he had already taken up intensely during the previous years.[126]

He also concerned himself with physics publications. Thus in 1953 he was one of the founders and first editorial board members of "Advances in Physics" (Fortschritte der Physik), which with its name and profile connected itself to the rich tradition of the first journal of the Physical Society from 1845 and planned to concentrate on the publication of reviews of current developments in physics. Given the scientific isolation and material needs of the GDR, this was a worthy task. Möglich had already done a great service a few years previously by helping the Annalen der Physik to reappear. The downfall of the Third Reich not only caused the entire political life of Germany to come to a halt, but scientific activities were also affected. The Allies not only temporarily closed all the universities and banned all political organizations from the Third Reich, but all other organizations were considered dissolved and had to apply for a special license from the Allies in order to re-establish themselves.

This was also the case for all magazines and journals in the land, whether of political or scientific character. The situation was additionally difficult for physics and other areas of the natural sciences because the Allied Contact Law Number 25 basically forbade research of military-technical relevance and initially placed it under a very tight control. Since during the Second World War in particular, many areas of scientific and especially physical research were carried out for military goals, key areas of physics like atomic and nuclear physics, high-frequency technology, and electronics as well as other research areas were affected by the Allied research ban. Under these conditions, and given the international isolation in which German science found itself after the horrors of the Third Reich, it was very problematic to reestablish a specialized physical journal.

Even when it was a matter of the oldest German journal with the richest tradition and highest international recognition, what should the pages be filled with, and how could international standards be guaranteed? According to Rompe, it was Max von Laue, who like no other during the postwar period tried to keep up the great traditions of German physics, who lobbied for keeping the Annalen going.[127] In Möglich he found an active and helpful collaborator. Möglich was, in comparison to the two other established and respected editors, Eduard Grüneisen in Marburg and the aged Max Planck in Göttingen, of special importance because he lived in the Soviet zone. Not only was the Leipzig publisher Johan Ambrosius Barth Verlag here, but during the early postwar period culture and science were valued highly by the Soviet occupation policies, so that the conditions and the interest in giving out such a license were more favorable than in the Western zones. Naturally Möglich's good connections to the occupation government and other responsible offices were also known. In the summer of 1946 their efforts finally were successful, and the SMAD gave the Leipzig publisher A. Meiner the license to begin publishing the *Annalen* again on August 1, 1946. Möglich edited the journal, whose first issue appeared at the beginning of 1947, in close collaboration with his co-editor, who always came from the western part of Germany. Thereby the "Germany as a whole" character of the journal was underlined, whereby this connecting link for the German physics community not only held during the Cold War, but also remained strong during the increased "distancing policy" of the later GDR. However, it was not possible completely to re-establish the international reputation of the journal and central role it had played. The language of physics had definitively become English, and the key role which the *Annalen* or *Zeitschrift für Physik* had played in the 1920s and 1930s in the physics community had now been definitely passed on to the *Physical Review* and other Anglo-Saxon journals.

Moreover, the shadow of his past also darkened Möglich's activity as editor. When a anniversary issue of the *Annalen* was being prepared in early 1949 for Max von Laue's seventieth birthday and Lise Meitner was asked for a contribution, she wrote a bitter letter to Eduard Grüneisen:

it was not easy for me to contribute to the memorial issue for Planck. You remember what happened on the occasion of Planck's 80th birthday as well as I do. But it was out of the question for me not to use the opportunity for the last time to express my honor and thanks to Planck. The editor wrote me a brief letter of thanks after the appearance of the Planck issue which was certainly very friendly. But he did not consider it necessary in any way to touch upon his behavior in 1938. If he had written a sentence to the effect that at the time he could not have acted differently or did not believe that he could have, then in my eyes the matter – as far as the past can retroactively be corrected in any case – would have been settled. But the attitudes, that incidents lose their significance by ignoring them, appears to me unfortunate with regard to the future of the world and I do not wish to support the editor of the *Annalen* in this inclination.[128]

Möglich's German colleagues, in the East as well as the West, had significantly fewer problems with his National Socialist past. In the SBZ/GDR, which subsequently handled its own anti-fascist founding myths so self-righteously, Möglich rose even further. After the Berlin Academy of Science resumed its activities in the summer of 1946, it began not only to care for its tradition as a learned society, but simultaneously to build itself up into a potent research academy.[129] In 1947 the corresponding expansion of the Academy's research potential made great strides, whereby in particular the former institutes of the

Kaiser Wilhelm Society were attached to the Academy. In the course of their restructuring, the group around Möglich, which up until now had researched solid state physics in the Berlin-Buch institute complex under the authority of the SMAD and "during this period through not insignificant acquisitions"[130] had experienced a considerable expansion, was transferred to the Academy.

Moreover, the noticeable expansion of the research potential became the subject of a critical article in the *New York Times*, which reported violations of the Control Council laws in the SBZ and suspected Möglich's institute of carrying out forbidden nuclear physics experiments for the Soviet Union.[131] It is improbable that Möglich at that time was actually involved in nuclear physics research or related work. Instead, the report should be seen as part of the encroaching Cold War. However, a completely different question arises in this context: why was Möglich so intensively and exclusively involved with problems of solid state physics in the postwar period and why did he set aside his biophysics research interests and give them to Walter Friedrich, who beginning in 1947 established the "Institute for Medicine and Biology" in Buch? The fact that, at the time solid state physics not only held great scientific interest but also had immense practical and economic significance, or that considerable personal tension existed between Möglich and Friedrich, is only part of the explanation. Perhaps Möglich once again followed the "spirit of the times." At that time, biophysical research, with its close connection to genetics was anything other than an apolitical field of study. Lyssenkoism had ideologically discredited modern genetics so strongly in the Soviet empire that it was barely tolerated and certainly in no way supported by GDR official science policy.[132]

After the institute was taken over by the academy, Möglich was given the assignment of expanding the existing research group further and establishing a special institute for solid state research, which would research comprehensively the physical properties of the solid state and their condensed phases. The intentions behind the founding of such an institute reflect the central significance of solid state physics for basic physics research, as well as its practical applications. The latter showed itself especially in the development of circuitry, which documented the important role of solid state physics for the scientific-technical developments and inventions, and thereby also for the national economy of the country. Both aspects were taken into account with the establishment of the Academy, and the institute experienced a great expansion in personnel as well as space within a few years. Thus the core group expanded from a small research group around Möglich in the first years to over thirty scientists in the 1950s, who were organized in six departments. Connected to this expansion in personnel was the establishment of its own institute building in the inner city of Berlin, which, however, could only be opened in the year of Möglich's death.

Möglich naturally showed himself to be an excellent scientific manager through the construction and direction of such a large institute, and he simultaneously stepped out of the niche of "pure science" and became a public figure. As such, he naturally was expected to comment on current science policy problems as well as on questions of daily politics. In corresponding statements for the daily press and other main media, Möglich thereby supported the "peace policy" of the Soviet Union, argued for German reunification in the GDR sense, praised the successes achieved in the GDR, and in particular, greeted the science policy measures and motions of the SED. In internal reports this earned him the "praise" that he identified himself "in open conferences, statements, and interventions as an unconditional supporter of the goals of the GDR, and especially of Socialism"[133] and was prepared "to speak of the successes of our young republic while teaching."[134] However, the students and close collaborators judged these more as lip service than conviction, since in his close circle of co-workers political themes were only very rarely touched upon and in personal conversations he seemed to distance himself cautiously from the GDR.[135]

Even when Möglich euphemistically noted in a c.v., probably from the summer of 1946, that his "stance had always been socialist,"[136] the judgment of the Head of Personnel at the Academy was probably more accurate that "his progressive stance has unfortunately still not resulted in a clear taking of sides...and for Professor Möglich, his own success seems to be more important than the collective success of the German Democratic Republic."[137] Möglich's political statements thus corresponded to the opportunistic or schizophrenic conduct so very common in the GDR. In this sense Möglich did use the usual propagandistic rhetoric, but his contributions were usually relatively sober and factually accurate. He certainly did not belong to the fanatics and propagandists of the worker and farmer state, but that was also neither expected nor demanded of him.

The so-called "bourgeois intelligentsia" were expected to raise the scientific-technological standards and thereby the performance and competitiveness of the country, both nationally and internationally. What was required was politically conformist conduct and, at least from the perspective of the outside, a documented solidarity with this state and its goals. Such corresponding good conduct was rewarded with the opportunity to make a professional career, approved social mobility and many privileges. This social contract is documented almost ideally in Friedrich Möglich's biography. His appointment at the University of Berlin in 1945–6 not only equipped him with the high social prestige of a German professor, but also brought him to one of the most prestigious and significant professorships in Germany, with Max Planck, Erwin Schrödinger, and Max von Laue among his predecessors. Furthermore the Academy provided Möglich with the opportunity to build up his own research institute, which enjoyed a significant research potential and thereby offered many possibilities for increasing his scientific and social prestige.

All of this would not have been posible in equal measure in the Western zones of occupation or the subsequent Federal German Republic, where the competition particularly in the area of theoretical physics was significantly stronger. Thus of all outstanding theoretical physicists after the war, along with Möglich only the former collaborator and student, respectively, of Werner Heisenberg, Friedrich Hund and Bernhard Kockel, had remained in the East, while in contrast Heisenberg himself, Pascual Jordan and other prominent members of the discipline lived in the Western zone. Also, in the West there would not have been so helpful a friend and colleague as Robert Rompe, who pulled the strings for his professional career and at the same time helped to neutralize his National Socialist past in such a comfortable manner.

Moreover, all of this was "gilded" by a contract which ensured a salary which lay far above the average earnings of the "normal GDR citizen,"[138] and at least in the early years could compete with the West and thereby ensured a high standard of living. There was also no shortage of public honors. In 1953 Möglich was honored for his scientific work with the National Prize, the highest scientific award of the GDR, and in the same year the President of the German Academy of Sciences appointed him a member of the Physics Section. He was still not a full member of the Academy, but his appointment can be seen as an attempt to prepare for such an election. In 1947 the physicist Karl Willy Wagner had already tried to initiate an election for Möglich,[139] but at that time failed. Unfortunately the official correspondence gives no information on the reasons for this failure. Möglich's National Socialist part could have played a role, for several academy members had been expelled because of their entanglements with National Socialism.

However, these additional appointments for Möglich represent more than a further recognition of his scientific performance. The academy leadership also created such consultant positions in order to give the activities of the learned society a broader basis and to connect it more closely to the work of the Academy Institute. Moreover, the appointment of Möglich and other "progressive scholars" was intended to enhance the strength of these forces in comparison to the "bourgeois" academy members and, not least, colleagues in the west. Möglich's appointment also honored the loyalty which he had shown in the previous years, especially in questions of science policy.

Möglich provided a valuable service by looking after contacts to the West for the Physical Society, and he also worked to win West German scientists for Academy institutes.[140] In addition, he participated in the boards overseeing different Academy institutes and, through his scientific positions, also tried to influence the research policy profile of the Academy. First and foremost here is the example of the so-called "Physics Memorandum" of 1952. In this memorandum the physicists, O. Hachenberg, F. Möglich, R. Rompe, and R. Seeliger developed guidelines for the further development of physical research at the Academy. They stated that, with regard to development, the GDR had fallen behind the international trend and called for recognition that "physics has its rightful place" in the Academy as well as the entire GDR economy, and to concentrate its development in selected important disciplines as well their connection with economic planning.[141]

Solid state physics, electronics, plasma physics, and naturally also nuclear physics were named in this regard. With regard to the latter, at about the same time (and thus long before the corresponding Control Council decrees had been rescinded) Möglich also made the politically extremely controversial proposal to begin immediate preparatory work for the construction of a nuclear reactor in the GDR.[142] However, Möglich otherwise treated nuclear physics as an unwanted child and potential competition for his own research area. Thus in different publications and statements he called for solid state physical research, not nuclear physics, to receive special support.

At the beginning of the 1950s, Möglich enjoyed great recognition in the scientific and public life of the GDR, but these years marked not only the high point of his professional career, but also the beginning of a turning away from the GDR and its socialist ordering of society. During these years, not only Möglich but also many other GDR citizens increasingly lost the hopes, illusions, and spirit of the years of reconstruction as conflicts and arguments in the GDR society increased. The workers revolt of 17 June 1953 was an obvious expression of this fact, as well as the continually increasing flight to the West, which included trained specialists and the intelligentsia in particular. Along with material considerations, an important role was played by the fact that the science and cultural politics of the SED may have adorned themselves on the outside with Bert Brecht and Ernst Bloch, but internally became more and more dogmatic and intolerant and in the official propaganda did not shy away from ideologically defaming important areas of the modern sciences, including genetics, psychology, and cybernetics. Moreover, the Cold War was reaching its high point and this led to an additional polarization.

All of this had an effect on Möglich, whose inner break with the GDR was marked by the events in Hungary in the autumn of 1956 when Soviet troops bloodily put down the Hungarian rebellion. At the same time, in the GDR the destalinization phase came to an early end with the arrests of W. Janka, G. Harich, and other critical intellectuals. From around this time, at least within his closest family circle, Möglich no longer concealed his rejection of the communist regime and purposefully prepared for his flight to the West; thus at least twice he traveled to confidential meetings in West Germany and Switzerland in order to explore the possibility of a job with the firm SEL, a large communication technology company.[143] Möglich had probably also been in contact during the previous years with American officials in West Berlin and had provided them with information on the Academy.[144]

Möglich's early death stopped his planned flight to the West. He died on 17 June 1957 in Berlin of liver failure, not yet fifty-five years old. As his obituary from the Physical Society put it, certainly with unintentional irony,[145] this came at a time when "a further new and fruitful epoch began during his lifetime." In general, the obituaries honored Möglich's scientific rank extensively and with the highest praise, characterizing him as "a man of unusual format"[146] and "world-renowned theoretician."[147] Here he was also portrayed as an "active friend of the working man."[148] In contrast, his position in the Third Reich was handled as briefly and superficially as possible; references were made merely to his "open opposition" and "long imprisonment," which led to a prohibition on "every activity at German universities." "He took sides, but also did not shy away, when he recognized that his position was false, from drastically correcting them."[149]

The apologia of these obituaries and Möglich's own behavior after 1945 show many parallels to the myths and legends which were propagated and developed after the war by West German physicists.[150] In both the East and the West, physicists believed that they had to justify themselves by asserting that they had resisted Hitler, either that they had denied him nuclear weapons, or protested against anti-Semitism. Nevertheless, they also claimed to have been as apolitical as possible during the Third Reich. However, the justifications differ on one point: whereas in the West the physicists were in their self-conception and public image "apolitical citizens" of the Federal Republic and therefore could continue their previous behavior in the West, in the East scholars like Möglich were portrayed as "progressive scientists" who had learned a lesson under National Socialism and therefore changed their ways. This rhetoric thereby became not only the matter of official propaganda, but the physicists also used it themselves; for example, in 1952 when Möglich professed his desire to the GDR State Secretary for Higher Education to "be able to work together with you for a long time to come for the reconstruction and introduction of Socialism in our common fatherland."[151]

Conclusion

Our analysis has showed that Möglich, together with Rompe and others in the GDR, systematically strove to be seen not only as an alienated former National Socialist but also as a victim of the Nazis and an anti-fascist representative of GDR science. This official portrayal contradicted historical facts, for Möglich's alienation from National Socialism, which began in 1935 and became complete only after the start of the war, had a very specific and particular cause: the effect of the Nuremberg laws on his relationship with his lover and his academic career. The letter he wrote from his Parisian exile hardly suggests that he opposed anti-Semitism in principle, rather that he refused to accept that the Countess Bubna was not Aryan enough. Nevertheless, there is no doubt that because of this affair, Möglich learned to hate the Nazis and eventually also became a consistent opponent of National Socialism, although the point in time when his opposition began can no longer be reconstructed exactly. However, his subsequent opposition cannot cover up the political connection or general opportunism he demonstrated earlier.

This is all the more clear because, after the war, Möglich was able to make a new career in the Soviet Zone and the GDR. Moreover, his political loyalty as well as his value as a scientist for this state made it possible that his National Socialist past could not only be forgotten but even reinterpreted as anti-fascist.

Only this last point distinguishes his biography from those scientists in the Western zones of occupation, in which the victorious powers were interested for their research and development work and therefore were "merely" denazified and quickly received a new chance at a career.[152] In both cases, this was not merely a matter of official propaganda and ideology, for as Möglich's behavior documents, he actively influenced the process through opportunism, personal ambition, and conscious silence with regard to a distortion of facts.

The same can be said of his two most important supporters, Max von Laue and Robert Rompe, who in striking continuity before and after the war tolerated Möglich's political convictions and pushed his scientific competence into the foreground. The Soviet Military Administration and East German state were willing to proclaim Möglich an anti-fascist because of his scientific ability and political loyalty. Indeed, all of the victorious powers sometimes turned a blind eye to the political past of the German scientists and engineers they employed after the war. Ironically, the only individual or organization that stuck to its principles was the NSDAP, which refused to tolerate a party comrade who did not accept the race laws of the Third Reich.

In 1935 Möglich was living proof that a scientist could be well respected both as a colleague by the German physics community and as a National Socialist by Hitler's followers. He underwent a political transformation and died more or less as a hero of the Socialist state and a victim of National Socialism. But it all might have ended very differently, if he had never fallen in love.

Notes

1 Max Planck Institute for the History of Science, Berlin, Germany.
2 Department of History, Union College, Schenectady, NY.
3 "Enge Verbindung von Theorie und Praxis. Zum Tode des Physikers Prof.Dr. Friedrich Möglich," *Berliner Zeitung* (June 19, 1957), 3.
4 Bl. 658, Bl. 331–404, Philosophische Fakultät, Archiv der Humboldt-Universität zu Berlin (HUA).
5 Friedrich Möglich, "Beugungserscheinungen an Körpern von ellipsoidischer Gestalt" *Annalen der Physik*, 83 (1927): 609–734.
6 Friedrich Möglich, "Beugung; Ponderomotorische Wirkungen der Strahlung," in E. Gehrcke (ed.), *Handbuch der physikalischen Optik* (Leipzig: Verlag J.A. Barth), 1927), v. 1, 499–660, 941–57.
7 Bl. 1244, Bl. 200–214, Philosophische Fakultät, HUA.
8 Friedrich Möglich "Quantentheorie schwingender Kontinua," *Annalen der Physik*, 2 (1929): 676–85.
9 Friedrich Möglich, "Zur Quantentheorie des rotierenden Elektrons," *Zeitschrift für Physik*, 48 (1928): 852–67.
10 Schrödinger's comments in his report were typical in this regard: "Möglich has exceptional experience in the mathematical treatment of old and new problems in theoretical physics, further a far-reaching knowledge of the current state of theoretical physics and an exceptional ability to familiarize himself with the ideas of others and to expand them; but little originality." Bl. 1244, Bl. 207, Philosophische Fakultät, HUA.

11 In interviews, his later students and collaborators all made this point: W. Brauer (April 18, 1996), B. Mühlschlegel (June 11, 1996) and B. Seraphin (August 18, 1996).
12 W. Pauli, "Die allgemeinen Prinzipien der Wellenmechanik," in Hans Geiger and K. Scheel (eds), *Handbuch der Physik*, 24, 1 (Berlin: Verlag Julius Springer, 1933), 235.
13 1244, Bl. 207, Philosophische Fakultät, HUA.
14 Friedrich Möglich to das Kreisgericht der NSDAP, Berlin (August 8, 1938) Möglich File, Berlin Document Center (BDC).
15 Interview with his son Peter-Michael Möglich, Berlin (January 4, 1996); also see the short reference in Möglich's c.v., that his father was a retired first lieutenant.
16 Robert Rompe, "Friedrich Möglich – sein Beitrag zum Aufbau der Physik in der DDR," *Annalen der Physik*, 47, 4 (1990): 320.
17 Naturally the two Nobel laureates Philipp Lenard and Johannes Stark should be mentioned in this regard, but so also should the physical chemist Peter-Adolf Thiessen.
18 "Urteilsbegründung in der Strafsache gegen Friedrich Möglich, Berlin" (July 6, 1937), 2, Möglich File, BDC.
19 Gau-Dozentenbundführer to das Gaugericht der NSDAP, Berlin (November 16, 1939), Möglich File, BDC.
20 Draft (no date) Bl. 14, Personalakte Friedrich (PA) Möglich, HUA.
21 Interview with Heinz Schmellenmeier (February 9, 1993).
22 Friedrich Möglich, "Über die Vollständigkeit der Gruppentheorie," *Sitzungsberichte der Preußischen Akademie der Wissenschaften* (1933): 639–41.
23 See Mark Walker, *Nazi Science: Myth, Truth, and the German Atomic Bomb* (New York: Plenum, 1995), 74ff; Dieter Hoffmann, "Johannes Stark – eine Persönlichkeit im Spannungsfeld von wissenschaftlicher Forschung und faschistischer Ideologie," *Philosophie und Naturwissenschaften in Vergangenheit und Gegenwart, Humboldt-Universität zu Berlin*, 22 (1982): 90–101.
24 See W. Moore, *Erwin Schrödinger* (Cambridge: Cambridge University Press, 1989), 265; among students the rumor even circulated that for a time Möglich had stood in for Goebbels in the Berliner Party leadership. Interview with Heinz Schmellenmeier, Berlin (February 9, 1993).
25 Gau-Dozentenbundführer to das Gaugericht der NSDAP, Berlin (November 16, 1939), Möglich File, BDC.
26 Robert Rompe, "Friedrich K.S. Möglich – 12.10.1902 bis 17.6.1957," *Annalen der Physik*, 20, 1 (1958): 2.
27 Robert Rompe, "Friedrich Möglich," 320.
28 Interview with P.-M. Möglich, Berlin (February 29, 1996).
29 Friedrich Möglich to Leyhausen, Paris (November 10, 1935), Möglich File, BDC.
30 Robert Rompe to den Rektor der Universität Berlin M. Dersch, Berlin (June 14, 1948), Bl. 31, PA Möglich, HUA.
31 Max von Laue, "Mein physikalischer Werdegang. Eine Selbstdarstellung," in Max von Laue, *Aufsätze und Vorträge* (Braunschweig: Verlag Friedr. Vieweg & Sohn, 1961), xxix.
32 Brief des Reichs- und Preußischen Minister für Wissenschaft, Erziehung und Volksbildung, Berlin (February 29, 1936), 445 PA, HUA.
33 Friedrich Möglich to Leyhausen, Paris (November 10, 1935), Möglich File, BDC.
34 C.V. (September 13, 1939) family papers (in the possession of P.M. Möglich).
35 See "Urteilsbegründung," 13, Möglich File, BDC.
36 Friedrich Möglich to Gaugericht Berlin (July 17, 1939), Möglich File, BDC.

37 "Gesetz zum Schutze des deutschen Blutes und der deutschen Ehre" vom 15.9.1935; *Reichsgesetzblatt*, 1935, Part I, 1146.
38 Gestapo to die Gauleitung Berlin der NSDAP, Berlin (January 18, 1937), Möglich File, BDC.
39 Anklageschrift (February 9, 1937), Möglich File, BDC.
40 Max von Laue to den Dekan der mathematisch-naturwissenschaftlichen Fakultät, Berlin (July 8, 1937) Bl. 32, PA Möglich, HUA.
41 Friedrich Möglich to den Prorektor der Universität, Berlin (October 27, 1937) Bl. 27–28, PA Möglich, HUA.
42 See the Möglich File, BDC.
43 "Eröffnungsbeschluß," Berlin (January 24, 1938), Möglich File, BDC.
44 "Beschluss" (June 10, 1938), Möglich File, BDC.
45 Friedrich Möglich to Gaugericht Berlin, (July 17, 1939), Möglich File, BDC.
46 Friedrich Möglich to Gaugericht Berlin (December 3, 1939), Möglich File, BDC.
47 Friedrich Möglich to Kreisgericht (August 8, 1938), Möglich File, BDC.
48 Communication from Prof. Dr. G. Richter, Berlin.
49 "Protokoll der Hauptverhandlung vor dem Gaugericht Berlin," (November 7, 1939), Möglich File, BDC.
50 Friedrich Möglich to Kreisgericht (August 8, 1938), Möglich File, BDC.
51 Friedrich Möglich to das Gaugericht Berlin, (November 5, 1939), Möglich File, BDC.
52 "Urteilsbegründung des Gaugerichts Berlin," (November 7, 1939), 6, Möglich File, BDC.
53 "Begründung" (July 3, 1940), Möglich File, BDC.
54 NSD-Dozentenbund to das Gaugericht Berlin, (November 16, 1939), Möglich File, BDC.
55 Friedrich Möglich to Leyhausen, Paris (November 10, 1935), Möglich File, BDC.
56 Oberstes Parteigericht to Reichssicherheitshauptamt, (January 22, 1940); Sicherheitsdienst to Oberstes Parteigericht (April 19, 1940), Möglich File, BDC.
57 Oberstes Parteigericht to Friedrich Möglich, (June 7, 1940), Möglich File, BDC.
58 Friedrich Möglich to Oberstes Parteigericht, Berlin (June 14, 1940), Möglich File, BDC.
59 Friedrich Möglich to Mathematisch-Naturwissenschaftliches Dekanat, Berlin (September 21, 1939) Bl.35, PA Möglich, HUA.
60 "Reichserziehungsministerium" (September 14, 1940) Bl.37, PA Möglich, HUA.
61 Interview with P.-M. Möglich, Berlin (January 4, 1996).
62 Robert Rompe, "Friedrich Möglich," 320.
63 Interview with P.-M. Möglich, Berlin (February 29, 1996).
64 Questionaire (March 18, 1948), No. 659, Akademieleitung Personalia, Archiv der Berlin-Brandenburgischen Akademie der Wissenschaften, (ALP, BBAA).
65 *Statistisches Jahrbuch des Deutschen Reiches 1941/42* (Berlin: 1943).
66 For a biography of Robert Rompe, see P. Nötzoldt, "Robert Rompe," in W. Schreier (ed.), *Biographien bedeutender Physiker* (Berlin: Verlag Volk und Wissen, 1984), 350–51; Robert Rompe, *Ausgewählte Vorträge und Aufsätze*, v. 1 und 2 (Berlin: Physikalischen Gesellschaft der DDR, 1980 and 1985); Dieter Hoffmann, *Das Müggelheim Buch* (Berlin: 1997), 168–70.
67 Robert Rompe, "Friedrich Möglich," 321.
68 Interview with Heinz Schmellenmeier, Berlin (February 9, 1993).
69 Interview with P.M. Möglich, Berlin (February 29, 1996).
70 Robert Rompe, "Friedrich K.S. Möglich," 3.
71 Friedrich Möglich, "Über das Masseverhältnis Elektron-Neutron," *Naturwissenschaften*, 26 (1938): 409–10.

72 Friedrich Möglich and Robert Rompe, "Einige Folgerungen aus der Existenz einer kleinsten Länge," Zeitschrift für Physik, 113 (1939): 740–50.

73 See D. Granin, Sie nannten ihn Ur (Berlin: Verlag Volk und Welt, 1988), 157ff; Richard Beyler, "Targeting the Organism," ISIS, 87, 2 (1996): 248–73.

74 Friedrich Möglich, Robert Rompe and Nikolas Timoféef-Ressovsky, "Bemerkungen zu physikalischen Modellvorstellungen über Energieausbreitungsmechanismen im Treffbereich bei strahlenbiologischen Vorgängen," Naturwissenschaften, 30 (1942): 409–19; Friedrich Möglich, Robert Rompe and Nikolas Timoféef-Ressovsky, "Über die Indeterminiertheit und die Verstärkereigenschaften in der Biologie," (manuscript submitted to the Naturwissenschaften, which could not appear because of the chaos at the end of the war).

75 Max von Laue and Friedrich Möglich, "Über das magnetische Feld in der Umgebung von Supraleitern," Sitzungsberichte der Preussischen Akademie der Wissenschaften, 16 (1933): 544–65.

76 See Dieter Hoffmann, "Wendepunkt in der Geschichte der Supraleitung. Vor 60 Jahren wurde der Meissner-Ochsenfeld-Effekt entdeckt," Physikalische Blätter, 49, 10 (1993): 899–902; P.F. Dahl, Superconductivity (New York: American Institute of Physics, 1992), 176ff, 195ff.

77 Friedrich Möglich and Robert Rompe, "Plasmaschwingungen als Ursache der Supraleitung," Annalen der Physik, 1 (1947): 27–40.

78 See P.F. Dahl, Superconductivity, 245ff.

79 P.F. Dahl, Superconductivity, 254ff.

80 Interviews with W. Brauer (April 18, 1996) and B. Seraphin (August 18, 1996).

81 For the history of modern solid state physics, see Lillian Hoddeson et al. (eds), Out of the Crystal Maze (New York: Oxford University Press, 1992).

82 See the list of his publications in J.C. Poggendorff's Biographisch-Literarisches Wörterbuch (Berlin: Akademie Verlag, 1959), v. VIIa, 318.

83 Friedrich Möglich and Robert Rompe, "Zur Entwicklung der "Photometalle," (confidential) study (October 16, 1940), Family Papers, P.-M. Möglich. Friedrich Möglich and Robert Rompe, "Zur Entwicklung der "Photometalle," (confidential) study (October 16, 1940), Family Papers, P.-M. Möglich.

84 Friedrich Möglich and Robert Rompe, "Über Energieumwandlung im Festkörper," Zeitschrift für Physik, 115 (1940): 707–28; Friedrich Möglich and Robert Rompe, "Zur Statistik der Vielfachstöße, Zeitschrift für Physik, 117 (1941): 119–24.

85 W. Weizel, "Gutachten Wissenschaftliche Bedeutung von Friedrich Möglich," Bonn (March 12, 1946) Bl. 13, PA Möglich, HUA.

86 Contract (August 29, 1941), Family Papers, P.-M. Möglich.

87 E. Swinne, Richard Gans. Hochschullehrer in Deutschland und Argentinien (Berlin: ERS Verlag, 1992), 112ff.

88 Interview with H. Schmellenmeier, Berlin (February 9, 1993).

89 Interview with P.-M. Möglich, Berlin (January 4, 1996).

90 John Gimbel, Science, Technology and Reparations: Exploitation and Plunder in Postwar Germany (Stanford, CA: Stanford University Press, 1990); Matthias Judt and Brughard Ciesla (eds), Technology Transfer out of Germany after 1945 (Amsterdam: Harwood Academic Press, 1996).

91 Ulrich Albrecht, Andreas Heinemann-Grüder, and Arnd Wellmann, Die Spezialisten. Deutsche Naturwissenschaftler und Techniker in der Sowjetunion nach 1945, (Berlin: Dietz Verlag, 1992).

92 Dieter Hoffmann (ed.), Operation Epsilon (Berlin: Rowohlt Berlin, 1993); Walker, Nazi Science, 207–41.

93 Linda Hunt, Secret Agenda: The United States Government, Nazi Scientists and Project Paperclip, 1945–1990 (New York: St. Martin's Press, 1991; Tom Bower, The

Paperclip Conspiracy: The Battle for the Spoils and Secrets of Nazi Germany (Boston: Little, Brown, 1987).

94 Interview with P.-M. Möglich, Berlin (January 4, 1996).

95 Granin, *Sie nannten ihn Ur.*

96 Nikolaus Riehl and Frederick Seitz, *Stalin's Captive: Nikolaus Riehl and the Soviet Race for the Bomb* (New York: American Chemical Society, 1996).

97 Interview with P.-M. Möglich, Berlin (January 4, 1996).

98 See Manfred Heinemann, "Der Wiederaufnbau der Kaiser-Wilhelm-Gesellschaft und die neugründung der Max-Planck-Gesellschaft (1945–1949)," in Rudolf Vierhaus and Bernhard vom Brocke (eds), *Forschung im Spannungsfeld von Politik und Gesellschaft. Geschichte und Struktur der Kaiser-Wilhelm-/Max-Planck-Gesellschaft* (Stuttgart: DVA, 1990).

99 Another example from physics was the Optical Laboratory of Ernst Lau and R. Ritschl (two former members of the Physikalisch-Technische Reichsanstalt) in Berlin-Karow, which was also directly subordinate to the Soviet authorities and did military-technical research and development for the Soviet Union. In the later 1940s it was brought into the Academy.

100 See Burghard Ciesla, "Die Transferfalle: Zum DDR-Flugzeugbau in den fünfziger Jahren," in Dieter Hoffmann and Kristie Macrakis (eds) *Naturwissenschaft und Technik in der DDR* (Berlin: Akademie Verlag, 1997), 193–212.

101 P. Nötzoldt, "Wissenschaft in Berlin – Anmerkungen zum ersten Nachkriegsjahr 1945/46," *Dahlemer Archivgespräche*, 1 (1996): 115–30.

102 Friedrich Möglich to Max von Laue, Berlin (February 5, 1946), Family Papers, P.-M. Möglich.

103 Max von Laue to Friedrich Möglich, Alswede (February 16, 1946) Bl.10, PA Möglich, HUA.

104 Interview with H. Schmellenmeier, Berlin (February 9, 1993).

105 Bl. 28–29, v. R-2 1307, Bundesarchiv, Abteilung Potsdam (BAP); we are grateful to Mitchell Ash for directing us to this document.

106 Mitchell Ash, "Verordnete Umbrüche – Konstruierte Kontinuitäten: Zur Entnazifizierung von Wissenschaftlern und Wissenschaften nach 1945," *Zeitschrift für Geschichtswissenschaften*, 43, (1995): 10, 903–923; also see Mitchell Ash, "Scientific Changes in Germany 1933, 1945 and 1990: Towards a Comparison," *Minerva*, 37 (1999): 329–54.

107 R. Jessen, *Akademische Elite und kommunistische Diktatur. Die ostdeutsche Hochschullehrerschaft in der Ulbricht-Ära* (Göttingen: Vandenhoeck & Ruprecht, 1999), 42ff.

108 H.U. Feige, "Aspekte der Hochschulpolitik der Sowjetischen Militäradministration in Deutschland (1945–1948)," *Deutschlandarchiv*, 25 (1992): 1169–80.

109 Chr. Gerthsen to Zentralverwaltung, Berlin (October 27, 1945) Bl.5, PA Möglich, HUA.

110 Stroux to Zentralverwaltung für Volksbildung, Berlin (November, 19, 1945), Bl. 52R, PA Möglich, HUA.

111 Rektor der Friedrich-Wilhelms-Universität to Friedrich Möglich, Berlin (January 12, 1946); Ernennungsurkunde des Präsidenten der DVV (January 29, 1946), Bl. 8; 9, PA Möglich, HUA.

112 Max von Laue to Robert Rompe, Alswede (March 9, 1946), Bl. 1, PA Möglich, HUA.

113 Th. Brugsch to Präsidenten Wandel, Berlin (April 16, 1946), Bl. 19, PA Möglich, HUA.

114 Notiz (June 14, 1948) Bl. 30, PA Möglich, HUA.

115 Robert Rompe to den Rektor der Universität Berlin, (June 14, 1948), Bl. 32, PA Möglich, HUA.

116 Personalia No. 659, AL, BBAA.
117 M. Strauß Papers, SAPMO IV2, 9.04, 295, Bl. 50ff; AL, BBAA; See also, Dieter Hoffmann, "Remigrierte Naturwissenschaftler in der DDR. Das Beispiel der Physiker Klaus Fuchs, Fritz Lange und Martin Strauss" (forthcoming).
118 BAP, R-2 959, Bl. 38, for the quotation see: Ash, "Verordnete Umbrüche," 913.
119 Vierhaus and vom Brocke, 401ff.
120 See. P. Nötzoldt, "From German Academy of Sciences to Socialist Research Academy," in Kristie Macrakis and Dieter Hoffmann (eds), *Science under Socialism* (Cambridge, MA: Harvard University Press, 1999), 140–57.
121 Max von Laue to Friedrich Möglich, Alsweede (February 16, 1946) Bl. 10, PA Möglich, HUA.
122 See R. Enderlein, F. Kaschluhn, "Das Institut für Theoretische Physik an der Humboldt-Universität zu Berlin und die theoretischen Bereiche der Sektion Physik von 1946 bis zur Gegenwart," *Wissenschaftliche Zeitschrift der Humboldt-Universität zu Berlin, Mathematisch-Naturwissenschaftliche Reihe* 32, 5 (1983): 595–600; Robert Rompe, "Friedrich Möglich," 322.
123 Interview with W. Brauer and B. Mühlschlegel.
124 Bl. 114f., PA Möglich, HUA.
125 See Dieter Hoffmann, "Die Physikalische Gesellschaft (in) der DDR," in Th. Mayer-Kuckuk (ed.), *150 Jahre Deutsche Physikalische Gesellschaft* (Weinheim: VCH, 1995), 157ff.
126 Bl. 37, PA Möglich, HUA.
127 Robert Rompe, "Friedrich Möglich," 319.
128 Lise Meitner to E. Grüneisen, Stockholm (March 28, 1949), Folder 5/6 , Lise Meitner Papers, Archives of Churchill College, Cambridge.
129 See Nötzoldt.
130 *Jahrbuch der Deutschen Akademie der Wissenschaften 1946/49* (Berlin: Akademie Verlag, 1950), 76.
131 "Arms, Atom Work in Soviet Zone Is Reported in Detail at Berlin," *New York Times* (July 13, 1946); also see Bl. 56f., PA Möglich, HUA.
132 See R. Hohlfeld, "Between Autonomy and State Control: Genetic and Biomedical Research," in Macrakis and Hoffmann.
133 Hausmitteilung des Staatssekretariats für Hochschulwesen (October 22, 1952) Bl. 58, PA Möglich, HUA.
134 Stellungnahme der Kaderabteilung (December 22, 1956) Bl. 113, PA Möglich, HUA.
135 Interview with W. Brauer and B. Mühlschlegel.
136 CV from August 20, 1946, Family Papers, P.-M. Möglich.
137 Gutachten des komm. Personalleiters der DAW (August 18, 1952), Personalia Nr. 659, AL, BBAA.
138 Personalia Nr. 659, AL, BBAA.
139 Personalia Nr. 677, Bl.5, AL, BBAA.
140 D.-G. v. der Schulenburg to Max von Laue, München 15.8.1951, MPGA, NL Laue, Nr. 1816, Bl.1.
141 "Denkschrift zur Entwicklung der Physik in der Akademie der Wissenschaften (1952)," *Adlershofer Splitter*, 4 (1998): 12–21.
142 Protokoll der Sitzung des wissenschaftlichen Kuratoriums des Instituts Miersdorf v. (October 20, 1952), AL 29, BBA.
143 Interview with P.-M. Möglich, Berlin (January 4, 1996).
144 According to B. Seraphin (Interview August 18, 1996), who was confronted with this information during his flight to West Berlin in the summer of 1953.
145 H. Simon, "Professor Möglich," *Mitteilungen der Physikalischen Gesellschaft in der DDR*, 6 (1957): 2.

146 Robert Rompe, "Friedrich Möglich".

147 H. Simon, "Nationalpreisträger Prof. Dr. Friedrich K. Sidney Möglich," *Zeitschrift für Physikalische Chemie*, 207 (1957): 157.

148 "Enge Verbindung von Theorie und Praxis. Zum Tode des Physikers Prof. Dr. Friedrich Möglich," *Berliner Zeitung* (June 19, 1957).

149 Robert Rompe, "Friedrich Möglich," 2.

150 Many studies are available for this subject, see for instance: Klaus Hentschel and Monika Renneberg, "Eine Akademische Karriere. Der Astronom Otto Heckmann im Dritten Reich," *Vierteljahresschrifte für Zeitgeschichte*, 43, 4 (1995): 581–610.

151 Friedrich Möglich to G. Harig (October 29, 1952) Bl. 59, PA Möglich, HUA.

152 The case of Wernher von Braun and the other German rocket researchers is certainly the best known; see Michael J. Neufeld, *The Rocket and the Reich* (New York: The Free Press, 1994).

INDEX